Heterogeneous Catalytic Oxidation

# Heterogeneous Catalytic Oxidation

## Fundamental and Technological Aspects of the Selective and Total Oxidation of Organic Compounds

B.K. HODNETT

*Department of Chemical and Environmental Sciences*
*and*
*The Materials and Surface Science Institute*
*University of Limerick, Ireland*

**JOHN WILEY & SONS, LTD**

Chichester • New York • Weinham • Brisbane • Singapore • Toronto

Baffins Lane, Chichester,
West Sussex PO19 1UD, England
National 01243 779777
International (+44) 1243 779777
e-mail (for orders and customer service enquiries): cs-books@wiley.co.uk
Visit our Home Page on http://www.wiley.co.uk
or http://www.wiley.com



*Other Wiley Editorial Offices*

John Wiley & Sons, Inc., 605 Third Avenue,
New York, NY 10158-0012, USA

WILEY-VCH Verlag GmbH, Pappelallee 3,
D-69469 Weinheim, Germany

Jacaranda Wiley Ltd, 33 Park Road, Milton,
Queensland 4064, Australia

John Wiley & Sons (Asia) Pte Ltd, 2 Clementi Loop #02-01,
Jin Xing Distripark, Singapore 129809

John Wiley & Sons (Canada) Ltd, 22 Worcester Road,
Rexdale, Ontario M9W 1L1, Canada

***Library of Congress Cataloging-in-Publication Data***

Hodnett, BK
Heterogeneous catalytic oxidation: fundamental and technological aspects of the selective and total oxidation of organic compounds/B.K. Hodnett.
p. cm.
Includes bibliographical references and index.
ISBN 0-471-48996-8 (alk. paper)
1. Oxidation. 2. Heterogeneous catalysis. I. Title.

QD281.O9 H63 2000
541.3′93–dc21

99-052447

***British Library Cataloguing in Publication Data***

A catalogue record for this book is available from the British Library

ISBN 0 471 48994 8

Typeset in 10/12pt Times by Keytec Typesetting Ltd, Bridport, Dorset
Printed and bound in Great Britain by Antony Rowe Ltd, Chippenham Wiltshire
This book is printed on acid-free paper responsibly manufactured from sustainable forestry, in which at least two trees are planted for each one used for paper production.

**To Eveline**

# Contents


**Preface** **xi**

**1 General Introduction** **1**
1.1 Presentation of Material 3
1.2 The Role and Importance of Heterogeneous Catalytic Oxidation Processes in the Chemical Industry 5
References to Chapter 1 7

**2 Introduction to Heterogeneous Catalysis** **8**
2.1 Macroscopic and Microscopic Structure of Heterogeneous Catalysts 9
2.2 The Five Basic Steps in Heterogeneous Catalysis 14
2.3 Testing of Heterogeneous Catalysts 15
2.4 Diffusion 25
2.5 Adsorption at the Gas–Solid Interface 29
2.6 Kinetics of Oxidation Reactions 44
2.7 Characterization of Oxidation Catalysts 50
References to Chapter 2 62

**3 General Principles of Heterogeneous Catalytic Oxidation** **65**
3.1 Generic Features of Catalytic Oxidation Reactions 66
3.2 The Mode of Activation of the Hydrocarbon and Discriminating Capacity of Active Sites in Oxidation Catalysis 72
3.3 Types of Active Oxygen Species Involved in Oxidation Catalysis 82
3.4 Structural Features of Oxidation Catalysts 90
References to Chapter 3 99

**4 Propene Oxidation to Acrolein: Allylic Oxidation over Catalysts with Highly Mobile Oxygen Anions** **102**
4.1 Introduction 103
4.2 The Composition of Multicomponent Bismuth Molybdate Catalysts for Propene Oxidation 105

4.3 The Surface Allyl Species 116
4.4 Source of Oxygen Inserted into Propene to Form Acrolein 119
4.5 Oxygen Transport within Bismuth Molybdate Catalysts 124
4.6 The Overall Reaction Mechanism 126
References to Chapter 4 129

**5 Oxidation Activation of Alkanes: Oxidation of *n*-butane to Maleic Anhydride 132**
5.1 Introduction 133
5.2 Technologies for the Production of Maleic Anhydride 135
5.3 The Preparation, Structure and Performance of Vanadium Phosphorus Oxide Catalysts 139
5.4 Activation of *n*-butane 149
5.5 The Nature of the Active Site and the Reaction Mechanism 153
5.6 Conclusions 156
References to Chapter 5 157

**6 Epoxidation of Alkenes: Reactivity of Electrophilic Oxygen Species 160**
6.1 Introduction 161
6.2 Technology for Ethylene Oxide Production 162
6.3 Catalyst Composition, Structure and Performance 164
6.4 Reaction Kinetics and Network 170
6.5 The Nature of the Oxygen Species on and under the Silver Surface 174
6.6 Reactivity of Oxygen Species on Silver 179
References to Chapter 6 187

**7 Catalytic Combustion 189**
7.1 Introduction 190
7.2 Environmental Applications of Catalytic Combustion 191
7.3 Principles of Catalytic Combustion 199
7.4 Choice of Materials for the Construction of Combustion Catalysts 201
7.5 Poisoning of Combustion Catalysts 216
7.6 Reactivity of the Substrate 218
7.7 Kinetics of Catalytic Combustion Reactions 223
7.8 Nature of the Active Species of Working Precious Metal Catalysts 227
References to Chapter 7 238

**8 Activation and Insertion of Nitrogen into Organic Compounds** **240**
8.1 Introduction 241
8.2 Technology for Propene Ammoxidation 243
8.3 Catalysts Used in Ammoxidation 249
8.4 The Ammoxidation Reaction Network 250
8.5 Mechanism of Nitrogen Insertion in Ammoxidation 254
8.6 Nitrogen Insertion in Ammoxidation 260
References to Chapter 8 262

**9 Catalytic Oxidation Reactions Involving Chlorine: Oxychlorination and the Oxidative Destruction of Chlorinated Hydrocarbons** **264**
9.1 Introduction 265
9.2 Vinyl Chloride by Oxychlorination 266
9.3 Technology for Vinyl Chloride Production 268
9.4 Oxychlorination Catalyst Composition, Preparation and Characterization 275
9.5 The Mechanism of Oxychlorination 276
9.6 Destruction of Chlorinated Hydrocarbons by Catalytic Oxidation 285
9.7 By-product Formation during the Catalytic Oxidation of Chlorinated Hydrocarbons 290
9.8 Polychlorinated Biphenyls, Dioxins and the Environment 295
References to Chapter 9 299

**10 Oxidation Reactions Catalysed by Zeolites and Structured Mesoporous Materials** **301**
10.1 Zeolites 302
10.2 Redox Molecular Sieves 313
10.3 Structured Mesoporous Silicates in Oxidation Reactions 328
References to Chapter 10 336

**Index** 339

# Preface

Heterogeneous catalytic oxidation is an important technological area which lies at the heart of a multitude of processes for the production of bulk chemicals, for pollution abatement and, increasingly, for the production of fine chemicals. In recent years this field has progressed sufficiently to allow a molecular level understanding of oxidation phenomena at catalyst surfaces. Hence, the modern researcher in this area must have a thorough understanding of fundamental chemical and engineering phenomana that range in scale from a typical reactor, constructed on the 1–10 m scale, to events occurring on active sites on the catalyst surface, constructed on the nanometre scale.

Over the years many superb experiments have been designed to probe the macroscopic and microscopic structure of heterogeneous oxidation catalysts and to investigate the mechanism of interaction with and transformation of hydrocarbons on these catalysts. This book attempts to describe some of these experiments and illustrates how the resultant data and its interpretation has led to our current understanding of heterogeneous catalytic oxidation.

The main body of the book centres around seven case studies (Chapters 4–10) each dealing with an important commercial process involving heterogeneous catalytic oxidation. Each study touches on a small number of generic features of oxidation catalysis and focuses in detail on a different fundamental or engineering aspect for each process. The choice of case studies was based on the author's own experience in this field. Other areas could equally validly have been included, but those offered are sufficient to illustrate all the major generic aspects of heterogeneous catalytic oxidation, which are introuced in Chapter 3.

The greatest benefit of presenting a detailed account of seven separate case studies in catalytic oxidation is that, taken together, they allow a clear overview of the entire field of heterogeneous catalytic oxidation. This overview in turn points to a small number of features common to all heterogeneous catalytic oxidation processes. Identification of these generic features lifts a lot of the mystery surrounding heterogeneous catalysis and enables researchers to view their field from a more logical and scientific perspective.

The book is written from the perspective that the newcomer to the field of heterogeneous catalysis is best introduced to the subject through an in-depth presentation and analysis of a very specific subsection of the field. Heterogeneous catalytic oxidation serves this purpose very well, with its wealth of classical experimentation, clear-cut experimental results and treatment of all chemical engineering and fundamental aspects of catalysis. Chapter 2 of this book has been written with the newcomer in mind. It is intended to be an introduction to the essential concepts and techniques of heterogeneous catalysis. This aspect of the book is not intended to be exhaustive and other texts, especially those dealing with characterization techniques, should be read in association with this chapter.

I wish to express my gratitude to a number of people who helped at various stages during the preparation of this book. A large part was written during a sabbatical leave with the group of Prof. RA van Santen at Eindhoven University of Technology, The Netherlands and I am particularly grateful to Pieter Stobbelaar for many discussions and for his patience in helping me with my basic computing skills. I also wish to thank my colleagues at the University of Limerick, in particular Seamus McMonagle, Tom O'Dwyer, Teresa Curtin, Finbar O'Regan and Aidan Doyle for reading and offering critical feedback on various parts of the book.

Most especially I wish to thank my wife Eveline and children, Anneke and Lieneke for their unending support and for putting up with my many absences from normal family life for the 18 months that this book was in preparation.

# 1

# General Introduction

1.1 Presentation of Material 3
1.2 The Role and Importance of Heterogeneous Catalytic Oxidation Processes in the Chemical Industry 5
References to Chapter 1 7

This book concentrates on those chemical processes in which an organic substrate is oxidized over a heterogeneous catalyst. It covers areas such as the production of bulk organic chemicals, the production of fine chemicals and environmental applications. For each of these areas the fundamental chemistry is explored as well as aspects of the industrial processes which they drive.

Selective oxidation and ammoxidation reactions and processes form a large part of the commercial processes covered here. These are reactions in which C—H bonds in the organic substrate are ruptured and oxygen or nitrogen is inserted to yield the corresponding aldehyde, ketone, carboxylic acid or nitrile. A large part of the technology currently used for the functionalization of hydrocarbons which in turn serve as feedstocks for plastic and polymer production is driven by these reactions. Acrolein, maleic anhydride and acrylonitrile production are examples of chemicals produced using technologies based on heterogeneous catalysts capable of rupturing a specific C—H bond in an organic substrate and assisting in the insertion of oxygen or nitrogen into the product. Some examples of these reactions are listed in Table 1.1.

A second class of oxidation reaction important in the production of bulk organic chemicals by heterogeneous catalysis are those in which a functional group is inserted into the molecule without rupture of a C—H bond. Reactions covered here include epoxidation of alkenes where an oxygen species is added across the double bond of an ethylene molecule and oxychlorination reactions, which involves the addition of chlorine to ethylene. This book explores fundamental and applied aspects of the epoxidation of ethylene over silver/alumina catalysts and the oxychlorination of ethylene over copper chloride/alumina catalysts.

Heterogeneous catalysts, very often in the form of zeolites and mesoporous materials, are now finding new applications for the production of fine

**Table 1.1** Typical oxidation, ammoxidation and oxychlorination reactions catalysed by heterogeneous catalysts

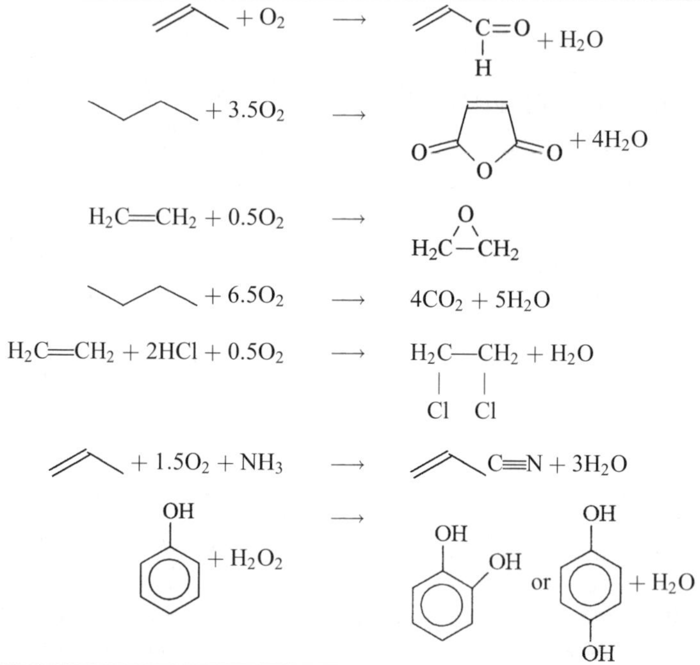


chemicals through oxidation reactions. This development is explored here with a review of the new materials that have been developed, principally transition metal substituted zeolites, zeotypes and mesoporous materials. An example of a reaction catalysed by one such material, namely titanium substituted into a form of silicalite, is the hydroxylation of phenol, shown in Table 1.1.

Finally applications of heterogeneous oxidation catalysts for the abatement of environmental problems will be explored. These include the introduction of catalyst based technologies for the modification of combustion processes so that less pollution is produced (for example low temperature catalytic combustion of natural gas below the threshold temperature for NOx production) or applications to situations where waste streams must be treated prior to discharge into the environment (for example the catalytic oxidation of a gaseous waste streams containing less than 1 vol% of a hydrocarbon, i.e. at a concentration that is too low to sustain normal thermal combustion). One aspect of the use of heterogeneous oxidation or redox catalysis not covered here is the so called deNOx problem, wherein a catalyst is used to reduce oxides of nitrogen (NOx) into nitrogen ($N_2$) through the reducing action of ammonia or a hydrocarbon. In that process the emphasis is on the reduction part of the redox cycle, namely the reduction of NOx, rather than the oxidation of the reductant, and for that reason deNOx chemistry and processes are not covered in this book.

## 1.1 PRESENTATION OF MATERIAL

This book is intended to cover the basics of heterogeneous catalytic oxidation as well as applied aspects of individual processes. Accordingly, this chapter serves as an introduction to the role of oxidation catalysis within the framework of the overall chemical industry. Fundamentals of heterogeneous catalysis, with a particular emphasis on the elementary processes and the most basic characterization methods important in oxidation catalysis, are covered in Chapter 2. Hence, mechanical aspects of heterogeneous catalysts, diffusional processes, adsorption of oxygen and hydrocarbon species on catalyst surfaces, the nature of these adsorbed species and kinetic models useful in describing catalytic oxidation processes will be explored. Methods for the testing and physicochemical analyses of catalyst materials will also be presented, although it should be pointed out that there will not be an elaborate presentation of surface analytical methods. The emphasis throughout this chapter is on those fundamental principles and basic characterization techniques essential for a full understanding of oxidation catalysis. Where a more detailed elaboration of a special characterization topic, beyond that presented in Chapter 2, is required, this will be presented within the relevant chapter. For a more general introduction to heterogeneous catalysis, the reader is referred to recent publications that cover in more detail fundamental principles of heterogeneous catalysis and modern characterization techniques (Thomas and Thomas, 1997; Niemantsverdriet, 1997; Farrauto and Bartholomew, 1997).

Chapter 3 serves as a general introduction to the topic of heterogeneous catalytic oxidation, and covers generic features of oxidation reactions. The reader is introduced to the concept of active sites, the nature and reactivity of oxygen species on heterogeneous oxidation catalysts, the concept of kinetic control of oxidation reactions, the mechanism of activation of hydrocarbons and the origins of selectivity in selective oxidation and ammoxidation reactions.

Chapters 4–10 are case studies of aspects of individual selective and total oxidation reactions, including the oxidation of propene to acrolein (Chapter 4), the oxidation of *n*-butane to maleic anhydride (Chapter 5), the epoxidation of ethylene (Chapter 6), catalytic combustion (Chapter 7), the ammoxidation of propene to acrylonitrile (Chapter 8), the oxychlorination of ethylene and the combustion of chlorinated volatile organic compounds (Chapter 9) and the application of heterogeneous oxidation catalysts for the production of fine chemicals (Chapter 10). It is not the intention to cover all commercial oxidation reactions that involve heterogeneous catalysts. Sufficient material is

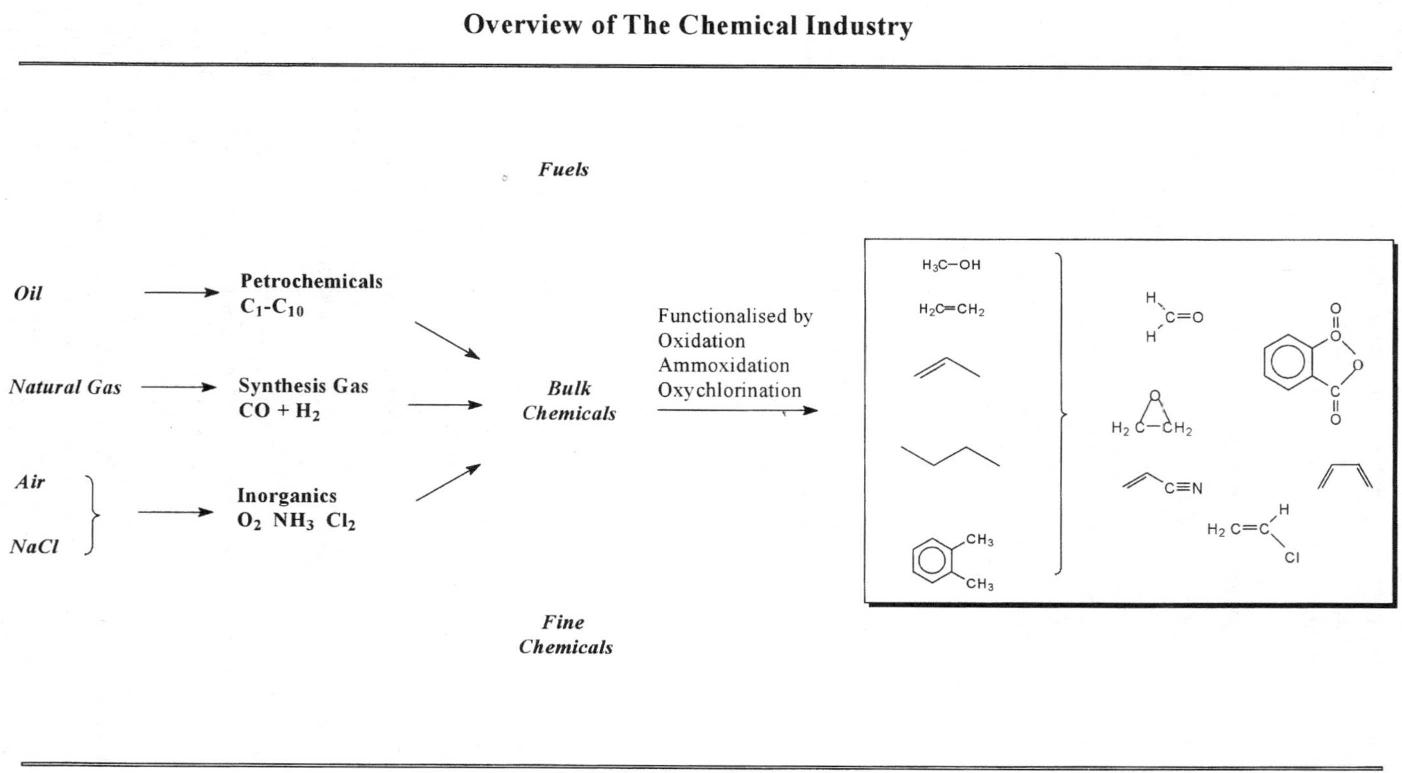


**Scheme 1.1** An overview of the role of oxidation processes in the chemical and petrochemicals industries.

presented in these chapters to elucidate the nature of heterogeneous oxidation catalysis and the general principles developed through these case studies can be applied throughout the field.

## 1.2 THE ROLE AND IMPORTANCE OF HETEROGENEOUS CATALYTIC OXIDATION PROCESSES IN THE CHEMICAL INDUSTRY

Oxidation processes that are driven by the heterogeneous catalysts can be divided into two groups, namely selective oxidation, ammoxidation and oxychlorination where the desired reaction product is not the most thermodynamically stable, and total oxidation reactions where the desired reaction products (simply carbon dioxide and water in the case of hydrocarbons) are the most stable thermodynamically. In general, the former type of reaction is practised for the production of bulk organic chemicals whereas the latter is practised for energy conversion or for pollution abatement.

Scheme 1.1 presents an overview of the chemical industry and the place therein of bulk organic chemicals produced by heterogeneous oxidation, ammoxidation and oxychlorination.

All major components for the chemical industry originate from a limited number of starting materials. Oil is the principle source of organics with natural gas as a source of synthesis gas (CO and $H_2$) becoming increasingly

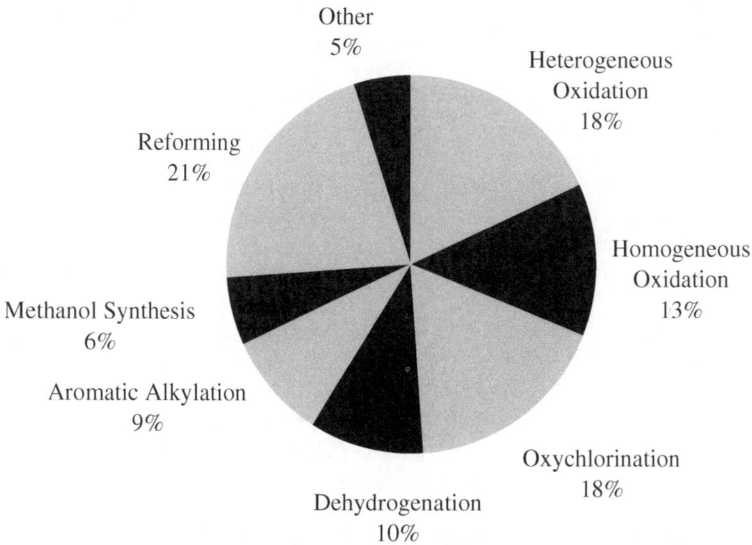


**Figure 1.1** Catalytic production of bulk organic chemicals in the United States in 1991 (Oyama *et al.*, 1993) (reproduced by permission of the American Chemical Society).

important. Ammonia is produced, through the Haber process, from the hydrogen component of synthesis gas and atmospheric nitrogen. Chlorine is sourced from the electrolysis of aqueous sodium chloride. Specifically, heterogeneous catalytic oxidation, ammoxidation and oxychlorination is concerned with those processes which add oxygen, nitrogen or chlorine or which remove hydrogen from $C_1$–$C_{10}$ hydrocarbons. Almost all the output from these processes is used as feedstock for polymer and plastics production.

About 66 % of organic chemicals production in the United States is carried out by catalytic means. Figure 1.1 presents a profile of the catalytic production of bulk organic chemicals in the United States for 1991. Heterogeneous oxidation, ammoxidation and oxychlorination forms nearly 40 %. In fact,

**Table 1.2** Production capacity and principal uses of the products of selective oxidation ammoxidation and oxychlorination processes
(Chemweek, 1997)

| *Chemical* | *World production/ktonnes yr$^{-1}$* | *Principal uses* |
|---|---|---|
| Formaldehyde | 18 000 | Urea resins (50 %)<br>Phenolic resins (10 %)<br>Polyacetal resins (10 %) |
| Ethylene oxide | 10 600 | Ethylene glycol (58 %)<br>Etholates (18 %)<br>Diethylene, triethylene glycol (7 %)<br>Ethanolamines (6 %) |
| Vinyl chloride monomer | 19 900 | Polyvinylchloride |
| Acrylonitrile | 4600 | Acrylic fibre (53 %)<br>ABS/SAN (30 %) |
| 1,3-Butadiene | 5700 | Styrene–butadiene rubber (29 %)<br>Butadiene rubber (27 %)<br>Adiponitrile (14 %)<br>Styrene–butadiene latex (12 %) |
| Maleic anhydride | 600 | Unsaturated polyester resins (63 %)<br>Lube oil additives (12 %)<br>Copolymers (6 %) |
| Phthalic anhydride | 2300 | Plasticizers (53 %)<br>Unsaturated polyester (26 %)<br>Alkyd resins (10 %) |

worldwide the catalytic oxidation, ammoxidation and oxychlorination business has a turnover of about 50 billion dollars per year (1998) and is fairly evenly distributed between America, Europe and Asia. The core of this technology is catalytic oxidation in which a hydrocarbon gas is mixed with air or oxygen (and with ammonia or hydrochloric acid for ammoxidation and oxychlorination, respectively) and passed over a solid catalyst usually at temperatures in the range 600–900 K.

There are about 15 specific selective oxidation, ammoxidation and oxychlorination reactions practiced commercially in about 1000 chemical plants globally. Some of the major commodity chemicals (average price of 0.6–1.0 $ per kilogram in 1997) produced by heterogeneous oxidation, ammoxidation and oxychlorination are presented in Table 1.2, with an indication of their main uses.

## REFERENCES TO CHAPTER 1

*Chemweek, 1997, Product Focus Reports* (http://www.chemweek.com/marketplace/prod_focus.html).

Farrauto RJ and Bartholomew CH (1997) *Fundamental of Industrial Catalytic Processes*. Blackie, London.

Niemantsverdriet JW (1997) *Spectroscopy in Catalysis. An Introduction*. VCH, Weinheim.

Oyama ST, Desikan AN and Hightower JW (1993) Research challenges in selective oxidation. *ACS Symposium Series 523, Catalytic Selective Oxidation*, Oyama ST and Hightower JW (Eds). American Chemical Society, Washington, pp. 1–14.

Thomas JM and Thomas WJ (1997) *Principles and Practice of Heterogeneous Catalysis*. VCH, Weinheim.

# 2

# Introduction to Heterogeneous Catalysis


2.1 Macroscopic and Microscopic Structure of Heterogeneous Catalysis 9
 2.1.1 Mechanical Properties of Catalysts 13
2.2 The Five Basic Steps in Heterogeneous Catalysis 14
2.3 Testing of Heterogeneous Catalysts 15
 2.3.1 Steady State Testing Procedures 17
 2.3.2 Transient Analysis Procedures 17
 2.3.3 Safety Considerations: Flammability of Hydrocarbon–Oxidant Mixtures 20
2.4 Diffusion 25
 2.4.1 External Diffusion 25
 2.4.2 Internal Diffusion 27
2.5 Adsorption at the Gas–Solid Interface 29
 2.5.1 Adsorption Isotherms 31
 2.5.2 Heats of Adsorption 36
 2.5.3 Nature of Adsorbed Species 42
2.6 Kinetics of Oxidation Reactions 44
 2.6.1 Mars and van Krevelen Kinetics 44
 2.6.2 The Eley–Rideal Mechanism 45
 2.6.3 The Langmuir–Hinshelwood Mechanism 48
 2.6.4 Power Rate Laws 49
2.7 Characterization of Oxidation Catalysts 50
 2.7.1 Catalyst Characterization by X-Ray Diffraction Methods 50
 2.7.2 Surface Area and Porosity Determination 53
 2.7.3 Selective Chemisorption Techniques 57
 2.7.4 Temperature Programming Methods 59
References to Chapter 2 62



Summary

Heterogeneous catalysis lies at the heart of many technologies for the production of petrochemicals, bulk chemicals and for pollution abatement. In this a chapter the newcomer to the field is introduced to the main macroscopic and microscopic structural features of heterogeneous catalysts, and a description of how oxidation catalysts are evaluated. The basic steps in all heterogeneous catalytic processes, namely diffusion, adsorption and surface reaction are also discussed. A number of kinetic rate laws, which find widespread application in oxidation catalysis are introduced as well as a small number of characterization techniques, necessary for a preliminary understanding of oxidation catalysis.

## 2.1 MACROSCOPIC AND MICROSCOPIC STRUCTURE OF HETEROGENEOUS CATALYSTS

During a typical heterogeneous catalytic oxidation reaction occurring at the gas–solid interface, a mixture of a hydrocarbon and oxygen is passed through a catalyst bed at an elevated temperature. A typical fixed bed reactor is shown schematically in Scheme 2.1 and this figure also shows the consumption of reactants and the generation of products as the gas mixture passes through the reactor. This apparently simple procedure involves a number of fundamental

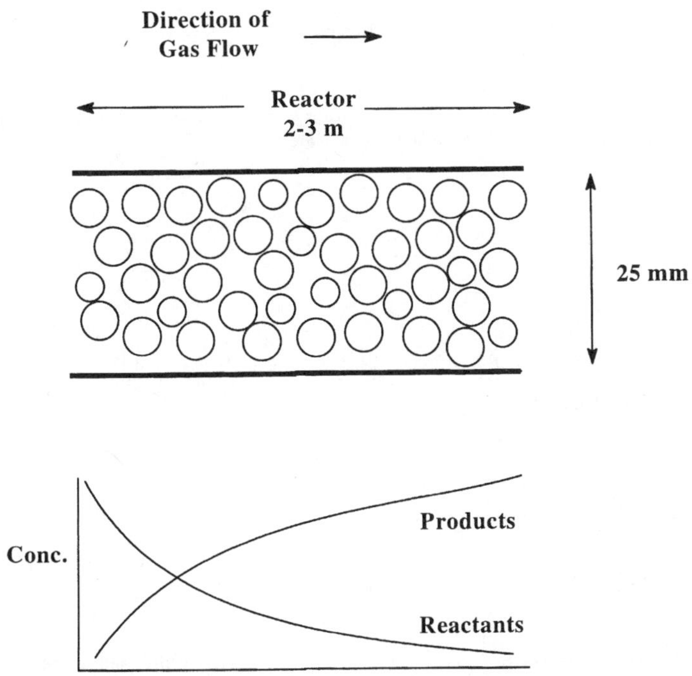


**Scheme 2.1** Schematic diagram of a fixed bed reactor packed with catalyst particles and an indication of the reactant and product profiles along its axial length.

engineering and chemical principles essential for a full understanding of heterogeneous catalytic oxidation. A necessary first step is to build up a picture of the composition of a typical heterogeneous catalyst.

The structure of heterogeneous catalysts can be understood from the perspective of a number of physical scales. Imagine that a catalytic process can be viewed under the microscope at increasing magnifications, such as is outlined schematically in Scheme 2.2.

The first scale to consider is that of an industrial reactor. A typical multi-tubular fixed bed reactor will comprise 30 000–60 000 individual tubes each about 25 mm in diameter and up to 6 m in length. A reactor of this size can accommodate up to 100 $m^3$ of catalyst. Other reactor configurations are also possible, such as fluidized bed and honeycomb reactors, and these will be

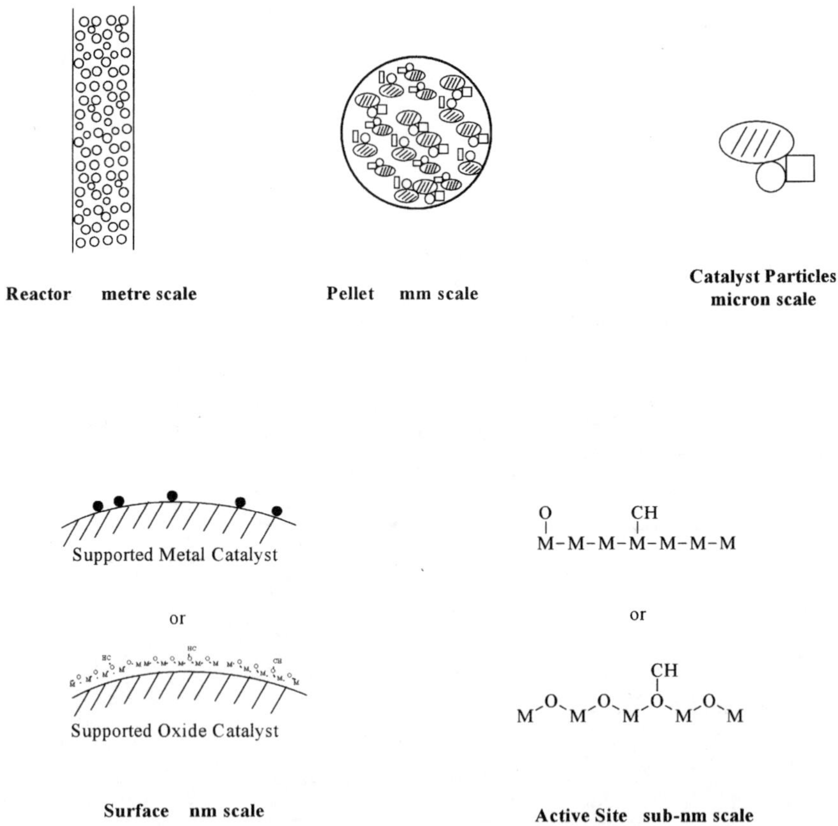


**Scheme 2.2** Illustration of dimensions important in concepts of heterogeneous catalysts.

dealt with in individual chapters of this book, but they are similar in size to the fixed bed reactors. A major engineering feature of these reactors is the sheer difficulty in pumping the required volume of gas through these reactors, and ensuring that it comes in intimate contact with the catalyst particles, without generating too great a pressure drop.

A fixed bed multi-tubular reactor may be filled with spherical pellets which appear on the 1–10 mm scale. Many other shapes are possible, such as cylinders and doughnuts. Spherical pellets are typically 1–5 mm in diameter, occasionally extending up to 10 mm. Uniform particle sizes are needed so that there is sufficient void volume between the pellets. When small particles are mixed with bigger ones there is a tendency for the smaller particles to plug the void volume that is needed for free passage of gas through the reactor bed.

The next scale shown in Scheme 2.2 concerns the composition of individual catalyst pellets. These are made up of catalyst particles in the form of a powder of typical dimension 1–50 μm. The powder is compacted to form the pellet and it may be necessary to add a binder if the natural adhesion of the powder particles is insufficient. The void volume between the particles now constitutes a type of porosity (usually macroporosity) and in addition, if the catalyst is for example a molecular sieve or a porous alumina, there may be some additional micro- or mesoporosity inside the catalyst particles. An essential feature of heterogeneous catalysis is that the reacting gas must diffuse into the interior of each catalyst pellet and be free to move around the outside, and in the case of porous materials, around the inside of the catalyst particles.

When each catalyst particle is looked at on a scale 1000 times smaller again, namely the nanometre scale, individual surface features begin to emerge (Scheme 2.2). For example, if the catalyst is a metal supported on an inorganic oxide, such as platinum supported on silicon oxide, the individual particles of the platinum metal will begin to appear on this scale. Typically, supported metal particles are about 1–10 nm in size, and will cover the surface of the oxide to a greater or lesser extent depending on how the catalyst was prepared.

Figure 2.1 presents electron micrographs of a number of oxidation catalysts. Palladium particles supported on $Al_2O_3$ (Fuchs *et al.*, 1991) and gold supported on $TiO_2$ (Haruta *et al.*, 1993) are shown in Figure 2.1. In each case the supported metal particles are in the range 2–10 nm. Unsupported metal oxide particles, often used in selective oxidation reactions, and also shown in Figure 2.1, tend to be larger than about 1 μm (Horowitz *et al.*, 1988).

When viewed on the final scale shown in Scheme 2.2, the sub-nanometre scale, individual active sites on the surface may be observed. Active sites are those areas of the surface where reactants can adsorb, where surface reactions occur and from which the final products of the reaction may desorb. This represents the atomic scale and the processes which occur on this scale are purely chemical.

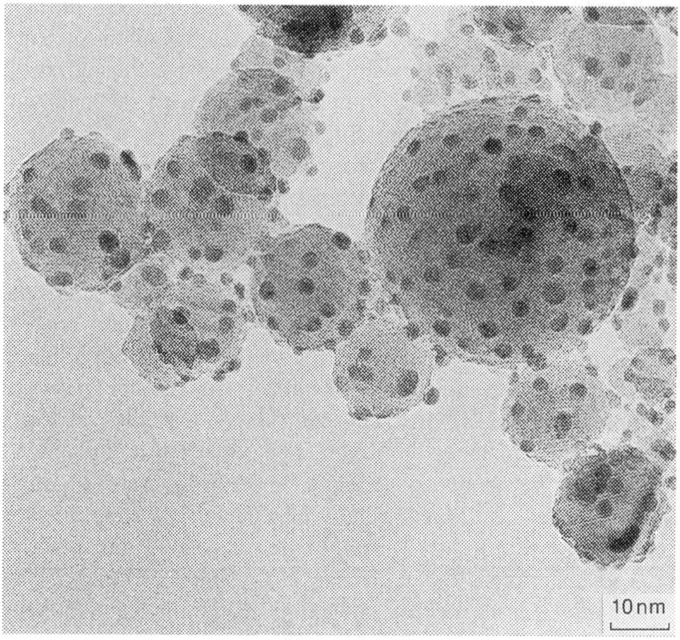


(a)

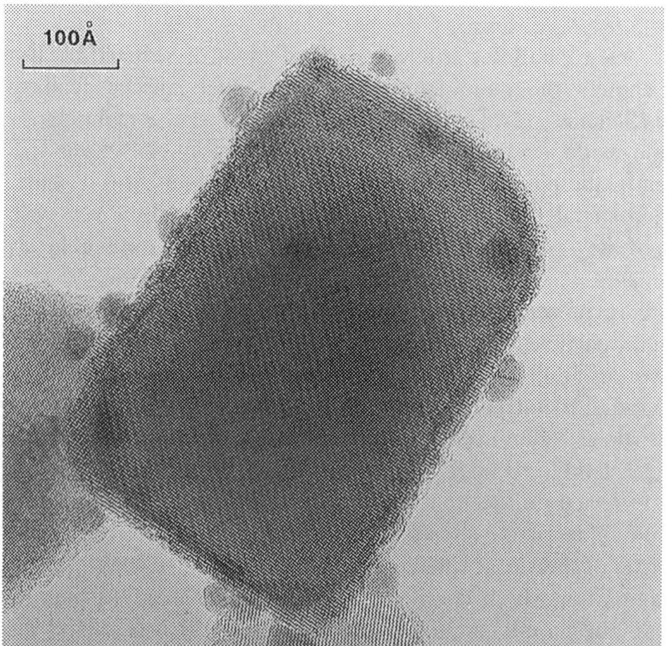


(b)

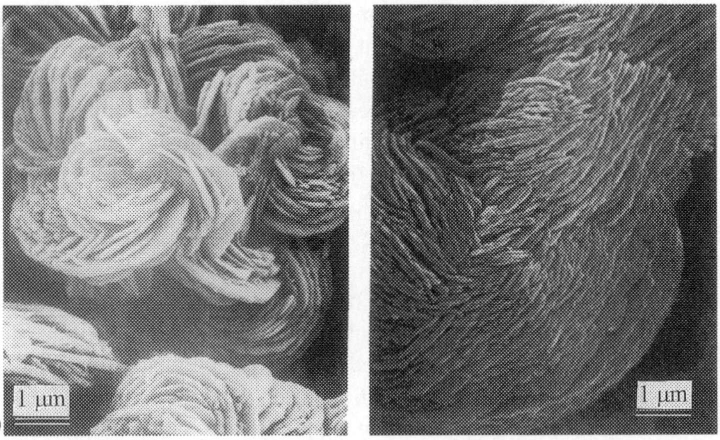


(c) (d)

**Figure 2.1** Transmission electron micrographs of (a) Pd supported on $Al_2O_3$ (Fuchs *et al.*, 1991) (reproduced by permission of the American Chemical Society), (b) Au supported on $TiO_2$ (Haruta *et al.*, 1993) (reproduced by permission of Academic Press), (c) and (d) scanning electron micrographs of an unsupported vanadium phosphorus oxide catalyst (Horowitz *et al.*, 1988) (reproduced by permission of Elsevier Science).

### 2.1.1 Mechanical properties of catalysts

All industrial catalysts are subjected to considerable thermal and mechanical strains, which can lead to their destruction as effectively as any chemical action. Catalysts are formulated as spheres or cylinders for use in fixed bed reactors and as powders of uniform particle size for use in fluidized and moving bed reactors. These materials must possess key mechanical properties which will allow them to operate without degradation in the reactor for which they have been designed.

Degradation of catalyst particles is manifested by the production of ‘fines’, namely any material that is less than about 2/3 the size of the original particles, but in practice tends to be as small as 1 μm. Fines can be parts of a support or small particles of the supported phase that become detached from the rest of the pellet. Fines can destroy the flow characteriztics of a reactor bed by clogging up interstitial spaces and valves.

Surprisingly, for fixed bed reactors, the strongest mechanical forces are experiences during transportation, loading, heating and cooling of catalysts. By contrast, mechanical forces during operation are much less important unless there is an upset in plant operation, such as a sudden pressure surge or change in temperature. For example, during heating, differing expansion coefficients for the reactor material (generally metals) and the various catalyst materials (oxides for the most part) can create some interesting effects. The combined effects of vibrations and expansion of the reactor may cause newly loaded catalyst particles to settle lower down in the reactor. When the system is cooled the catalyst particles may not be sufficiently fluid to relax back into their original position, and may be caught in a large lateral strain. Heating fresh catalysts can result in a large pressure buildup within pellets due water vaporization and stream generation. This phenomenon can lead to cracking of pellets. Supported phases can become detached from supports for a variety of reasons including loss of adhesion due to differing rates of expansion or drastic changes in molar volume as the supported phase is transformed from a precursor to an active form.

Study of the mechanics of granular media has not advanced sufficiently to allow a meaningful theoretical treatment of the phenomena described. Instead, catalyst manufacturers have developed their own empirical tests to assess whether their catalyst will withstand the strains described above. These tests cover resistance to crushing, abrasion and attrition. Crushing tests can be carried out on individual particles (and repeated many times to arrive at a statistically significant result) or on collections of particles. The apparatus in question is rather simple and involves applying a measured load to the sample being examined. A good rule of thumb is that crush strength needed to break a catalyst particle increases as the square of the particle diameter $d_p$.

Attrition is usually associated with fluidized bed operation. The most straightforward test is to place the sample to be examined in a small scale fluidized bed reactor operated with an appropriate flow of fluidizing gas. Any fines that are produced are elutriated through the top of the bed and collected. The percentage fines produced is determined and the suitability of the catalysts for eventual use on the industrial scale evaluated. There are other tests available also which rely on shaking a sample of the catalyst material and measuring the production of fines.

## 2.2 THE FIVE BASIC STEPS IN HETEROGENEOUS CATALYSIS

The fundamental processes occurring when a gas mixture is passed through a catalytic reactor may be described as follows:

Step 1 Diffusion of the reactants to the active site. This terms covers boundary layer diffusion and pore volume diffusion.
Step 2 Adsorption of at least one reactant on the catalyst surface.
Step 3 Surface reaction.
Step 4 Desorption of the products.
Step 5 Diffusion away from the active site.

The rates for steps 1 and 5 are determined by diffusion coefficients and porosity, whereas steps 2–4 are determined by chemical phenomena occurring at the molecular scale. Any one of these steps can represent the slow step of the overall process. Below, Steps 1 and 5 will be treated together as will steps 2 and 4.

Before looking at these steps in detail a concise description of the reactor and analytical system most often used to test oxidation catalysts will be presented.

## 2.3 TESTING OF HETEROGENEOUS CATALYSTS

The reaction rate per unit volume associated with many industrial catalytic processes falls within the limits $10^{-6}$–$10^{-5}$ mol cm$^{-3}$ s$^{-1}$. At reaction rates below the lower limit the process it drives is uneconomical, whereas operation above the limit leads to heat and mass transfer problems. This observation was made by Weisz and is presented in Scheme 2.3, which shows the range of reaction rates common to many industrial processes and compares these with the corresponding rates in biological and geological systems (Weisz, 1982).

Oxidation catalysts are often evaluated in a test rig such as that shown in Figure 2.2. The rig consists of three principal parts: (i) gas delivery system, (ii) reactor and (iii) the analytical system.

The individual gas flows are controlled by mass flow controllers. Gas mixtures are made up to the required composition and diluted, as appropriate, with helium or nitrogen. Normally the total gas pressure is 1 atm.

The reactor is normally designed to accommodate 0.1–5 g of catalyst and operates in the range 400–1100 K. The simplest reactors are usually straight quartz tubes with internal diameters of 6–10 mm where the catalyst sample is held in place between two plugs of quartz wool. In some circumstances quartz diluent is also added to the catalyst bed.

The analytical system usually comprises GC and sometime MS analysis. Regular calibration is essential for accurate work. An essential feature of the testing rig is the ability to analyse the feed gas mixture before it contacts the catalyst. To do this, a means of bypassing the reactor is incorporated into the gas delivery system.

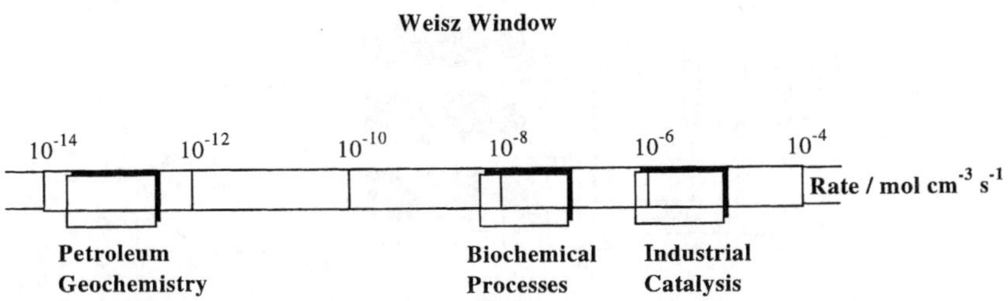


**Scheme 2.3** Weisz window for the operation of industrial catalysts (Weisz, 1983) (reproduced by permission of the American Chemical Society).

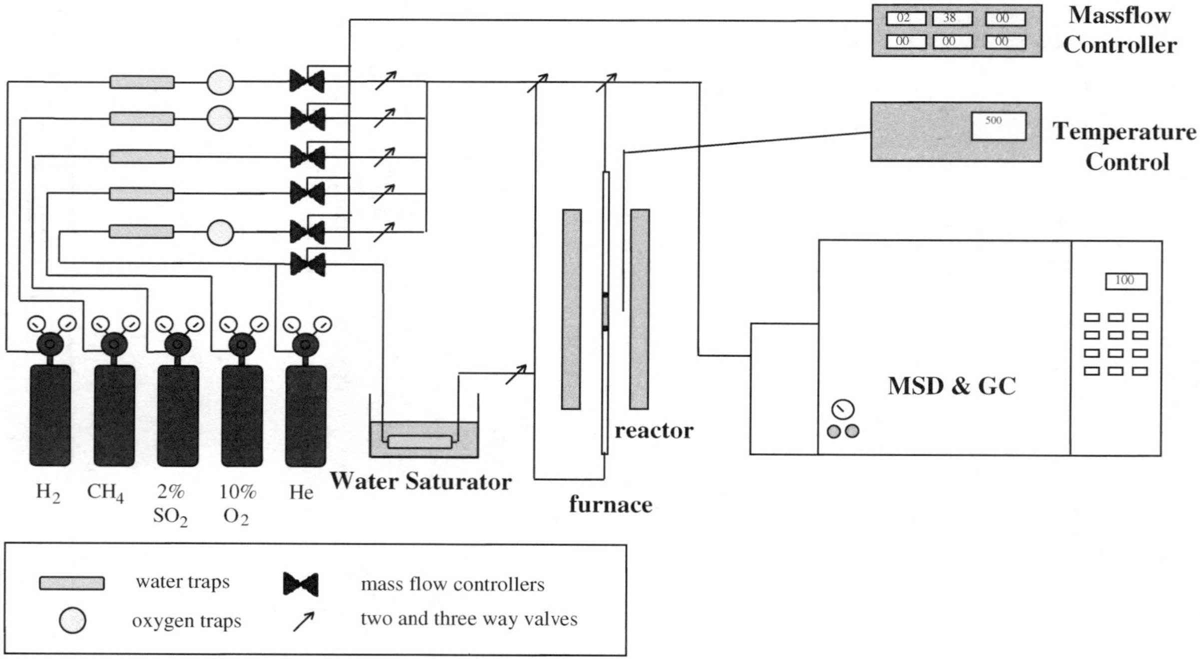


**Figure 2.2** Diagram of a typical catalyst testing facility for oxidation reactions.

For reasons of experimental flexibility, additional gas lines are sometimes incorporated into the design and a facility to introduce water vapour is usually included also.

### 2.3.1 Steady state testing procedures

The important parameters to follow during catalyst testing include conversion of reactants, selectivity to a specified product and yield of a specified product. It may also be necessary to determine reaction rates. Exact definitions of each of these terms is given in Table 2.1. Testing often involves increasing the reaction temperature and observing changes in conversion, selectivity and yield. A more thorough evaluation of the performance of a selective oxidation catalyst in particular is obtained by operating at a fixed reactor temperature and varying the function $W/F$, namely mass of catalyst divided by the total gas flow. In this way the conversion can be varied at a fixed temperature and in selective oxidations this results in a variation in selectivity. It should be realized that the most important property of selective oxidation catalysts is that they should exhibit a high selectivity to the desired product. A typical selectivity–conversion plot is shown in Figure 2.3(A) for two catalysts tested at the same reaction temperature and the corresponding yield–conversion plot is shown in Figure 2.3(B). It can be seen that for every value of conversion one catalyst is superior in terms of the selectivity and yield.

### 2.3.2 Transient analysis procedures

Transient techniques have found widespread application in catalysis research and a large number of individual techniques have been developed. The most advanced of these is temporal analysis of products (TAP). This technique was developed in the mid-1980s and brought for the first time sub-millisecond resolution to transient experiments over catalyst surfaces (Gleaves *et al.*, 1988). The technique involves injecting a gas pulse of very short duration

**Table 2.1** Definitions commonly used to describe catalyst properties

| | |
|---|---|
| % Conversion of a specified reactant: | $X_R = (F_R\|_{in} - F_R\|_{out})/F_R\|_{in}$ |
| % Selectivity to a specified product: | $S_P = F_P\|_{out}/(F_R\|_{in} - F_R\|_{out})$ |
| % Yield of a specified product: | $Y_P = S_P X_R = F_P\|_{out}/F_R\|_{in}$ |
| Space time e.g. (kg s mol$^{-1}$) | $W/F_R\|_{in}$ |
| Reaction rate, e.g. for a specified product (mol kg$^{-1}$ s$^{-1}$) | $F_P\|_{out}/W$ |

$F_R|_{in}$: molar flow of reactant at the inlet of the reactor mol s$^{-1}$.
$F_R|_{out}$: molar flow of reactant at the outlet of the reactor mol s$^{-1}$.
$F_P|_{out}$: molar flow of product at the outlet of the reactor mol s$^{-1}$.
$W$: mass of catalyst kg.

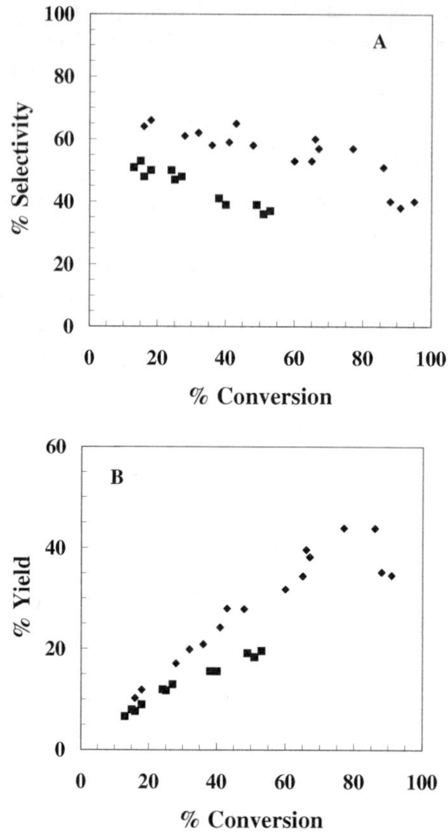


**Figure 2.3** (A) Selectivity to and (B) yield of maleic anhydride as a function of *n*-butane conversion over VPO catalysts with (◆) and without (■) promoters (Horowitz *et al.*,1988) (reproduced by permission of Elsevier Science).

(1–10 ms) into one end of a small cylindrical reactor and continuously evacuating the other end. Following reaction, the exit gas pulse, dissipated to a width of ca 500 ms, travels as a molecular beam through the differentially pumped vacuum system. Analysis of the reactor effluent is by mass spectroscopy. The ultimate resolution in terms of the width of the injected gas pulse can be as little as 10 μs. The reactor and analytical arrangement are presented in Figure 2.4. The main components of the apparatus are the special high speed valves used to deliver the gas pulses, the catalytic reactor, the differential pumping system and the mass analysis system.

Typically two types of experiment can be performed using this arrangement. In the first of these a pulse of reactants is passed through the reactor and the time at which the various reactants and products emerge from the

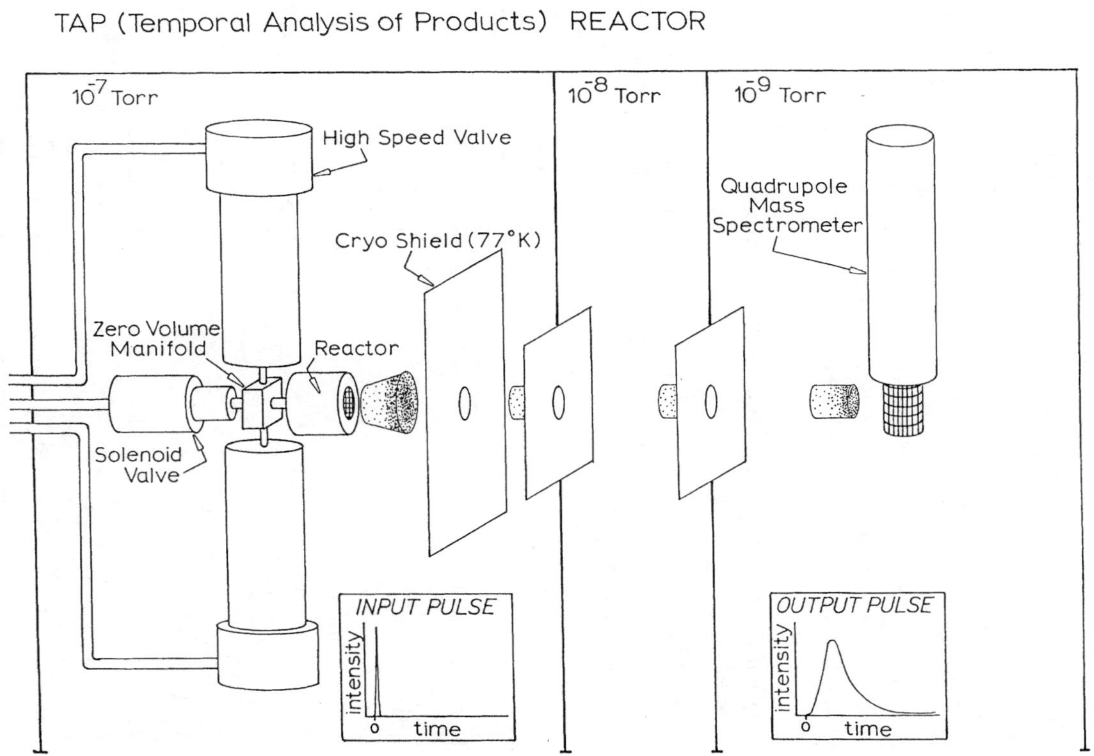


**Figure 2.4** Temporal analysis of products (TAP) system showing movement of a gas pulse from the microreactor to the mass detector (Gleaves *et al.*, 1988) (reproduced by permission of Marcel Dekker).

reactor is monitored. The essential logic is that for a sequential reaction with $A \rightarrow B \rightarrow C \rightarrow D$ the reactant should emerge from the reactor first, followed by the intermediate B, then C and finally the product D, because D requires longer than C to form and C in turn longer than B. This is demonstrated in Figure 2.5 for the oxidation of propene. In this case unconverted propene emerges first from the reactor, followed by acrolein, then carbon dioxide and water.

The second type of experiment is called a pump–probe experiment designed to determine the lifetime of reactive surface intermediates. Here a pulse of reactant A is injected, followed at selected intervals by a pulse of the second reactant B. For short lived intermediates formed from A the rate of reaction falls off rapidly when the interval between pulses is increased as a consequence of the degradation of short lived surface intermediate.

### 2.3.3 Safety considerations: flammability of hydrocarbon-oxidant mixtures

Of central importance in the safe testing and operation of oxidation catalysts is a knowledge and respect for the existence of flammability limits for combustible–oxidant mixtures.

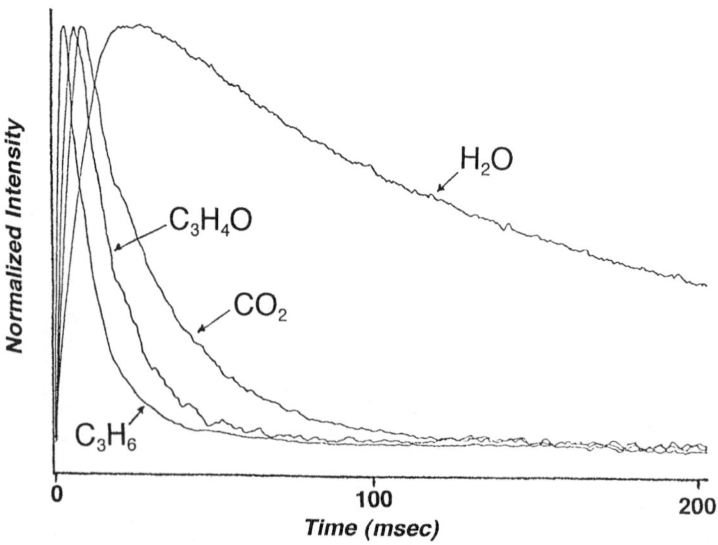


**Figure 2.5** TAP experiment showing the emergence of reactants and products during the reaction of propene and oxygen over $Bi_2MoO_6$ at 723 K (Gleaves *et al.*, 1988) (reproduced by permission of Marcel Dekker).

*Flammability limits in air*

Homogeneous combustible gas–air or vapour–air mixtures are flammable, namely they can propagate flame freely, only within a limited range of compositions. Typically, trace amounts of methane in air oxidize readily when in contact with a hot surface, but a flame will propagate from the ignition source at ambient temperature and pressure only if the surrounding mixture contains at least 5 but less than 15 vol% methane. The more dilute mixture is called the lower flammability limit (LFL) and the more concentrated is referred to as the upper flammability limit (UFL). In practice each combustible has flammability limits, but these limits are affected by temperature, pressure, direction of flame propagation and the surroundings (wall effects) (Zabetakis, 1965).

Table 2.2 presents a list of UFLs and LFLs for a range of gases and vapours in air at 298 K. As a general rule the LFL for most hydrocarbons is in the range 45–50 mg per litre of air. Expressed in volume per cent terms this means that methane at 5 % has the highest LFL of all the gaseous hydrocarbons whereas the C-4 hydrocarbons have LFLs below 2 %. (see Table 2.2). Generally useful rules of thumbs have been developed to relate flammability limits at 298 K to each other and to the stoichiometric composition ($C_{st}$):

$$\mathrm{LFL}_{298} = 0.55C_{st} \tag{2.1}$$

$$\mathrm{UFL}_{298} = 4.8{C_{st}}^{0.5} \tag{2.2}$$

$$\mathrm{UFL}_{298} = 6.5{\mathrm{LFL}_{298}}^{0.5} \tag{2.3}$$

Generally the flammability limits widen as temperature and pressure are increased, and become narrower as inert gases, such as carbon dioxide or nitrogen are added.

*The influence of temperature on flammability limits*

The influence of temperature on the LFL of a range of hydrocarbons is presented in Figure 2.6. The straight lines in this figure are given by:

$$\mathrm{LFL}_T/\mathrm{LFL}_{298} = 1 - 0.000784\ (T - 298\ \mathrm{K}) \tag{2.4}$$

The corresponding expression for the upper limit is:

$$\mathrm{UFL}_T/\mathrm{UFL}_{298} = 1 + 0.000721\ (T - 298\ \mathrm{K}) \tag{2.5}$$

For the lower limits when the ratio $\mathrm{LFL}_T/\mathrm{LFL}_{298}$ is plotted against temperature a single straight line results (Figure 2.7) which can be used to calculate

**Table 2.2** Summary of limits of flammability at 298 K and autoignition temperatures of individual gases and vapours in air at atmospheric pressure (Zabetakis, 1965)

| *Combustible* | *Limits of flammability/vol* | | *Autoignition temperature (AIT)/K* |
|---|---|---|---|
| | $LFL_{298}$ | $UFL_{298}$ | |
| Hydrogen | 4 | 75 | 673 |
| Deuterium | 4.9 | 75 | — |
| Ammonia | 15 | 28 | — |
| Carbon monoxide | 12.5 | 74 | — |
| Methane | 5 | 15 | 813 |
| Ethane | 3 | 12.4 | 788 |
| Ethene | 2.7 | 36 | 763 |
| Ethylene oxide | 3.6 | 100 | — |
| Acetylene | 2.5 | 100 | 578 |
| Vinyl chloride | 3.6 | 33 | — |
| Propane | 2.1 | 95 | 723 |
| Propene | 2.4 | 11 | 733 |
| Acrolein | 2.8 | 31 | 508 |
| Acrylonitrile | 3.0 | — | — |
| Acetone | 2.6 | 13 | 738 |
| 1,3-Butadiene | 2.0 | 12 | 693 |
| 1-Butene | 1.6 | 10 | 658 |
| 2-Butene | 1.7 | 9.7 | 598 |
| *n*-Butane | 1.8 | 8.4 | 678 |
| Isobutane | 1.8 | 8.4 | 733 |
| Isobutene | 1.8 | 9.6 | 738 |
| Benzene[a] | 1.3 | 7.9 | 873 |
| Toluene[a] | 1.2 | 7.1 | 753 |
| *n*-Nonane | 0.85 | — | 478 |
| 2,2,3,3-Tetramethylpentane | 0.80 | — | 703 |

[a] At 373 K.

values of $LFL_T$ for any hydrocarbon, provided the $LFL_{298}$ value is known. For example, for methane the ratio of $LFL_T/LFL_{298}$ at 873 K is 0.55 from Figure 2.7. The $LFL_{298}$ is 5 %, hence the $LFL_{873}$ is 5.0 vol% × 0.55 = 2.75 vol%.

*Mixtures of combustibles*

When homogeneous mixtures of combustibles in air are considered the LFL of the mixture can be calculated according to Le Chatelier's law, namely:

$$\mathrm{LFL}_{298} = 100 \Big/ \sum (C_i/L_i), \qquad \sum C_i = 100 \tag{2.6}$$

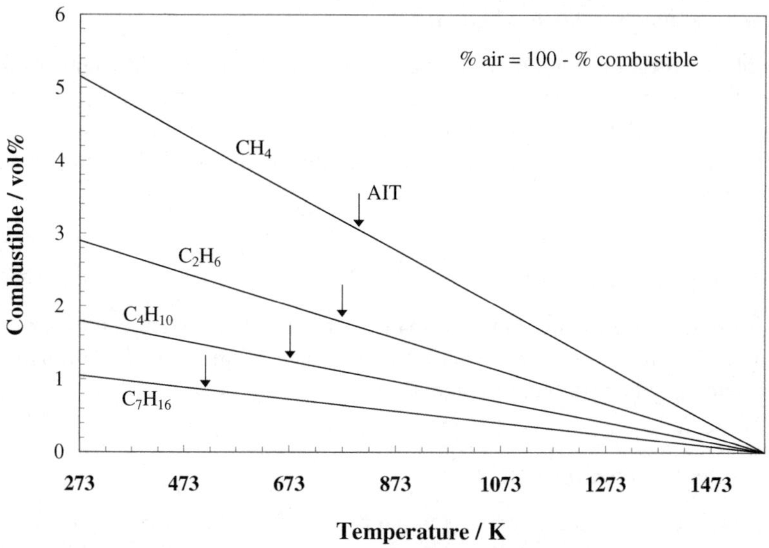


**Figure 2.6** The influence of temperature on the LFL for alkanes in air at 1 bar (Zabetakis, 1965).

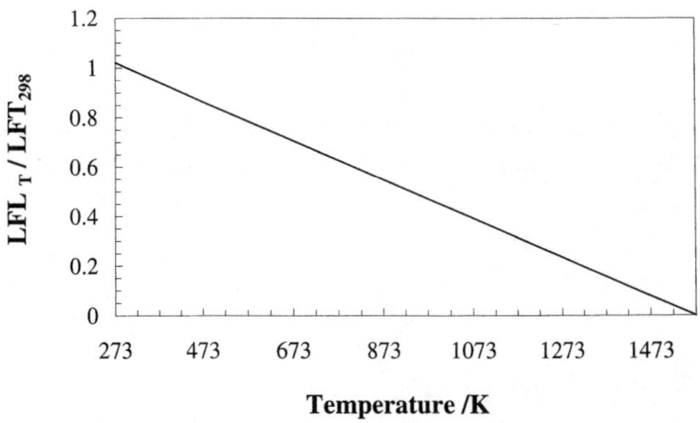


**Figure 2.7** Influence of temperature on the ratio $LFL_T/LFL_{298}$ for alkanes in air at 1 bar (Zabetakis, 1965).

where $L_i$ is the LFL of component $i$. Accordingly a mixture comprising 60 parts by volume methane, 25 parts by volume ethane and 15 parts by volume propane has a $LFL_{298}$ in air of

$$LFL_{298} = 100/(60/5 + 25/3 + 15/2.1) = 3.6 \text{ vol\%} \tag{2.7}$$

*Influence of inerts in flammability limits*

The introduction of an inert component significantly changes the flammability limits to a point where for many combustibles a flammable mixture is never possible when the inert component exceeds 50 vol%. This is illustrated in Figure 2.8 for propane–air mixtures to which have been added carbon dioxide or nitrogen.

*Influence of pressure*

The influence of pressure on the propane–air–nitrogen system can be see also in Figure 2.8 and shows that even in the presence of an inert the flammability range widens when pressure is increased.

*Autoignition temperature (AIT)*

The AIT is defined as the lowest temperature at which ignition can occur and values are presented in Table 2.2 for a range of combustibles. AIT may be correlated with molecular structure and for alkanes the average chain length of a hydrocarbon ($L_{ave}$) is important. The longer the average chain length the lower is the AIT. The $L_{ave}$ is defined as:

$$L_{ave} = 2\sum g_i N_i / M(M-1) \tag{2.8}$$

where $g_i$ is the number of possible chains within the hydrocarbon in question each containing $N_i$ carbon atoms and $M$ is the number of $CH_3$ groups.

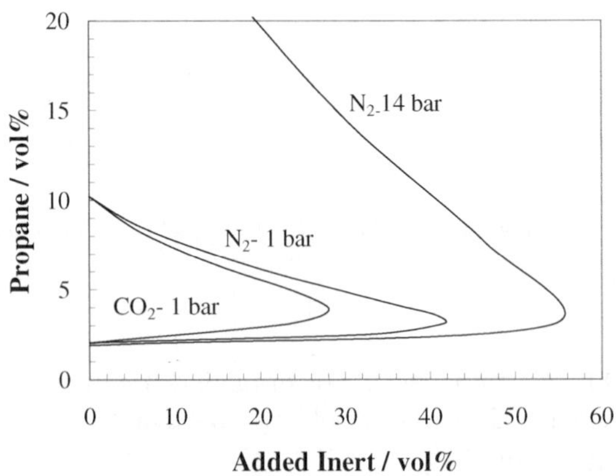


**Figure 2.8** Flammability limits for $C_3H_8$–$CO_2$–air and $C_3H_8$–$N_2$–air mixtures at 298 K and 1 bar and $C_3H_8$-$N_2$-Air at 14 bar (Zabetakis, 1965).

Comparing *n*-nonane and 2,2,3,3-tetramethylpentane, each having 9 carbon atoms, the former has two methyl groups and the latter has six; the normal alkane has just one chain of nine carbon atoms whereas the branched alkane has four chains of three carbon atoms, eight chains of four carbon atoms and three chains of five carbon atoms. Thus *n*-nonane has an $L_{ave}$ of nine whereas 2,2,3,3-tetramethylpentane has a $L_{ave}$ of 3.9 (Zabetakis, 1965).

## 2.4 DIFFUSION

Two diffusion processes need to be considered in heterogeneous catalysis, namely diffusion across the boundary layer that surrounds the catalyst pellet, referred to below as external diffusion, and diffusion inside the pore volume (internal diffusion). In the appropriate experimental conditions either of these processes can be rate limiting. Two situations may be envisaged, the first in which there is no significant concentration gradient through the boundary layer and throughout the pore volume. This situation arises when the diffusion steps are significantly faster than the chemical steps of adsorption, surface reaction and desorption. The second situation arises where there are significant concentration gradients. These gradients arise because the chemical steps are much faster than the diffusion steps so that, for example, reactants are consumed at a faster rate than they can be replenished by diffusion. Experimental studies that seek true chemical kinetics must be carried out in conditions that avoid the latter.

### 2.4.1 External diffusion

A boundary layer forms at a gas–solid interface owing to the accumulation of a stationary layer composed of reactants, products and diluents. The reactants and products must pass through this layer to contact the active catalytic surface. This layer offers resistance to flow thereby generating a concentration gradient across the layer. Diffusion through this layer is governed by Fick's laws of diffusion, so that

$$F = D\frac{\mathrm{d}c}{\mathrm{d}x} \tag{2.9}$$

where $D$ is the diffusion coefficient for mass transport through the boundary layer and $\mathrm{d}C/\mathrm{d}x$ the concentration gradient though a layer of thickness $x$.

Scheme 2.4 shows a boundary layer surrounding a catalyst pellet, and the types of concentration gradients, from steep to flat, that can be generated across this layer. The flow rate through this layer is determined by the diffusion coefficient, the concentration gradient and the external surface area of the catalyst particle:

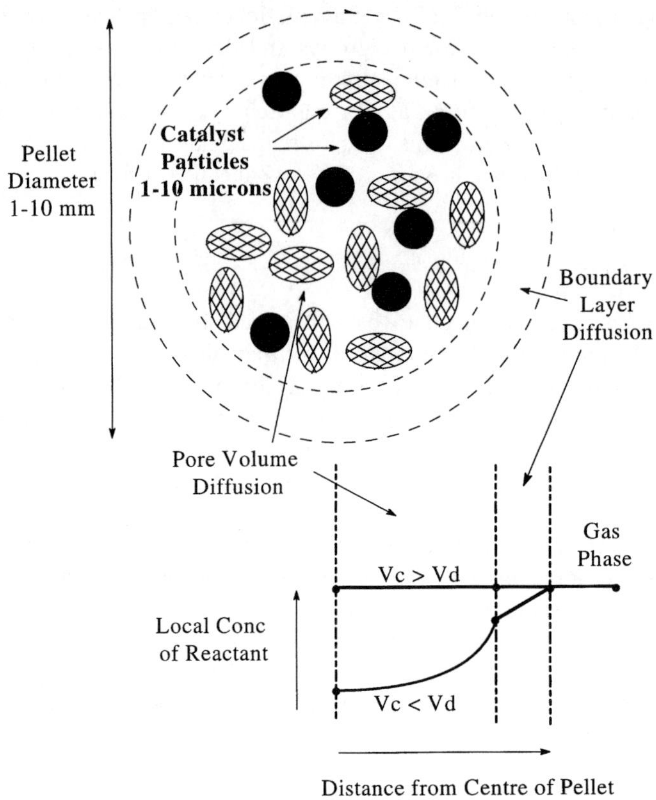


**Scheme 2.4** Test for external diffusion and schematic diagram of concentration gradient at the boundary layer. A decline in the percentage conversion at low superficial gas velocities is an indication that the slow step in the reaction conditions is transport of material across the boundary layer.

$$r = k_s A_s C_h \tag{2.10}$$

In fact for spherical particles the rate of transfer per gram of catalyst ($V$) can be expressed as:

$$V = 2.2 B_1 \frac{\mu^{0.6}}{\Sigma \rho_g d_p^{1.4}} C_h \tag{2.11}$$

where

$V$ is the transfer rate per unit of catalyst mass,
$B_1$ is a constant characteriztic of the gas,

$\Sigma$ is the void fraction of the catalyst bed,
$\mu$ is the superficial linear flow rate through the catalyst bed,
$C_h$ is the concentration of reacting gas,
$\rho_g$ is the density of the catalyst pellet,
$d_p$ is the pellet diameter.

This equation demonstrates a dependence of the rate of transfer across the external film boundary on the linear velocity of the reacting gas. An experiment can be readily devised to determine if the observed rate is limited by external diffusion. In this experiment the catalyst volume is changed over a factor of about 5. As the catalyst volume is increased the mass flow of gas is changed accordingly. As a result the contact time between catalyst and reactants remains constant. However, the linear velocity increases by the same factor as the mass of catalyst used. According to equation (2.11) the rate of diffusion across the external film boundary should be proportional to $\mu^{0.6}$ if the experiment is limited by external diffusion. As the linear velocity decreases the percentage conversion of reactants should decrease if external diffusion is the slow step in the overall process. This phenomenon arises because at very low superficial gas velocities a significant reactant concentration gradient develops, especially over those catalyst particles that are placed lower in the bed.

### 2.4.2 Internal diffusion

When the reactants have traversed the boundary layer they must then diffuse through the pore volume to arrive at an active site on the surface of the catalyst. Pore diffusion has additional complications brought about by the pore diameter. Resistance to flow in the pore volume comes about from the material already inside the pores. Two types of diffusion are identified, namely Knudsen diffusion where the average mean free path of the diffusing species is of the same order of magnitude as the pore diameter, and adsorbed phase diffusion, where gases condense inside the pores. Internal diffusion can lead to concentration gradients inside pores.

The most effective way of showing internal or pore diffusion effects in heterogeneous catalysis is through use of an effectiveness factor. This is defined as the ratio of the observed rate to the intrinsic rate, where the latter is the rate in the absence of diffusional limitations.

For pore volume diffusion the effectiveness factors is defined as:

$$\eta_1 = \frac{R_1}{R_1*} = \frac{3}{\phi_1}\left(\frac{1}{\tanh \phi_1} - \frac{1}{\phi_1}\right) \tag{2.12}$$

where $R_1$ is the observed rate and $R_1{}^*$ the rate in the absence of internal diffusion effects. $\phi_1$ is the Thiele modulus defined as:

$$\phi_1 = \frac{d\phi}{2}\left(\frac{k_i \rho_g S}{D_e}\right)^{0.5} \tag{2.13}$$

where

$k_i$ is the intrinsic rate constant,
$d_\phi$ is the diameter of a spherical pellet,
$\rho_g$ is the density of the pellet,
$S$ is the specific surface area,
$D_e$ is the coefficient of pore diffusion.

In general the effectiveness factor decreases as the Thiele modulus increases as shown in Figure 2.9 for isothermal conditions. Pore diffusion becomes a factor when the Thiele modulus is high, and this happens with large particles, high intrinsic rate constants, high surface areas, or low diffusion coefficients. Conversely diffusional effects do not distort the chemical kinetics when small particle sizes, low reactivities, low surface areas and high diffusion coefficients are encountered.

The shape of the effectiveness factor–Thiele modulus curve can be distorted by other contributing factors. In particular in the context of oxidation reactions that are almost always exothermic, heat transfer effect can come into play. In particular the combination of a highly exothermic reactions with low thermal conductivities of catalyst materials can give rise to the types of

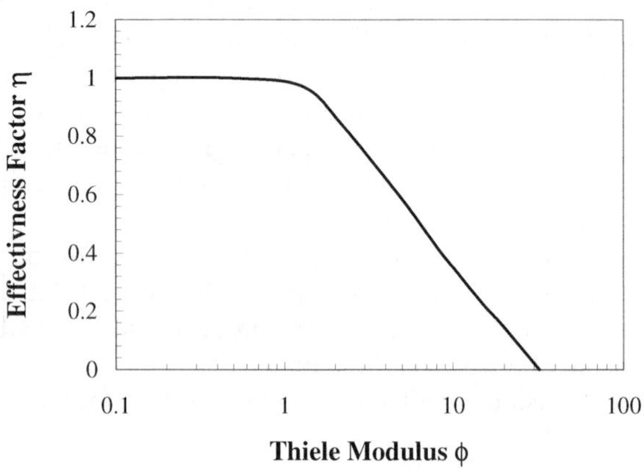


**Figure 2.9** Plot of effectiveness factor versus Thiele modulus for isothermal reaction conditions.

distortion whereby effectiveness factors considerably in excess of unity can occur, especially for low values of the Thiele modulus.

In practice, estimating effectiveness factors is probably only practical for large scale reactors where an abundance of parameters can be determined. For laboratory studies that seek accurate chemical kinetics it is essential to work in condition where the measured rate constants are not being falsified by the intrusion of diffusional effects. A simple experiments can be carried out to determine if pore diffusional effects are present. The Thiele modulus increases as the catalyst particle size increases, so pushing the effectiveness factor (Figure 2.9) towards lower values. The simplest test to apply is to measure the reaction rate with a range of particle sizes. A factor of 5 is usually sufficient to determine if internal diffusional effects are intruding. If diffusional effects are important the reaction rate per unit mass of catalyst should be less with the larger particles. In general the intrinsic rate constant is determined by carrying out the reaction being investigated with a finely divided powder (Le Page *et al.*, 1987).

## 2.5 ADSORPTION AT THE GAS–SOLID INTERFACE

The process whereby gaseous molecules (adsorbates) become attached to a solid surface (adsorbent) is known as adsorption. This terms also covers the adhesion of liquids on solid surface and gases on liquid surfaces.

Full characterization of an adsorption process requires an understanding of the extent to which the adsorbent is covered, the strength of interaction between the adsorbate and adsorbent and an understanding of the chemical nature of the adsorbed species. The extent of adsorption is usually characterized by measuring an adsorption isotherm, namely the amount of adsorption per unit mass of adsorbent is determined as a function of the applied pressure at a fixed temperature. The strength of interaction is determined from the heat of adsorption. The chemical nature of the adsorbing species is determined from physicochemical characterization.

In general adsorption types can be divided into two catagories, physical adsorption (physisorption) and chemical adsorption (chemisorption).

Physisorption is a process whereby gas molecules adsorb weakly on a solid surface without the formation of a chemical bond with the surface; the process tends to be very weakly exothermic, with heats of adsorption close to the corresponding heats of condensation. The extent of physisorption can be very large in some circumstances amounting to multiple layers of adsorbate. Because the bond with the surface is very weak, the adsorbed species readily desorbs resulting in a short average residence time on the surface. Realistically, physisorption is distinguished from elastic collisions with the surface on the basis of residence time. A typical time for a molecular vibration is $10^{-13}$ s. Physisorption is possible only where the residence time is sufficiently long to allow several molecular vibrations during the time the molecule spends on the surface. Typical residence times for physisorption are in the range $10^{-12}-10^{-6}$ s. The energetics of the physisorption process is shown in Scheme 2.5 in terms of the energy of a diatomic molecule as it approaches a solid surface onto which physisorption is possible. A shallow minimum in energy is observed at the point where physisorption occurs (Gasser, 1985).

Chemisorption involves the formation of a distinct chemical bond between the adsorbate and the solid surface; during chemisorption molecules can dissociate and heats of adsorption are similar to the bond energies for chemical bonds. Charged species may also develop of which $O^-$ and $O_2^-$ are especially important in oxidation reactions. However, the extent of adsorption tends to be limited to a small fraction of a monolayer. Residence times for chemisorbed species can be very long owing to the strong bonding between the surface and the adsorbate. Generally chemisorption is an activated process and Scheme 2.5 illustrates the point at which activation from the physisorbed to the chemisorbed state is required. Note also the much closer approach to the surface and the deeper energy well associated with chemisorption. Chemisorption is then usually divided into associative and dissociative. In the former the adsorbing molecule remains intact, whereas at least one bond in the adsorbing molecule ruptures during dissociative chemisorption.

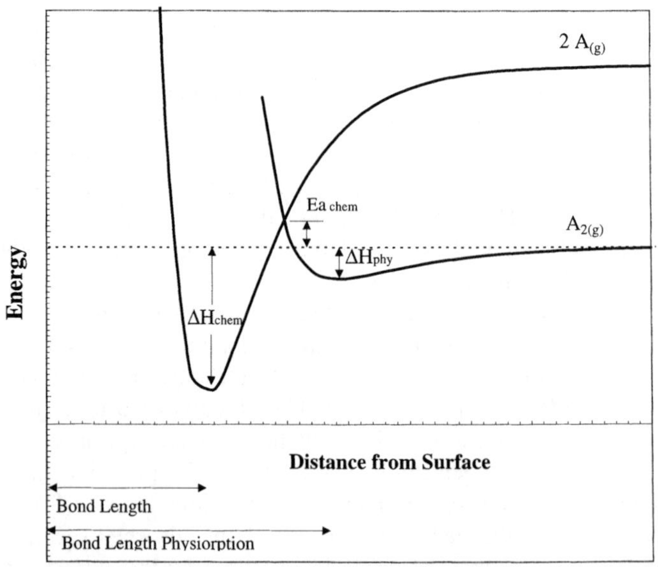


**Scheme 2.5** Potential energy diagram for the approach of a diatomic molecule towards a surface. The depths of the physisorbed and chemisorbed minima give the corresponding heats of adsorption and the point of crossover between the chemisorption and physisorption curves determine the activation energy for chemisorption.

Some of the main characteriztics of the two types of adsorption are summarized in Table 2.3.

### 2.5.1 Adsorption isotherms

The extent to which an adsorbent will cover a particular surface depends strongly on the temperature and pressure at which the measurement is made; generally the amount adsorbed decreases with increasing temperature but increases with increasing pressure for physisorption and chemisorption. In most experimental studies the temperature is held constant and the amount of adsorption is measured as a function of the gas phase pressure. This measurement is called an isotherm. Experimentally determined adsorption isotherms generally fall into five catagories, each identified from its distinctive shape (Gregg and Sing, 1982). These are outlined in Figure 2.10. Several of the

**Table 2.3** Characteriztics of physisorption and chemisorption

| | *Physisorption* | *Chemisorption* |
|---|---|---|
| Fractional coverage | $\theta > 1$ | $\theta \ll 1$ |
| Heat of adsorption/kJ mol$^{-1}$ | 0 to (−80) | (−100) to (−400) |
| Residence time/s | $10^{-6}$–$10^{-12}$ | $10^{-6}$–$10^{8}$ |

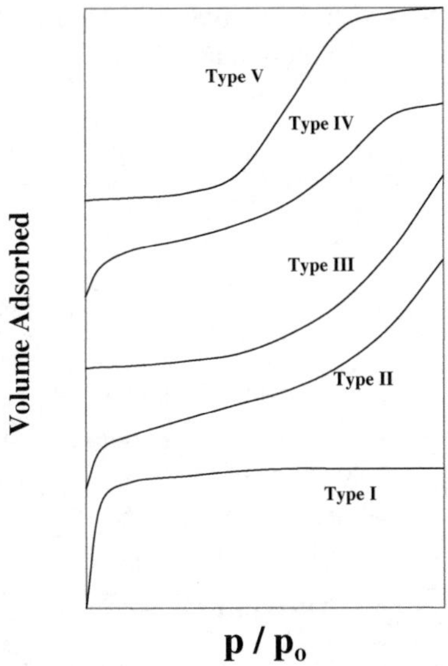


**Figure 2.10** The five types of adsorption isotherms.

isotherms also exhibit hysteresis whereby the shape of the isotherms obtained is different depending on whether the isotherm is established by progressive addition or removal of gas from the system. Generally, hysteresis is observed at low temperatures where physisorption is predominant and is associated with the filling and emptying of pores as the gas pressure over the sample is varied (Lower equilibrium pressures are required to empty pores, owing to capillary effects, so that the upper branch of the hysteresis loop is associated with the removal of gas from the system.) Generally, the five types of isotherm have been observed most often over solids of varying porosity. Type II isotherms form the basis of the famous BET model for surface area determination and porosity analysis elaborated on fully in Section 2.7.2.

Figures 2.11 shows some typical experimental isotherms for $N_2$, $C_3H_8$ and $O_2$ (Valenzuela and Myers, 1989) on a range of high surface area microporous materials. A common feature is a marked reduction in amount adsorbed as the temperature is increased. In addition, many of the isotherms seem to be heading for a limit in terms of the amount that can be adsorbed at any given temperature. Generally, the phenomenon of multiple layers of adsorbate only occurs at the gas–solid interface at temperatures close to the condensation temperature for the adsorbate and at high gas pressures.

One of the most widely used theoretical descriptions of adsorption was developed by Langmuir, to describe the situation where a gas adsorbs onto a fully uniform solid surface (Laidler, 1965). In this theory each adsorption site is regarded as identical, there is no interaction between adjacent sites and there is an equilibrium between the gas phase and adsorbed species. The situation is outlined schematically in line 1 of Table 2.4 for associative adsorption of oxygen onto a surface composed of uniform adsorption sites. Because all adsorption sites are regarded as identical and because of the lack

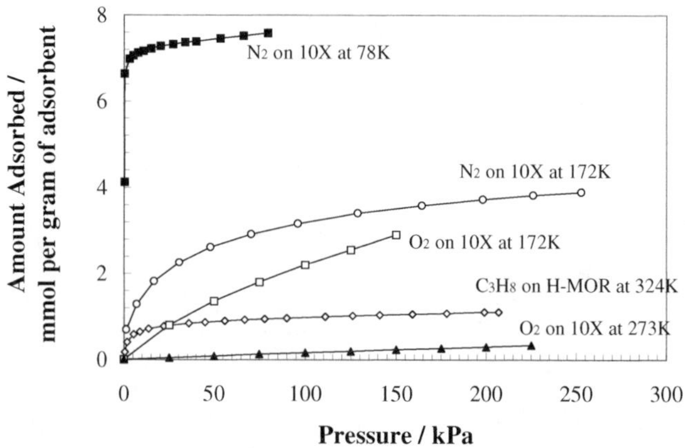


**Figure 2.11** Adsorption isotherm for $N_2$ and $O_2$ on 10X Linde molecular sieve and $C_3H_8$ on H-Mordenite at the temperatures indicated (Valenzuela and Myers, 1989).

**Table 2.4** Variations of the Langmuir adsorption isotherm

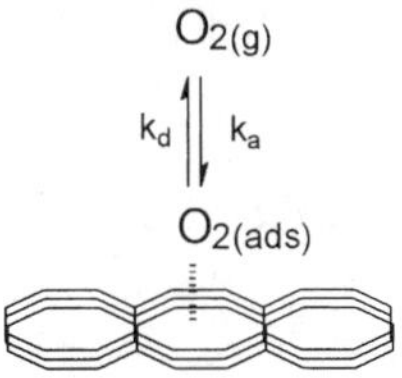


$$\theta_{O_2} = K_{O_2} p_{O_2}/(1 + K_{O_2} p_{O_2})$$

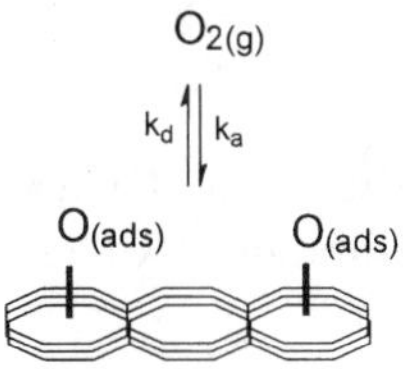


$$\theta_{O_2} = \mathrm{K_O}^{0.5} p_{O_2}{}^{0.5}/(1 + \mathrm{K_O}^{0.5} p_{O_2}{}^{0.5})$$

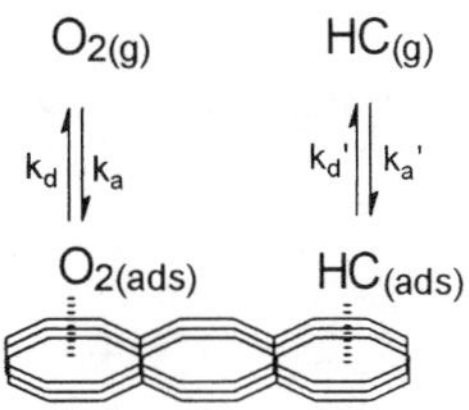


$$\theta_{O_2} = K_{O_2} \mathrm{p}_{O_2}/(1 + K_{O_2} p_{O_2} + K_{HC} p_{HC})$$
$$\theta_{HC} = K_{HC} p_{HC}/(1 + K_{O2} p_{O_2} + K_{HC} p_{HC})$$

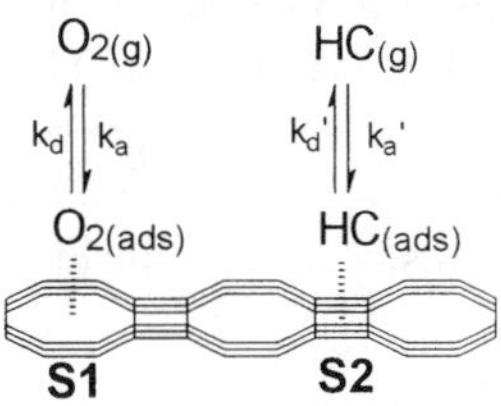


$$\theta_{O_2} = K_{O2} p_{O_2}/(1 + K_{O_2} p_{O_2})$$
$$\theta_{HC} = K_{HC} p_{HC}/(1 + K_{HC} p_{HC})$$

of interaction between adjacent sites the theory predicts that the heat of adsorption does not vary as the fractional coverages ($\theta$) changes. In reality, heats of adsorption vary widely with fractional coverage (see Section 2.5.2).

According to the Langmuir theory two types of surface site can be distinguished, namely those occupied by an adsorbed species and those unoccupied. The fraction of occupied sites is given by $\theta$ leaving $1 - \theta$ as the fraction of unoccupied sites. At equilibrium the rates of adsorption and desorption are equal, as are the heats of adsorption and desorption. The rate of adsorption is determined by the rate coefficient for adsorption, the partial pressure of adsorbate and the fraction of unoccupied sites, whereas the rate of desorption is determined by the rate coefficient for desorption and the fraction of occupied sites. Thus for an diatomic adsorbate $A_2$.

Rate of adsorption = Rate of desorption

$$k_a p_{A_2}(1 - \theta) = k_d \theta \tag{2.14}$$

or

$$\theta_{A_2} = K_{A_2} p_{A_2} / (1 + K_{A_2} p_{A_2}) \tag{2.15}$$

where $K = k_a / k_d$ (see Table 2.4).

Experimentally the term $\theta$ is defined as the amount of adsorption ($V$) at a particular pressure, divided by the maximum amount at saturation coverage ($V_m$), namely

$$\theta = V / V_m \tag{2.16}$$

Substitution and rearrangement gives the linear form of the Langmuir adsorption isotherm:

$$\frac{p_{A_2}}{V} = \frac{p_{A_2}}{V_m} + \frac{1}{KV} \tag{2.17}$$

The Langmuir description of dissociative chemisorption, is shown schematically in line 2 of Table 2.4 for a surface with uniform adsorption sites. The final expression for surface coverage by the adsorbing species is somewhat different because the adsorption step requires the presence of two unoccupied sites, and the desorption step requires two occupied sites. The equilibrium may be expressed as:

Rate of adsorption = Rate of desorption

$$k_a p_{A_2}(1 - \theta)^2 = k_d \theta^2 \tag{2.18}$$

and

$$\theta_A = K_A^{0.5} p_{A_2}^{0.5} / (1 + K_A^{0.5} p_{A_2}^{0.5}) \tag{2.19}$$

Many variations of the basic Langmuir model for adorption are possible. Two more are shown in Table 2.4. Line 3 shows competitive associative adsorption of a hydrocarbon and oxygen on the same type of surface site. The fractional coverage by one is inhibited by the other with an additional $Kp$ term appearing in the denominator of the expression. Line 4 of Table 2.4 shows the situation for a surface with two types of adsorption sites, one of which (S2) accommodates the adsorbing hydrocarbon and the other (S1) accommodates the oxygen, both adsorbing associatively.

Although there are many limitations associated with the Langmuir treatment of adsorption it still forms the basis for many of the kinetic expressions currently used to describe heterogeneous reactions. These will be introduced in Section 2.6.

A particular problem associated with oxygen in catalysis is its tendency not just to physisorb or chemisorb, but also to react with the underlying solid. If this solid is a metal then a metal oxide is produced, and if the underlying solid is an oxide then the oxygen may cause it to be further oxidized or it may exchange with oxygen of the lattice. These processes are outlined in Scheme 2.6 for the interaction of oxygen with palladium metal supported on silicon oxide. The scheme shows physisorption of oxygen onto the palladium metal and onto the silica support, associative and dissociative chemisorption onto the palladium metal and physisorption of oxygen onto palladium oxide previously formed by reaction of chemisorbed oxygen with the palladium metal.

Experimentally it may be very difficult to distinguish between these very different processes. A detailed study of the adsorption of oxygen onto a catalyst comprising 2.5 wt% Pd supported on silica presents a useful example (Leon *et al.*, 1991). Depending on experimental conditions employed, oxygen can physisorb, chemisorb or react with the palladium metal to generate palladium oxide. All three processes may proceed simultaneously. At 143 K this system presents a standard adsorption isotherm (see Figure 2.12), with upwards of 450 μmol of oxygen adsorbed per gram of catalyst. Most of this can be easily removed from the sample by lowering the equilibrium gas pressure. There remains approximately the 50 μmol of oxygen which is strongly bound to the sample for all applied pressures at 143 K. This is referred to as irreversibly adsorbed oxygen (chemisorbed), whereas the read-

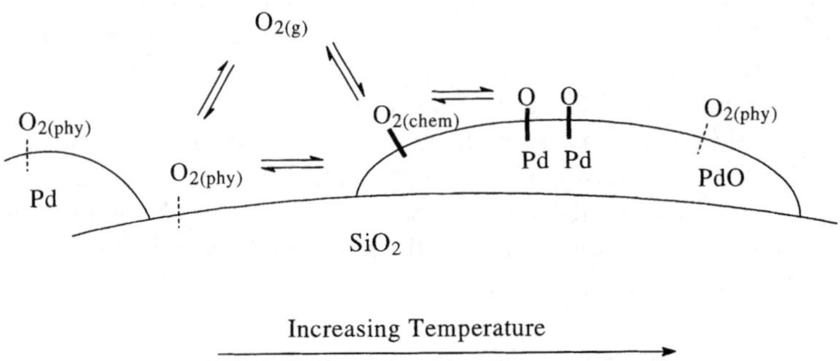


**Scheme 2.6** Schematic representation of the modes of interaction and reaction of oxygen with palladium metal supported on silica. As the temperature increases the mode of interaction can change from physisorption to associative chemisorption to dissociative chemisorption and finally to oxidation of the palladium metal.

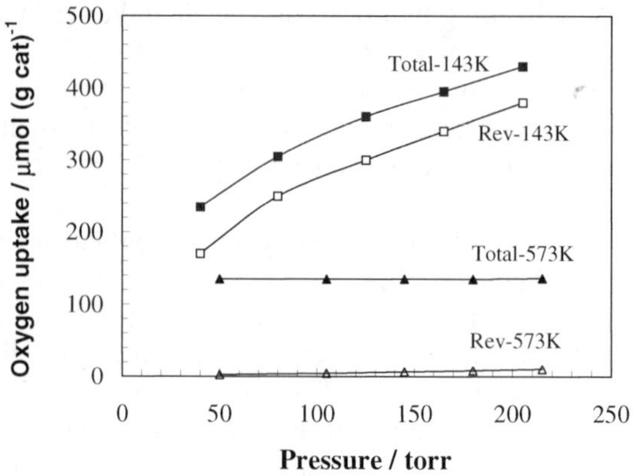


**Figure 2.12** Oxygen adsorption isotherms for adsorption onto 2.5% $Pd/SiO_2$ at the temperatures indicated. Each isotherm is presented as the total amount of oxygen adsorbed and as the amount reversibly adsorbed. By difference the amount irreversibly adsorbed can be calculated (Leon *et al.*, 1991) (reproduced by permission by Elsevier Science).

ily removed oxygen is referred to as reversibly adsorbed (physisorbed). The chemisorbed oxygen is probably associated with the palladium metal, whereas much of the physisorbed oxygen may be attached to the silica surface.

When the same experiment is repeated at 573 K the situation is completely different. Now the uptake of oxygen is less, as expected, but is nearly all of the chemisorbed kind. This experiment was repeated at a number of temperatures and the amounts of chemisorbed and physisorbed oxygen were determined for a single equilibrium pressure (200 Torr). In this way an adsorption isobar (defined as the amount adsorbed at constant pressure) was generated for both types of adsorbed oxygen. This is presented in Figure 2.13 and shows that the amount of physisorbed oxygen rapidly falls to zero just above 250 K, whereas the chemisorbed oxygen slowly increases as the temperature rises. The final measured value for the O/Pd ratio was close to 1, namely the value expected for bulk PdO (Leon *et al.*, 1991).

The overall process may be explained as follows: below 150 K physisorption predominates, with oxygen adsorbing weakly onto the palladium metal or onto the silica support. Above 250 K physisorption is nearly eliminated, but now the oxygen species can enter into the bulk of the palladium particles and form palladium oxide.

### 2.5.2 Heats of adsorption

Adsorption is a spontaneous process, namely one in which $\Delta G$ is negative. Furthermore adsorption is nearly always accompanied by a decrease in entropy, so it follows from

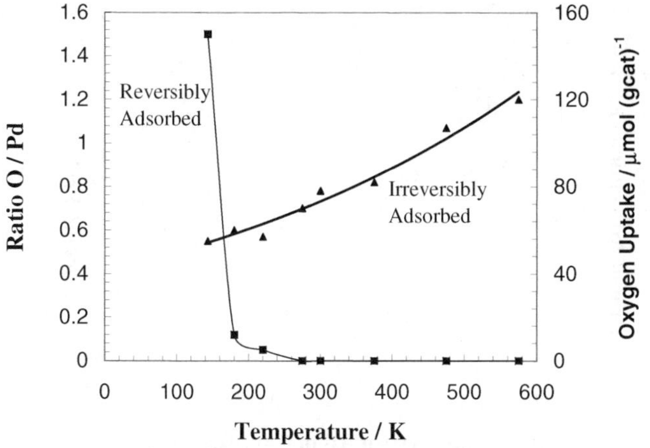


**Figure 2.13** Isobars for reversible and irreversible oxygen uptake at 27 kPa on 2.5 % $Pd/SiO_2$ (Leon *et al.*, 1991) (reproduced by permission by Elsevier Science).

$$\Delta G = \Delta H - T\Delta S \tag{2.20}$$

that adsorption processes are nearly always exothermic (Thomas and Thomas, 1997).

Many cases have now been documented where the value for the heat of adsorption varies appreciably with the amount of adsorption. Heats of adsorption can be measured on single crystal surface directly by microcalorimetry whereby a small amount of gas is exposed to the surface and the heat released for a given amount of adsorption is measured. In this way a profile can be built up of how the heat of adsorption varies with surface coverage.

Differential heats of adsorption for oxygen onto the (110) surfaces of platinum metal are shown in Figure 2.14 (Wartnaby *et al.*, 1996). The data illustrate that heats of adsorption are highest over the clean metal surface and fall off to a lower constant value as the surface coverage is increased. Non-uniform adsorption sites must play a role here as must interaction between adjacent sites. The high initial heats of adsorption can be associated with specific interactions between the most reactive surface sites onto which the first aliquots of adsorbate attach preferentially. Thereafter, heats of adsorption diminish and at high coverages repulsive interactions between adjacent sites must contribute to a reduction in heat of adsorption.

Meaningful comparisons of heats of adsorption can only be made when account is taken of the amount of adsorption that has occurred. These values of heats of adsorption are called isosteric heats of adsorption to indicate that the values quoted are for the same surface coverage. A point to note is that

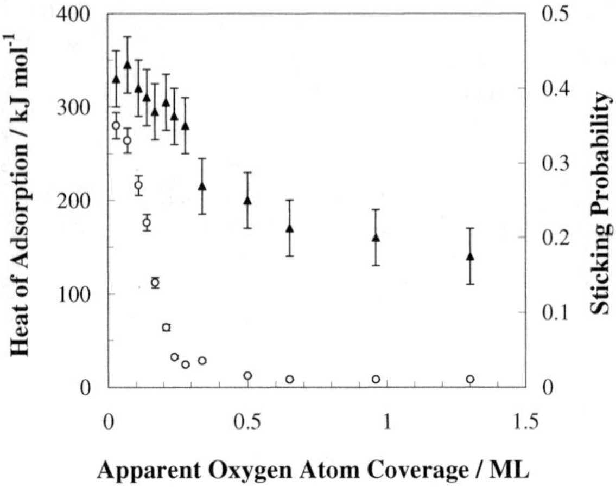


**Figure 2.14** Differential heats of adsorption (▲) and sticking probabilities (○) for oxygen on initially clean Pt(110) (Wartnaby *et al.*, 1996) (reproduced by permission of the American Chemical Society).

most studies of adsorption are carried out at much lower temperatures than those encountered in oxidation catalysis.

Some typical values for initial and final heats of adsorption are listed in Table 2.5. For the most part these values were determined by differential microcalorimatry or through use of the Claperyon–Clausius equation. The list is not exhaustive, but intended to illustrate the range of heats of adsorption possible and differences between initial values measured on clean surfaces and the corresponding values close to saturation coverage. Physisorbed species can be identified from the rather low range of heats of adsorption and the high coverages possible.

A second feature readily measured in microcalorimetric work is variations in sticking coefficient with coverage. The sticking coefficient is defined as:

$$s = \text{rate of gas adsorption/rate of collision with the surface}$$

and measures the probability that a particular collision will produce an adsorbed species. Figure 2.14 illustrates how sticking coefficients for oxygen on Pt(110) vary with surface coverage, being generally high at low coverages and tending towards zero even before 50 % coverage is achieved. In this regard many systems follow the Kisliuk equation developed to describe a role for a precursor species (not formally equivalent to a physisorbed species) in equilibrium with the gas phase. Chemisorbed species form from the precursor (Gasser, 1985):

$$\text{Gas} \rightleftharpoons \text{Precursor} \rightarrow \text{Chemisorbed species} \tag{2.21}$$

There are three possible routes whereby the precursor state can advance, namely by chemisorption, desorption or by migration to another site. The overall probabilites for chemisorption, desorption and migration from the precursor state depend on the fractional coverage by chemisorbed species. The probabilities of chemisorption, desorption and migration determine the final form of the Kisliuk equation. For single site adsorption the equation takes the following form:

$$\frac{s}{s_0} = \left(1 + \frac{\theta K}{1 - \theta}\right)^{-1} \tag{2.22}$$

where $s$ is the sticking probability, $s_0$ the sticking probability at zero surface coverage, $\theta$ is the fractional coverage and $K$ is the equilibrium constant.

For dual site adsorption the Kisliuk equation is:

$$\frac{s}{s_0} = \frac{(1 - \theta)^2}{1 - \theta(1 - K_1) + \theta^2 S_0} \tag{2.23}$$

Figure 2.15 shows the predicted shape of the plot of relative sticking coefficients as a function of coverage for a number of different values of $K$ for single site adsorption.

It should be noted that only the relative ratio of $k_d^*/k_a$, namely the value of $K$, determines the shape of the plot of sticking coefficient versus fractional coverage. As $K$ decreases chemisorption becomes more favoured, particularly at high coverages.

The proposal of a pre-equilibrium state for chemisorption is particularly useful in explaining the diminishing amounts of chemisorption observed as the adsorption temperature is increased. The chemisorptive state is fed by a pool of precursor species, which are only weakly held on the surface. Increasing the temperature favours desorption more than chemisorption, hence a reduced amount of chemisorption as the temperature is increased (Gasser, 1985).

Heats of adsorption can also be useful in estimating the residence time of a particular species on the surface. The theoretical background simply states that desorption can be achieved via a molecular vibration sufficiently energetic to overcome the bonding with the surface. Two approaches have been used to estimate surface residence times. The first of these was developed by Frenkel (Thomas and Thomas, 1997):

$$\tau = \tau_0 \exp(\Delta H/RT) \tag{2.24}$$

**Table 2.5** Heats of Adsorption

| *Adsorbent* | *Adsorbate* | *Temperature range/K* | *Method*[a] | $\Delta H_{ads}$ *Initial kJ mol*$^{-1}$ | $\Delta H_{ads}$ *final kJ mol*$^{-1}$ | *Final amount adsorbed mmol (g adsorbent)*$^{-1}$ | *Reference* |
|---|---|---|---|---|---|---|---|
| H-Mordenite | $C_3H_8$ | 283–324 | CC | 60 | 30 | 1.25–1.1 | Valenzuela and Myers, 1989 |
| 10X-Linde | $O_2$ | 172–273 | CC | 15 | 13 | 3.0–0.3 | Valenzuela and Myers, 198 |
| 10X-Linde | $N_2$ | 78–273 | CC | 5 | 8 | 7.6–0.8 | Valenzuela and Myers, 198 |
| 10X-Linde | CO | 172–273 | CC | 15 | 20 | 4.8–1.4 | Valenzuela and Myers, 198 |
| 0.5% Pd/$Al_2O_3$ | $H_2$ | 323 | Cal | 110 | 30 | $\theta = 0.95$ | Cordona-Martinez and Dumesic, 1992 |
| 1% Pt/$SiO_2$ | $H_2$ | 300 | Cal | 83 | 30 | 0.02 | Cordona-Martinez and Dumesic, 1992 |
| 1% Pt/Na-Y | $H_2$ | 390 | Cal | 145 | 40 | 0.019 | Cordona-Martinez and Dumesic, 1992 |
| Pt(111) | $O_2$ | 300 | Cal | 340 | 120 | $\theta = 0.7$ | Yeo *et al.*, 1997 |
| Pt(110) | $O_2$ | 300 | Cal | 335 | 140 | $\theta = 1.2$ | Wartnaby *et al.*, 1996 |
| 5% Rh/C | $O_2$ | 300 | Cal | 375 | 20 | 0.023 | Cordona-Martinez and Dumesic, 1992 |
| Pt | $C_2H_4$ | 300 | Cal | 148 | 100 | 0.016 | Cordona-Martinez and Dumesic, 1992 |
| Pt | $C_3H_6$ | 300 | Cal | 177 | 100 | 0.016 | Cordona-Martinez and Dumesic, 1992 |
| Pt | $CH_4$ | 300 | Cal | 151 | 50 | 0.004 | Cordona-Martinez and Dumesic, 1992 |
| Pt | $C_2H_6$ | 300 | Cal | 220 | 100 | 0.005 | Cordona-Martinez and Dumesic, 1992 |

| | | | | | | | |
|---|---|---|---|---|---|---|---|
| H-Y | $NH_3$ | 298 | Cal | 105 | 60 | 4.0 | Cordona-Martinez and Dumesic, 1992 |
| H-Y | Pyridine | 303 | Cal | 130 | 120 | 1.0 | Cordona-Martinez and Dumesic, 1992 |
| H-Y | $CO_2$ | 298 | Cal | 80 | 10 | 0.06 | Cordona-Martinez and Dumesic, 1992 |
| ZSM-5 | $NH_3$ | 473 | Cal | 121 | 65 | 0.45 | Cordona-Martinez and Dumesic, 1992 |
| ZSM-5 | Pyridine | 473 | Cal | 235 | 95 | 0.48 | Cordona-Martinez and Dumesic, 1992 |
| ZSM-5 | $CO_2$ | 298 | Cal | 96 | 2 | 0.09 | Cordona-Martinez and Dumesic, 1992 |
| $SiO_2$ | Pyridine | 473 | Cal | 95 | 92 | 0.03 | Cordona-Martinez and Dumesic, 1992 |
| $Al_2O_3$ | $H_2O$ | 373 | Cal | 220 | 30 | 2.0 | Cordona-Martinez and Dumesic, 1992 |

* Cal refers to microcalorimetry and CC to the Clapeyron–Clausius equation:

$$\left(\frac{\delta \ln p}{\delta T}\right)_\theta = \frac{\Delta H_{\mathrm{ST}}}{RT^2} \text{ or } \left(\ln \frac{p_1}{p_2}\right)_\theta = -\frac{\Delta H_{ST}}{R}\left(\frac{1}{T_1} - \frac{1}{T_2}\right)$$

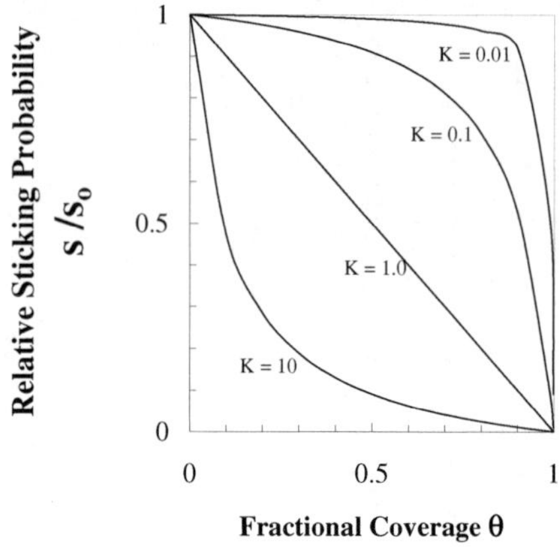


**Figure 2.15** Relative sticking probability as a function of surface coverage for a number of different values of $K$ in the Kisliuk expression (Gasser, 1985) (reproduced by permission of Oxford University Press).

and the second developed by Ertl and co-workers is:

$$\tau = \tau_0 \exp(-E_d/RT) \tag{2.25}$$

The two equations are very similar; the first takes the exponential of the heat of adsorption and the second uses the activation energy for desorption. In each case $\tau_0$ is the time for a single molecular vibration on the surface and the exponential term determines the probability that a particular vibration will have sufficient energy to overcome the attractive forces with the solid surface.

Enormous variations in residence times are possible according to these equations. Figure 2.16 shows variations in residence times for one physisorbed and one chemisorbed species at different temperatures. The residence time for a chemisorbed species with $\tau_0$ of $3 \times 10^{-13}$ s and a heat of adsorption of 125 kJ mol$^{-1}$ is approximately $1 \times 10^{11}$ s at 273 K and about $10^{-6}$ s at 973 K. For a physisorbed species the lifetime varies from $10^{-10}$ to $10^{-13}$ s for the same temperature range. The latter is less than the time for a single molecular vibration, and so should be regarded as a collision rather than an adsorption process. Alternatively, Figure 2.17 illustrates how residence time at 473 K varies with heat of adsorption. Physisorbed species feature residence times less than $10^{-6}$ s at 473 K and chemisorbed species have residence times greater than $10^{-1}$ s.

### 2.5.3 Nature of adsorbed species

Pioneering work in the 1970s has led to a very useful understanding of the nature of adsorbed species present on oxide surfaces. In this work the use of

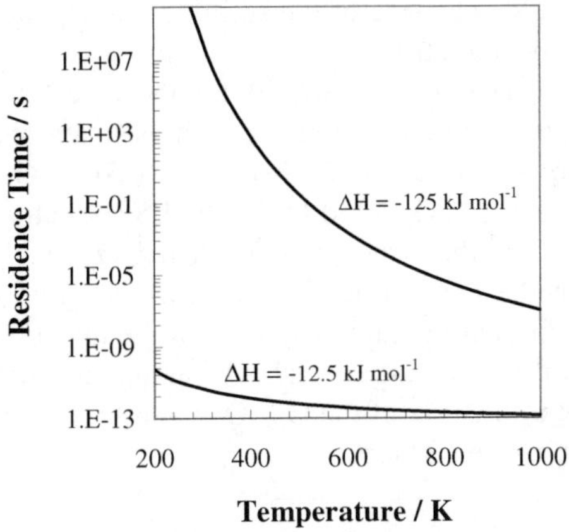


**Figure 2.16** Variation in residence times for one physisorbed species and one chemisorbed species as a function of temperature according to $\tau = \tau_0 \exp(\Delta H/RT)$, with $\tau_0$ equal to $3 \times 10^{-13}$ s.

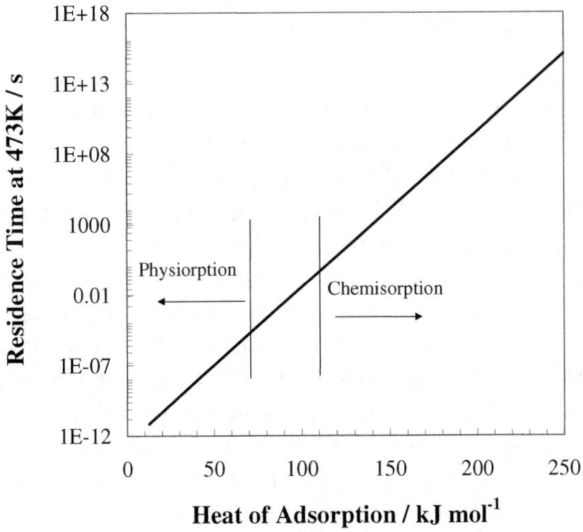


**Figure 2.17** Variation in residence time at 273 K as a function of heat of adsorption according to $\tau = \tau_o \exp(\Delta H/RT)$, with $\tau_o$ equal to $3 \times 10^{-13}$ s.

electron spin resonance (ESR) was particularly important in determining the structures of the species involved. This techniques is used to detect the presence of molecules with unpaired electrons, namely paramagnetic species. When placed in a magnetic field the energy level of the electron splits depending on whether the unpaired electron alligns parallel or anti-parallel with the magnetic field. The size of the splitting corresponds to the energy of a photon of microwave radiation.

The technique is particularly powerful for the detection of paramagnetic oxygen species on diamagnetic materials such as magnesium oxide. A wide range of chemisorbed species have been detected in this way, the most important of which are $O^-$, $O_2^-$ and $O_3^-$. The ESR spectra in themselves are not very informative of the structure but when $^{17}O_2$ or $N_2{}^{17}O$ is used as a source of oxygen the basic splitting is complemented by a wealth of hyperfine structure due to additional interactions between the unparied electron and the uneven spin (5/2) of the $^{17}O$ nucleus. This technique in particular has allowed the unequivocal determination of the $O^-$, $O_2^-$ and $O_3^-$ (Lunsford, 1973) on a wide range of metal oxide surfaces.

## 2.6 KINETICS OF OXIDATION REACTIONS

There are five basic types of kinetic rate laws commonly used to describe heterogeneous oxidations. These are listed in Table 2.6.

### 2.6.1 Mars and van Krevelen kinetics

In 1954 Mars and van Krevelen published the results of a study of the oxidation of benzene, toluene, naphthalene and anthracene over vanadium oxide catalysts (Mars and van Krevelen, 1954). The study consisted of an examination of the influence of hydrocarbon and oxygen partial pressures on the reaction rates measured in a fluidized bed reactor. This reactor type was selected because it presented a uniform temperature profile throughout the catalyst bed, even for very exothermic reactions. The mechanism proposed was that the hydrocarbon extracts lattice oxygen from the surface layer of the catalyst, thereby generating a reduced catalyst. The extent of the reduction generated during steady state operation of the catalyst varies and is not a crucial factor in the determination of the kinetics. However, an important part of the Mars and van Krevelen mechanism is that the lattice oxygen is replenished by gas phase oxygen in a reoxidation process. The mechanism may be visualized as follows:

$$HC_{(g)} + \beta O_L \overset{k_{red}}{\rightarrow} P + \beta\square \quad r_{red} = k_{red}\, p_R \theta \tag{2.25}$$

$$\square + \tfrac{1}{2}O_2 \overset{k_{ox}}{\rightarrow} O_L \quad r_{ox} = k_{ox}\, p_{O_2}^n (1-\theta) \tag{2.26}$$

where $\beta$ is the stoichiometric requirement in lattice oxygen to generate the reaction product, P, $O_L$ represents a lattice oxygen anion, $\square$ represents the corresponding vacancy and R is the component to be oxidized. At a steady state the rates of these reactions are equal, hence

$$r_{red} = r_{ox}/\beta \tag{2.27}$$

so

$$\beta k_{red} p_R \theta = k_{ox} p_{O_2}^n (1 - \theta) \tag{2.28}$$

then

$$\theta = \frac{k_{ox} p_{O_2}^n}{\beta k_{red} p_R + k_{ox} p_{O_2}^n} \tag{2.29}$$

Substituting into equation (2.25) gives

$$\frac{1}{r_{red}} = \frac{1}{k_{red} p_R} + \frac{\beta}{k_{ox} p_{O_2}^n} \tag{2.30}$$

This is the rate equation shown in Table 2.6 for the Mars and van Krevelen mechanism.

This mechanism has come to be the most widely applied in heterogeneous oxidation catalysis and some typical examples of oxidation reactions that have been reported to proceed via this mechanism are presented in Table 2.6. Many more will be encountered in later chapters. Ironically proof of the operation of a Mars and van Krevelen mechanism has come to rely much more heavily on techniques, such as isotopic studies, which distinguish between the incorporation of gas phase and lattice oxygen into reaction products, than on traditional kinetic methods. One of the reasons for this is the very large amount of experimental data needed to distinguish experimentally between the Mars and van Krevelen mechanism and the other mechanisms commonly encountered in heterogeneous catalysis.

### 2.6.2 The Eley–Rideal mechanism

The Eley–Rideal mechanism envisages a simple reaction between an adsorbed species and a second species reacting without adsorbing. This mechanism is shown in Scheme 2.7. This means that the gas phase reactant, such a carbon monoxide molecule as shown in Scheme 2.7, does not reside on the catalyst surface for longer than about $10^{-13}$ s, namely less than the time required to define a physisorbed state. In that time a simple reaction with an adsorbed species, $O_2$ or O appropriately charged in the case of oxidations, can occur according to this mechanism. Since the mechanism derives from the Langmuir adsorption isotherm it is constrained to describing energetically and chemically uniform adsorption and reaction sites. A feature is that the adsorbing species must be in equilibrium with the species from which it is

**Table 2.6** Rate laws for heterogeneous catalysis[a] (Mozzanega *et al.*, 1979) (reproduced by permission of the American Chemical Society)

| *Mechanism* | *Kinetic equation* | *Linear transform* | *Examples* |
|---|---|---|---|
| (1) Mars–Van Krevelen (surface oxido-reduction) | $r = \frac{k_{ox} k_{red} p_{O_2}{}^n p_R}{k_{ox} p_{O_2}{}^n + k_{red} p_R}$ | $\frac{1}{r} = \frac{1}{k_{red} p_R} + \frac{1}{k_{ox} p_{O_2}{}^n}$ | $1 - C_4H_8 + 0.5O_2 \xrightarrow[623\ K]{VPO} C_4H_6 + H_2O$ [Brkle and Trifiro, 1979]<br>$C_2H_4 + O_2 \xrightarrow[523\ K]{Pd/V_2O_5} CH_3COOH$ [Seoane *et al.*, 1980] |
| (2) Eley–Rideal (reaction between adsorbed $O_2$ and gaseous or weakly adsorbed R) | $r = \frac{kK^n_{O_2} p_{O_2}{}^n p_R}{1 + K_{O_2}{}^n p_{O_2}{}^n}$ | $\frac{1}{r} = \frac{1}{kp_R}\left(1 + \frac{1}{K_{O_2}{}^n p_{O_2}{}^n}\right)$ | $CH_4 + O_2 \xrightarrow[968\ K]{Fe_2(MoO_4)_3} HCHO + H_2O$ [Otsuka *et al.*, 1993] |
| (3) Langmuir–Hinshelwood (uniform surface with one type of site) | $r = \frac{kK_{O_2}{}^n K_R p_{O_2}{}^n p_R}{(1 + K_{O_2}{}^n p_{O_2}{}^n + K_R p_R)^2}$ | no linear transform; $r$ passes through a maximum equal to $k/4$ | $C_4H_6 + 2.50_2 \xrightarrow[623\ K]{VPO} C_4H_2O_3 + 2H_2O$ [Brklc and Trifiro, 1979] |

| | | | |
|---|---|---|---|
| (4) Langmuir–Hinshelwood (uniform surface with two types of sites) | $r = \frac{kK_{O_2}{}^n K_R p_{O_2}{}^n p_R}{(1 + K_{O_2}{}^n p_{O_2}{}^n)(1 + K_R p_R)}$ | $\frac{1}{r} = \frac{1}{kK_{O_2}{}^n K_R}\left(K_{O_2}{}^n + \frac{1}{p_{O_2}{}^n}\right)\left(K_R + \frac{1}{p_R}\right)$ | $C_2H_4 + 0.5O_2 \xrightarrow[493\ K]{Ag/Al_2O_3} C_2H_4O$ [Klugherz and Harriott, 1971] $2NH_3 + 1.5O_2 \xrightarrow[298\ K]{Ti} N_2 + 3H_2O$ [Mozzanega *et al.*, 1979] |
| (5) Power Law | $r = kp_{O_2}{}^n p_R^m$ | | $C_3H_6 + O_2 \xrightarrow[723\ K]{\beta-Bi_2Mo_2O_a} C_3H_4O + H_2O$ [Monnier and Keulks, 1981] $CH_4 + N_2O \xrightarrow[853\ K]{Mo/SiO_2} HCHO + H_2O + N_2$ [Liu *et al.*, 1984] |

[a] $r$ is the reaction rate; $k$, the reaction rate constant; $k_{ox}$, $k_{red}$, rate constant of oxidation or reduction; $n = 1$ or $\frac{1}{2}$ for $O_2$ molecular or dissociative adsorption; R, component to oxidize; $K_{O_2}$, $K_R$, adsorption coefficient of $O_x$ or R; $a$, stoichiometric coefficient.

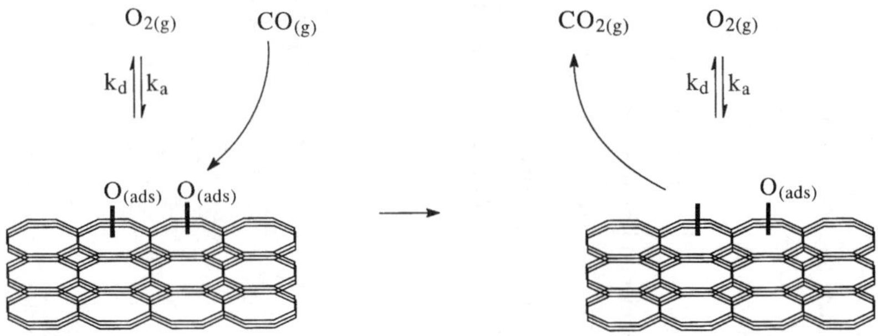


**Scheme 2.7** Schematic diagram of the Eley–Rideal mechanism for the reaction of an oxygen species, adsorbed on a uniform surface, with CO reacting from the gas phase.

derived in the gas phase. Scheme 2.7 shows the equilibrium dissociative adsorption of oxygen on a uniform surface governed by an equilibrium constant. The fractional coverage of the surface by atomic oxygen will be given according to the Langmuir adsorption isotherm as

$$O_{2(g)} \overset{K_{O_2}}{\rightleftharpoons} 2O_{(ads)} \qquad \theta_0 = \frac{K_{O_2} p_{O_2}{}^{0.5}}{1 + K_{O_2} p_{O_2}{}^{0.5}} \tag{2.31}$$

and subsequent reaction with a hydrocarbon species from the gas phase is given by

$$HC_{(g)} + O_{(ads)} \overset{k}{\rightarrow} P \qquad r = k\theta_0 P_R \tag{2.32}$$

The overall reaction rate increases as the surface coverage by oxygen increases and as the gas phase pressure of carbon monoxide increases. A more elaborate version of this expression is given in Table 2.6 with some appropriate examples of oxidation reactions that have been associated with this mechanism. Many minor variations in this basic mechanism are possible, such as when the hydrocarbon competes with oxygen species for adsorption sites, but does not react from the adsorbed state. The simplest Eley–Rideal reaction mechanism predicts a first order dependence on hydrocarbon partial pressure; first order in oxygen partial pressure at low partial pressures changing progressively to zero order at high pressures.

### 2.6.3 The Langmuir–Hinshelwood mechanism

This mechanism envisages that the slowest of the five steps involved in a catalysed reaction is the reaction between two adsorbed species on a uniform catalyst surface. Adsorption can be on a single type of surface site, shown in

Scheme 2.8 (line 3 in Table 2.6), or each adsorbing species can adsorb on its own type of surface site (line 4 Table 2.6), but each adsorbed species is in thermodynamic equilibrium with the corresponding gas phase species, according to the appropriate Langmuir adsorption isotherm. The overall reaction can be described by the following equations:

$$O_{2(g)} \overset{K_{O_2}}{\rightleftharpoons} 2O_{(ads)} \qquad \theta_0 = \frac{K_{O_2} p_{O_2}{}^{0.5}}{1 + K_{O_2} p_{O_2}{}^{0.5}} \tag{2.33}$$

$$HC_{(g)} \overset{K_R}{\rightleftharpoons} HC_{(ads)} \qquad \theta_{HC} = \frac{k_{RC}\, p_R}{1 + K_R\, p_R} \tag{2.34}$$

$$HC_{(ads)} + O_{(ads)} \underset{slow}{\overset{k}{\rightarrow}} P \qquad r = k\theta_0\theta_R \tag{2.35}$$

The kinetic expression predicts reaction orders in both reactants varying from one to zero as the partial pressures increases. When this type of kinetics is observed with one type of adsorption site a characteristic feature is the appearance of a maximum in rate in experiments where the partial pressure of one reactant is increased as the partial pressure of the second reactant is held constant. This phenomenon is associated with the inhibition of the adsorption of one reactant by too high a partial pressure of the second reactant.

### 2.6.4 Power rate laws

The final type of kinetic expression encountered in heterogeneous oxidation is the power rate law type. Some examples of this type of expression are indicated in Table 2.6 and they will be encountered in more detail in later chapters. In general, they tend to present a rather oversimplified rate expression and they can be difficult to interpret when fractional exponents are encountered.

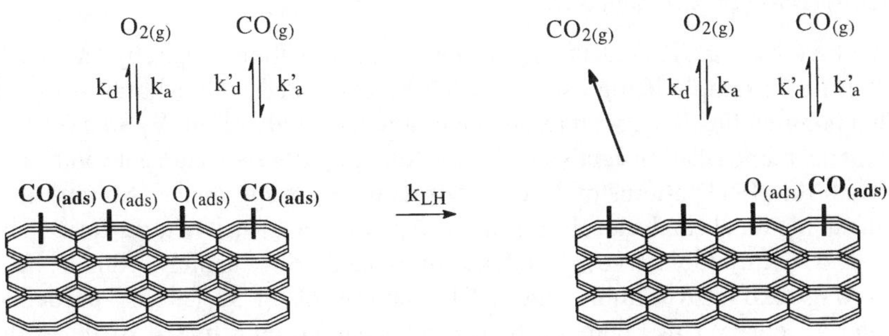


**Scheme 2.8** Schematic diagram of the Langmuir–Hinshelwood mechanism for the reaction of oxygen and hydrocarbon species adsorbed on a single type of uniform surface site.

## 2.7 CHARACTERIZATION OF OXIDATION CATALYSTS

In this section those basic characterization techniques that are widely used in studies of oxidation catalysis will be presented. It is not intended to present an exhaustive description of characterization methods, simply to introduce the most basic characterization techniques required for Chapters 3–10.

### 2.7.1 Catalyst characterization by X-ray diffraction methods

Many of the materials encountered in heterogeneous catalysis are crystalline, but the crystal quality is often poor, especially in the cases where preparation methods are used which produce small particles. Nevertheless, X-ray diffraction (XRD) is a standard method for the characterization of catalysts and generally has been used to determine the crystalline phase(s), including solid solutions, that are present and to measure the particle size and shape. With appropriate facilities it is also possible to follow by *in situ* XRD changes that occur during activation and testing of catalysts.

Monochromatic X-rays, incident on a crystalline solid, are diffracted owing to the crystal structure of the solid. For a maximum to occur in the diffraction pattern at a particular angle of incidence $\theta$ (with respect to lattice planes ($hkl$), the Bragg equation must be satisfied:

$$n\lambda = 2d_{hkl} \sin \theta_{hkl} \tag{2.36}$$

where:

$d_{hkl}$ = interplanar distance between ($hkl$) planes,
$n$ = order of diffraction,
$\lambda$ = wavelength of incident X-rays (e.g. 1.542 Å for copper K$\alpha$ X-rays).

*Identification of phases by XRD*

The first use of XRD data is for phase identification, whereby the XRD pattern of a sample being examined can be compared with a library of data. This point is illustrated with reference to a recent study of Sn/V/Sb oxides as propane ammoxidation catalysts (Albonetti *et al.*, 1998). Figure 2.18 indicates that all the XRD patterns of seven catalysts with varying Sn : V: Sb ratios are almost identical, and closely resemble that of $SnO_2$ in its rutile form. The catalyst with the largest amount of added Sb, namely, Sn : V : Sb= 1.0 : 0.05 : 2.0 also exhibits the XRD pattern of $Sb_6O_{13}$. This particular material (curve 4 in Figure 2.18) can be regarded as a mixed oxide in the sense that it is a mixture of $Sb_6O_{13}$ and another oxide with an XRD pattern very similar to that of $SnO_2$. What is this second oxide? Careful examination of Figure 2.18 reveals that as the Sn : V: Sb ratio changes the diffraction angle

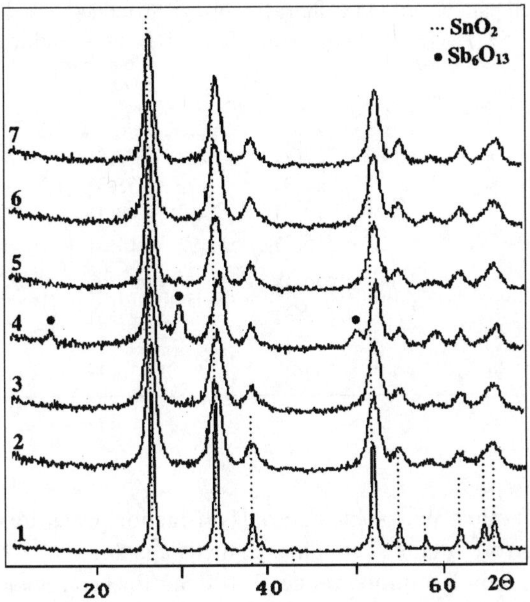


**Figure 2.18** XRD patterns of Sn–V–Sb oxides. (a) With increasing antimony amounts: curve 1, (Sn : V : Sb = 1.0 : 0.05 : 0.0); curve 2, (Sn : V : Sb = 1.0 : 0.05 : 0.5); curve 3, (Sn : V : Sb = 1.0 : 0.05 : 1.0); curve 4, (Sn : V : Sb = 1.0 : 0.05 : 2.0). (b) With increasing vanadium amounts: curve 5, (Sn : V : Sb = 1.0 : 0 : 1.0); curve 3, (Sn : V : Sb = 1.0 : 0.05 : 1.0); curve 6, (Sn : V : Sb = 1.0 : 0.1 : 1.0); curve 7, (Sn : V : Sb = 1.0 : 0.2 : 1.0) (Albonetti *et al.*, 1998) (reproduced by permission of Baltzer Scientific Publishers).

also changes very slightly. Rietveld analysis uses all the diffraction lines of a particular phase to determine the unit cell constants and volume and accordingly is very accurate. These data are presented in Table 2.7. This analysis reveals that there is a small but systematic change in the unit cell parameters and volume as the composition changes. This behaviour is characteristic of solid solution formation (see Section 4.2.2) so we can conclude that these materials (curves 1–6 in Figure 2.18) are solid solutions of structure similar to $SnO_2$.

*Particle size measurement by XRD*

Good quality crystalline material of dimension 0.1 mm is made up of approximately 100 000 crystal planes (assuming that $d_{hkl} = 1$ nm). When an X-ray collides with these planes close to the Bragg angle, a diffraction intensity is not observed because the X-rays reflected from the multitude of parallel planes

**Table 2.7** Unit cell parameters and volume for Sn–V–Sb oxides (Albonetti *et al.*, 1998) (reproduced by permission of Baltzer Scientific Publishers)

| *Composition Sn : V : Sb* | $a/nm$ | $c/nm$ | $V/nm^3$ |
|---|---|---|---|
| 1 : 0.05 : 0 | 4.727(2) | 3.177(2) | 70.99 |
| 1 : 0.05 : 0.5 | 4.724(3) | 3.155(3) | 70.31 |
| 1 : 0.05 : 1 | 4.712(3) | 3.147(3) | 69.87 |
| 1 : 0.05 : 2 | 4.715(3) | 3.114(3) | 69.22 |
| 1 : 0 : 1 | 4.720(3) | 3.134(3) | 69.82 |
| 1 : 0.05 : 1 | 4.712(3) | 3.147(3) | 69.89 |
| 1 : 0.01 : 1 | 4.705(3) | 3.145(3) | 69.62 |
| 1 : 0.2 : 1 | 4.702(4) | 3.136(3) | 69.33 |

annihilate one another. With such materials diffraction peaks are produced only within a very narrow window very close to the Bragg angle.

Particle sizes are often in the range 1–100 nm for heterogeneous catalysts. In such circumstances the number of stacked planes per particle is in the range 5–100. This results in the broadening of the XRD peaks because there are insufficient reflecting planes to annihilate phase reflections at angles that are fairly far removed from the Bragg angle.

This effect has been quantified as the Scherrer equation:

$$t_{hkl} = K\lambda / B_{hkl} \cos \theta_{hkl} \tag{2.37}$$

where

$t_{hkl}$ is the particle size measured from X-rays diffracted from the ($hkl$) planes,
$B$ is the peak width measured at half height measured in Radians,
$\theta$ is the Bragg angle for the reflection in question, and
K is a constant which depends to some extent on the particles shape. For spherical particles $K = 0.9$.

Scheme 2.9 demonstrates how particle size and shape can be determined by XRD analysis. A sample was analysed by XRD and exhibited the pattern shown, namely one very narrow line and one rather broad one. Conflicting information on the particle size? In this case the apparent conflict is caused by the shape of the particle. X-rays reflected from the (100) planes 'see' a large particle and a narrow line results. X-rays reflected from the (001) planes 'see' a small particle and a broad line results.

When measuring particle size using the XRD line broadening it is impor-

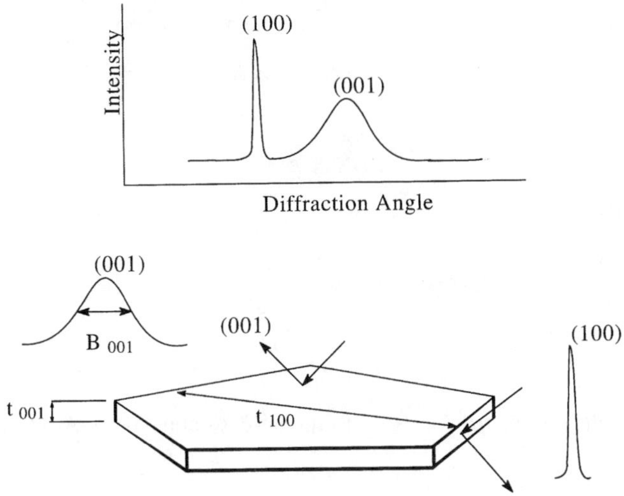


**Scheme 2.9** Schematic diagram of X-ray line broadening effects.

tant to realize that the not all the broadening is due to the particle size effect. Near perfect crystals should produce extremely narrow XRD lines, but even here some broadening occurs because of instrumental factors. Instrumental contribution to line broadening is substracted out by running the XRD of a good quality polycrystalline sample of the same material.

### 2.7.2 Surface area and porosity determination

*Surface area determination*

Brunauer, Emmet and Teller developed the now famous BET equation for the determination of the surface area of a solid (Brunauer *et al.*, 1938). The method is based on non-specific physisorption of a gas ($N_2$ or Ar) onto a solid close to the condensation temperature of the adsorbing gas. Adsorption is characterized by an isotherm which represents the equilibrium amount of gas adsorbed on a solid at a given temperature as a function of pressure (Richardson, 1989). Figure 2.19 is a typical type II isotherm which forms the basis for the BET analysis. This figure shows formation of a near monolayer of adsorbate at low pressures, followed by multilayer formation which accelerates at high partial pressures. Hysteresis is observed also at $p/p_0$ above 0.8, namely differing behaviour for whether the isotherm is determined by progressive addition or removal of adsorbate gas to the system (see Section 2.5.1)

Point A in Figure 2.19 corresponds to the stage at which there is approximately monolayer coverage. There is a difficulty in determining this

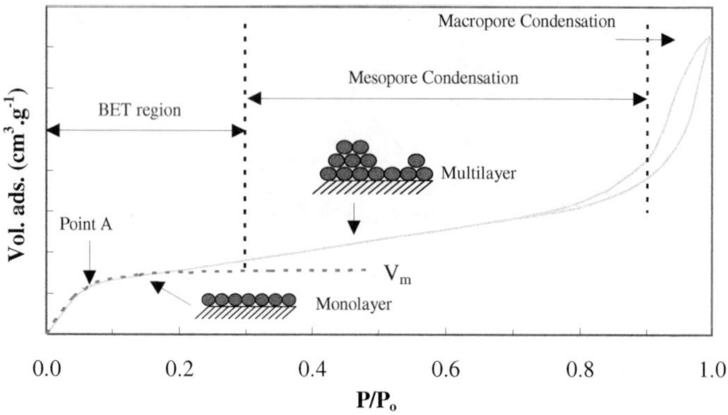


**Figure 2.19** Typical isotherm for $N_2$ adsorption–desorption on $\gamma$-$Al_2O_3$ (Macken, 1996).

point exactly as multilayer physisorption can occur before monolayer coverage is complete. The BET equation extends the Langmuir isotherm (equation (2.17)) to multilayer adsorption (Smith, 1970).

$$\frac{p}{V(p_0 - p)} = \frac{1}{V_m C} + \frac{(C-1)p}{C V_m p_0} \tag{2.38}$$

where

$V$ is the volume, reduced to standard conditions (STP) of gas adsorbed per unit mass of adsorbent at a given pressure, $p$ and constant temperature,
$p_0$ is the saturation pressure at the measurement temperature,
$V_m$ is the volume of gas adsorbed at STP per unit mass of adsorbent, when the surface is covered by a unimoleculer layer of adsorbate, and
$C$ is a constant, related to the free energy of adsorption.

According to equation (2.38) a plot of $p/V(p_o - p)$ versus $p/p_o$ should yield a straight line. The surface area is then calculated using:

$$S_{BET} = \frac{V_m A_m N_a}{V_{mol}} \tag{2.39}$$

where:

$N_a$ is Avogadro's number ($6.023 \times 10^{23}$),
$V_{mol}$ is the molar volume of absorbate gas at STP ($22.4\ \mathrm{l\,mol^{-1}}$), and
$A_m$ is the cross-sectional area of adsorbed gas ($A_m(N_2) = 0.162\ \mathrm{nm}^2$).

When nitrogen is the adsorbing gas this reduces to:

$$S_{\mathrm{BET}} = 4.353\,V_{\mathrm{m}} \tag{2.40}$$

The BET equation is applicable within the range of relative pressures $0.05 < p/p_o < 0.3$. At higher relative pressures, the BET equation is usually inaccurate owing to the effects of capillary condensation in the smallest micropores (Baiker, 1985). Figure 2.20 shows analysis of the data from Figure 2.19 in the range $0 < p/p_o < 0.3$. The surface area in this case is 128 $m^2g^{-1}$.

*Pore size distribution*

The International Union of Pure and Applied Chemistry (IUPAC) have classified pores into several size ranges (Ternan, 1994). Pores with diameters less than 2 nm are termed micropores, pores with diameters greater then 2 nm but less than 50 nm are termed 'transitional' or mesopores, while pores with diameters greater than 50 nm are termed macropores. In addition to the determination of the surface area of porous solids it is essential that the pore shape and pore size distributions are known to describe completely the textural properties of porous solids.

Traditionally pores with diameters less than 50 nm have been determined by nitrogen adsorption–desorption isotherms, while pore diameters greater than 100 nm have been measured with mercury porosimeters (Smith, 1970).

Two separate approaches are required to determine micropore and mesopore size distributions (Storck *et al.*, 1998). For micropores Mikhail *et al.*, (1968) devised a method for the determination of micropore area distributions and hence micropore volume distributions (the MP method) by manipulation of isotherm adsorption data (Figure 2.21). It is essentially a modification of Lippens and de Boer's, *t*-method where the variation of the thickness ($t$) of adsorbed $N_2$ layers with relative pressure ($p/p_o$) on a porous solid is com-

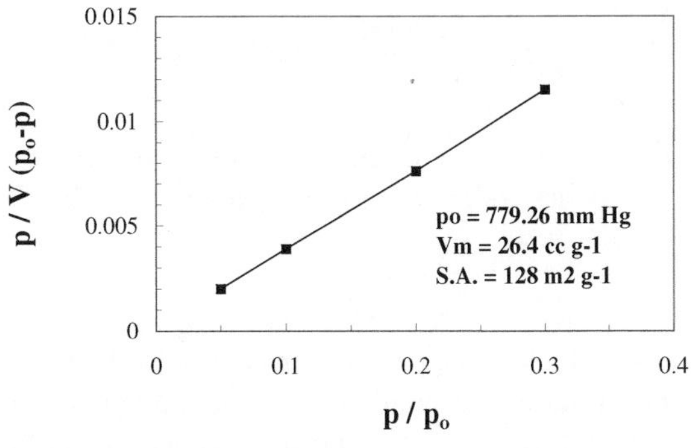


**Figure 2.20** BET plot for surface area determination for $\gamma$-$Al_2O_3$ (Macken, 1996).

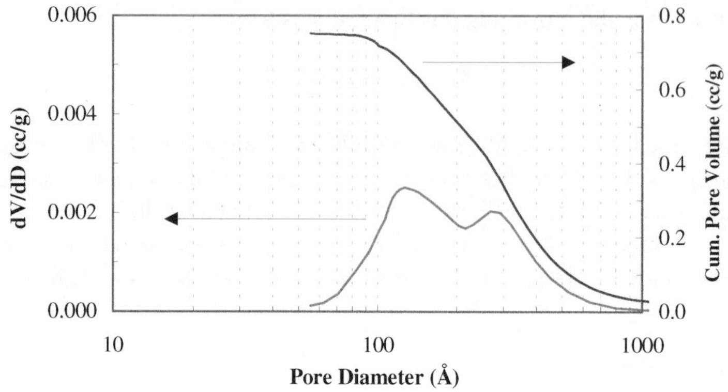


**Figure 2.21** BJH analysis of data from Figure 2.19 showing differential and cumulative mesopore volume distributions for $\gamma$-$Al_2O_3$ (Macken, 1996).

pared with a standard *t*-curve from a non-porous solid. The thickness of the adsorbed $N_2$ layers is obtained by dividing the liquid volume of $N_2$ adsorbed by the BET surface area:

$$t = \frac{V_{\text{liq}}}{S_{\text{BET}}} \times 10^3 (\text{nm}) \tag{2.41}$$

At higher relative pressures (higher values of t) deviations from a straight line due to micropore capillary condensation are used to evaluate micropore shape, size and volume distributions (Lowell, 1979).

Information on mesopores comes from the isotherm at $p/p_o$ values above 0.25. The isotherm in Figure 2.19 shows a splitting into an adsorption and a desorption branch. This hysteresis contains valuable information on mesopore geometry and the phenomenon is governed by the Kelvin equation (Lowell, 1979). The Kelvin equation with $N_2$ as the adsorbate is:

$$r_k = \frac{0.415}{\log(p_0/p)} (\text{nm}) \tag{2.42}$$

The Kelvin radius ($r_k$) indicates the radius into which condensation occurs at the required relative pressure. The actual pore radius is given by:

$$r_p = r_k + t \tag{2.43}$$

The term *t* refers to the thickness of a multilayer of adsorbed $N_2$ existing on the pore wall when evaporation or condensation occurs which is of the same depth as the adsorbed $N_2$ layers on a non-porous surface. It can be calculated from the Halsey equation (Halsey, 1948);

$$t = 3.54 \left[ \frac{5}{2.303 \log(p/p_0)} \right]^{1/2} \tag{2.44}$$

Numerical integration methods such as the BJH method (Barrett *et al.*, 1951) can be used to evaluate meso-pore area and volume distribution utilizing the Halsey equation. Figure 2.21 presents a BJH treatment of the isotherm shown in Figure 2.19, in terms of cumulative mesopore volume and size distribution.

### 2.7.3 Selective chemisorption techniques

The vast majority of heterogeneous catalysts comprise more than a single component. Examples include supported precious metals ($Pt/Al_2O_3$), supported oxides ($V_2O_5/TiO_2$), mixtures of oxides and solid solutions (multicomponent bismuth molybdates) and sulfides ($MoS_2/Al_2O_3$). A full description of these catalysts requires characterization of the individual components. A range of physicochemical techniques, such as XRD, X-ray photoelectron spectroscopy, electron microscopy and *in situ* infrared spectroscopy can be applied to this analysis. Selective chemisorption techniques, many of which predate the physicochemical methods, are still widely used and in many cases give more quantitative information especially on the dispersion of individual components of a multicomponent catalyst.

An illustration of selective chemisorption techniques is shown in Scheme 2.10 for the chemisorption of hydrogen on platinum and for carbon monoxide on palladium. In each case the adsorbing gas is capable of chemisorbing on the supported metal whereas physisorption occurs on the underlying support, which is most often an oxide. It is essential that the catalyst is treated in an appropriate way prior to attempting a selective chemisorption experiment. The exact type of pretreatment depends on the catalyst being examined and the adsorbent that is to be used. For example, prior to hydrogen chemisorption, the catalysts must be reduced at an elevated temperature, usually with hydrogen, to ensure that the metal is in the metallic state prior to exposure to hydrogen in the selective chemisorption experiment.

Selective chemisorption can be operated in pulse or continuous mode. Both are shown in Scheme 2.10. In the pulse mode a series of pulses is introduced, each with about $10^{18}$ molecules of adsorbate per gram of pretreated catalyst. The size of the pulse as recorded when the reactor is bypassed is compared with its size after passage through the reactor. The difference is assigned to chemisorption. The separation between pulses is usually 1–5 min. During this time the catalyst is being flushed by an inert gas and it is assumed that physisorbed species are removed between pulses by this flushing action. Nearly all of the adsorbate can be removed through chemisorption from the first pulses administered; thereafter the pulse size tends to revert to the value observed with an empty reactor. For experimental accuracy and convenience

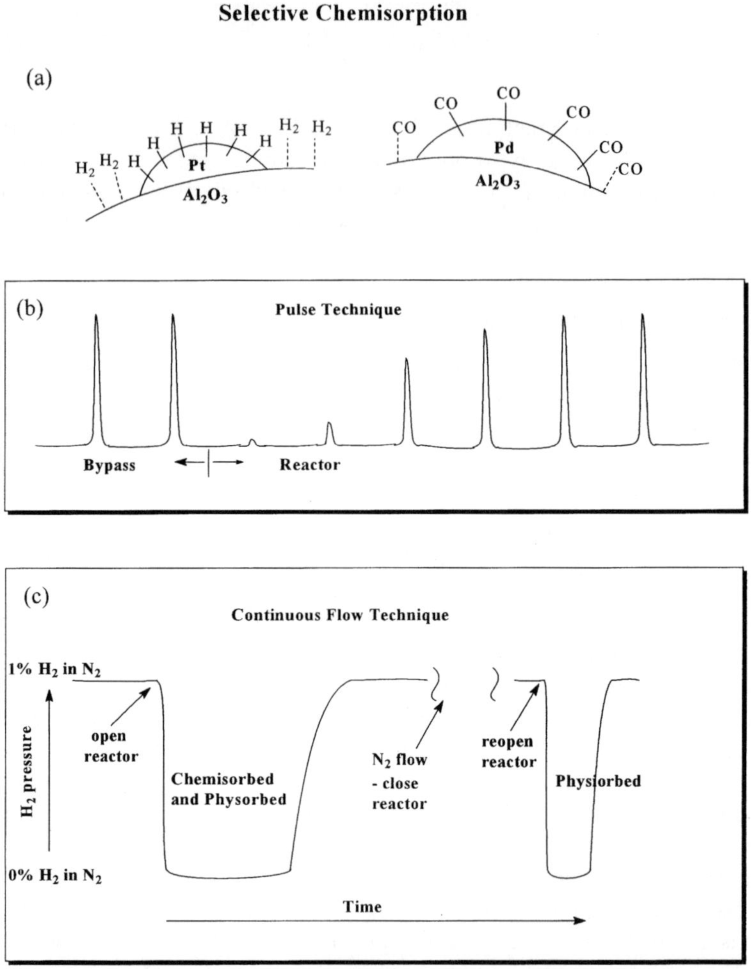


**Scheme 2.10** (a) Illustration of selective chemisorption of $H_2$ and CO on Pt and Pd respectively, with physisorption occurring on the alumina support. (b) Appearance of pulses of adsorbate in bypass mode and following opening of a reactor containing a suitably pretreated catalyst. (c) Continuous flow technique to evaluate selective chemisorption.

chemisorption is usually complete after 5–20 pulses. With fewer pulses experimental inaccuracies arise and a larger number can be cumbersome.

With the continuous flow technique a dilute stream of adsorbent in an inert gas (e.g. 1% $H_2$ in $N_2$) is passed over the suitably pretreated catalyst. Adsorption commences when the reactor is opened to allow the adsorbate stream to pass over the catalyst. The adsorbate concentration in the gas phase reaching the detector is a measure of the amount of chemisorption and physisorption occurring. At this point the adsorbate flow is stopped and replaced by a flow of pure carrier gas. This action is intended to remove all physisorbed species from the catalyst surface. Thereafter, the adsorbate flow is re-established and once again passed over the catalyst, which still retains its chemisorbed species. Adsorption restarts and, this time, is due exclusively to physisorption. Calibration and integration of the adsorption events gives the amount physisorbed, the amounts physisorbed and chemisorbed and by difference the amount chemisorbed.

The final part of the technique involves a knowledge of the adsorption stoichiometry and the area occupied by each chemisorbed species. Simple stoichiometries are common. For example the CO : Pd ratio is usually taken as 1 : 1 whereas the $H_2$ : Pt ratio is usually 0.5 : 1 reflecting the disociative chemisorption of hydrogen on this surface.

The area occupied by each chemisorbing species is usually related to the area occupied by each surface adsorbent species. For example, with Pt the surface area occupied by each surface Pt atom is taken as 0.64 $nm^2$, namely the area occupied by each surface Pt atom on the (111) cleavage plane.

Table 2.8 summarizes a number of selective chemisorption techniques and their applications to a range of catalysts. The table also lists the nature of the interaction, where known, between the chemisorbed species and a specific component in the catalyst. Finally, the adsorption cross-section (column 5), defined as the surface area occupied by each chemisorbed species, is usually determined by the surface crystallographic structure of the adsorbent.

### 2.7.4 Temperature programming methods

Temperature programmed desorption (TPD) and reduction (TPR) have become powerful tools for the characterization of catalysts (Hurst *et al.*, 1982; Bhatia *et al.*, 1990). In these experiments, an inert gas in the case of TPD and a reductant in the case of TPR, is continuously passed over a catalyst bed, while the temperature is increased. A TPD profile is obtained by monitoring the gases which are desorbed from the surface as the temperature is increased. The temperature of desorption is an indication of the strength of binding of the adsorbate. A TPR profile is obtained by monitoring the consumption of the reductant or the evolution of the product gases, as a function of temperature. Again the temperature at which a reduction event occurs is an indication of the ease with which a particular event occurs. The area under each peak indicates the amounts desorbed or consumed (Huang *et al.*, 1988).

Figure 2.22 shows a typical apparatus that can be used for temperature programming methods. The apparatus as shown is set up for TPR with hydrogen as the reducing agent. Consumed hydrogen is determined using a thermal conductivity cell. When desorption products are to be determined this method needs to be augmented with a mass spectrometer.

**Table 2.8** Selective chemisorption techniques

| *Adsorbate* | *Adsorbent* | *Typical application* | *Nature of interaction* | *Adsorption cross sectional area/nm² (atom adsorbent)⁻¹* | *Reference* |
|---|---|---|---|---|---|
| $H_2$ | $Pt_{(s)}$ | $Pt/Al_2O_3$ | $Pt_s$–H | 0.084[a] | Lematire *et al.*, 1984 |
| $H_2$ | $Ni_{(s)}$ | $Ni/SiO_2$ | $Ni_s$–H | 0.068[a] | Lematire *et al.*, 1984 |
| $N_2O$ | $Ru_{(s)}$ | $Ru/Al_2O_3$ | $Ru_s + xN_2O \rightarrow Ru_sO_x + xN_2$ | 0.091[b] | Berndt and Müller, 1999 |
| $N_2O$ | $Cu_{(s)}$ | $Cu/SiO_2$ | $Cu_s + N_2O \rightarrow Ru_2O + xN_2$ | 0.20[b] | Chinchen *et al.*, 1987 |
| $O_2$ | $VOx_{(s)}$ | $V_2O_5/SiO_2$ | [c] | 0.34[b] | Oyama *et al.*, 1989 |
| $O_2$ | $MoS_{2(s)}$ | $MoS_2/Al_2O$ | [c] | 1.44[b] | Bodrer and Bartholemew, 1983 |

[a] $nm^2$ (atom hydrogen)$^{-1}$. [b] $nm^2$ (atom oxygen)$^1$. [c] Unspecified.

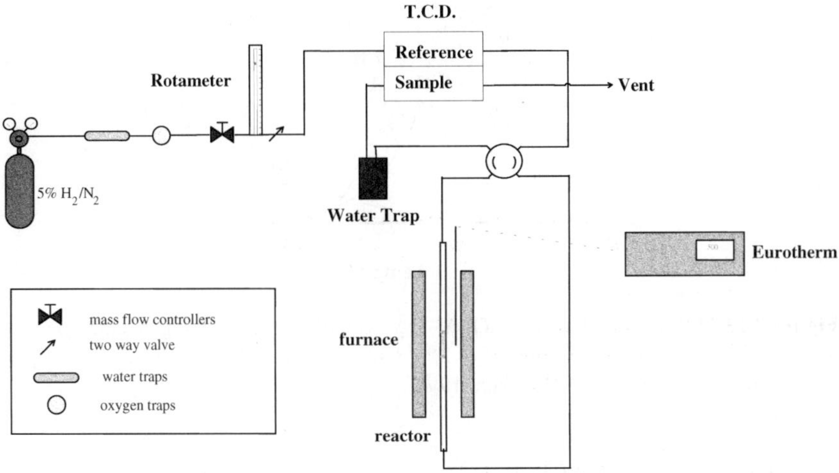


**Figure 2.22** Experimental test rig for TPR in 5 % $H_2$–$N_2$ reductant with a thermal conductivity detector. A mass selective detector (not shown) can also be used to monitor reduction products (Macken, 1996).

The selection of appropriate experimental conditions and their effect on a TPR profile have received a lot of attention in the literature. To avoid high distortions in the TPR profiles Malet and Caballero (1988) defined the parameter $P$:

$$P = \beta S_0 / F C_0 \tag{2.45}$$

where

$\beta$ = heating rate (K $\text{min}^{-1}$),
$S_0$ = initial amount of reducible species (μmol),
$F$ = flowrate ($\text{cm}^3 \text{min}^{-1}$),
$C_0$ = initial reductant concentration (μmol $\text{cm}^{-3}$).

This parameter should be kept below 20 K.

Figure 2.23 shows a typical TPR profile for the reduction of $CuSO_4$ supported on $Al_2O_3$ using a variety of reducing agents. A mass spectrometer was used to monitor the evolution of $SO_2$ from the catalyst during reduction. The influence of the reducing agent can be clearly seen with hydrogen the strongest and methane the weakest reducing agent used (Macken, 1996).

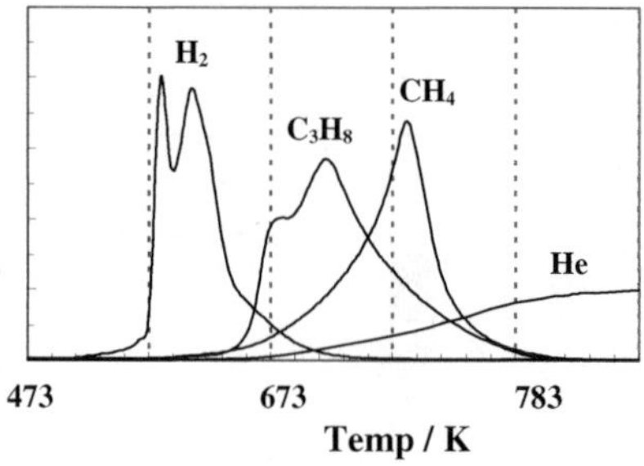


**Figure 2.23** TPR of sulfated $4.3CuO/Al_2O_3$ with different reducing agents ($\beta = 10$ K $min^{-1}$, $S_o = 27$ μmol (50 mg of 4.3% $CuO/Al_2O_3$), $F = 20$ $cm^3$ $min^{-1}$, $C_o = 2.232$ μmol $cm^{-3}$, $P = 6$ K) (Macken, 1996).

## REFERENCES TO CHAPTER 2

Albonetti S, Blanchard G, Burattin P, Masetti S, Tagliani A and Trifiro F (1998) A new ternary mixed oxide catalyst for ammoxidation of propane: Sn/V/Sb. *Catal. Lett.* **50**: 17–23.

Baiker A (1985) Experimental methods for the characterisation of catalysts. Part I. *Int. Chem. Eng.* **25**(1): 16–29.

Barrett EP, Joyner LG and Halenda PP (1951) The determination of pore volume and area distributions in porous substances: 1. Computations from nitrogen isotherms. *J. Am. Chem. Soc.* **73**: 373–380.

Berndt H and Müller U (1999) Determination of the surface area of dispersed ruthenium by reactive nitrous oxide chemisorption. *Appl. Catal. A: General* **180**: 63–69.

Bhatia S, Beltramini J and Do DD (1990) Temperature programmed analysis and its applications in catalytic systems. *Catal. Today* **7**(3): 309–435.

Bodrer TA and Bartholomew CH (1983) Oxygen chemisorption on $MoS_2$ and commercial hydrotreating catalysts. *J. Catal.* **84**: 145–155.

Brklc D and Trifiro F (1979) Kinetic analysis of 1-butene oxidation to maleic anhydride with a polyfunctional catalyst. *Ind. Eng. Chem. Prod. Res. Dev.* **18**: 333–339.

Brunauer S, Emmet PH and Teller E (1938) Adsorption of gases in multimolecular layers. *J. Am. Chem. Soc.* **60**: 309–319.

Cordona-Martinez N and Dumesic JA (1992) Application of adsorption microcalorimetry to the study of heterogeneous catalysis. *Adv. Cataly.* **38**: 149–244.

Chinchen GC, Hay CM, Vandervell HD and Waugh KC (1987) The measurement of copper surface areas by reactive frontal chromatography. *J. Catal.* **103**: 79–86.

Fuchs G, Neiman D and Poppa H, (1991) Three-dimensional transmission electron microscopy observations of supported palladium particles. *Langmuir*, **7**: 2853–2859.

Gasser RPH (1985) *An Introduction to Chemisorption and Catalysis by Metals.* Clarendon Press, Oxford.

Gleaves JT, Ebner JN and Kuechler TC (1988) Temporal analysis of products (TAP) – A unique catalyst evaluation system with submillisecond time resolution. *Catal. Rev. Sci. Eng.* **30**: 49–116.

Gregg SJ and Sing KSW (1982) *Adsorption, Surface Area and Porosity.* Academic Press, London.

Haruta M, Tsubota S, Kobayashi T, Kageyama H, Genet MJ and Delmon B (1993) Low temperature oxidation of CO over gold supported on $TiO_2$, $\alpha$-$Fe_2O_3$ and $Co_3O_4$. *J. Catal.* **144**: 175–192.

Halsey GD (1948) Physical adsorption on non-uniform surfaces. *J. Chem. Phys.* **16**: 931–937.

Horowitz HS, Blackstone CM, Sleight AW and Teufer G (1988) V–P–O catalysts for oxidation of butane to maleic anhydride. Influence of $(VO)_2H_4P_2O_9$ precursor morphology on catalytic properties. *Appl. Catal.* **38**: 193–210.

Huang YJ, Xue J and Schwarz JA (1988) Analysis of temperature-programmed reduction profiles from metal supported catalysts. *J. Catal.* **111**: 59–66.

Hurst NW, Gentry SJ and Jones A (1982) Temperature programmed reduction. *Catal. Rev. Sci. Eng.* **24**(2): 233–309.

Klugherz PD and Harriott P (1971) Kinetics of ethylene oxidation on a supported silver catalyst. *AIChE J.* **17**: 856–866

Laidler KJ (1965) *Chemical Kinetics.* McGraw-Hill, New York.

Lemaitre JL, Menon PG and Delannay F (1984) The measurement of catalyst dispersion. In *Characterization of Heterogeneous Catalysts*, Delannay F. (Ed). Marcel Dekker, New York, pp. 299–365.

Leon Y, Leon CA and Vannice MA (1991) Adsorption and catalytic properties of $Pd/SiO_2$, $Cu/SiO_2$ and $Pd\text{-}Cu/SiO_2$ systems. *Appl. Catal.* **69**: 269–290.

Le Page J-F, Cosyns J, Courty P, Freund E, Franck J-P, Jacquin Y, Judguin B, Marcilly C, Martino G, Miquel J, Montarnal R, Sugier A and van Landeghem H (1987) *Applied Heteroegeneous Catalysis.* Editions Technip, Paris.

Liu H-F, Liu R-S, Liew KW, Johnson RE and Lunsford, JH (1984) Partial oxidation of methane by nitrous oxide over molybdenum on silica. *J. Am. Chem. Soc.* **106**: 4117–4121.

Lowell S (1979) *Introduction to Powder Surface Area.* John Wiley & Sons, New York.

Lunsford JH (1973) ESR of adsorbed oxygen species. *Catal. Rev.* **8**: 135–157.

Macken C, (1996) Thesis, University of Limerick.

Malet M and Caballero A (1988) The selection of experimental conditions in temperature programmed reduction experiments. *J. Chem. Soc., Faraday Trans.* **84**: 2369–2375.

Mars P and van Krevelen DW (1954) Oxidations carried out by means of vanadium oxide catalysts. *Chem. Eng. Sci. Spec. Suppl.* **3**: 41–59.

Mikhail RH, Brunauer S and Bodor EE (1968) Investigations of a complete pore structure analysis—1. Analysis of micropores. *J. Colloid Interface Sci.* **26**: 45–53.

Monnier JR and Keulks GW (1981) The catalytic oxidation of propylene: IX The kinetics and mechanism over $\beta\text{-}Bi_2Mo_2O_9$. *J. Catal.* **68**: 51–66.

Mozzanega H, Hermann J-M and Pichat P (1979) $NH_3$ oxidation over UV-irridiated $TiO_2$ at room temperature. *J. Phys. Chem.* **83**: 2251–2255.

Otsuka, Wang Y, Yamanaka I and Morikawa A (1993) Kinetic study of the partial oxidation of methane over $Fe_2(MoO_3)_4$ catalyst. *J. Chem. Soc., Faraday Trans.* **89**: 4225–4230.

Oyama ST, Went GT, Lewis KB, Bell AT and Somorjai GA (1989) Oxygen chemisorption and laser Raman spectroscopy of unsupported and silica-supported vanadium oxide catalysts. *J. Phys. Chem.* **93**: 6786–6790.

Richardson JT (1989) *Principles of Catalyst Developement*, 1st ed. Plenum, New York.

Seoane JL, Boutry P and Montarnal R (1980) Ethylene oxidation to acetic acid with $Pd\text{–}V_2O_5$ catalysts. II Kinetics of the catalytic reaction. *J. Catal.* **63**: 191–200.

Smith JM (1970) *Chemical Engineering Kinetics*, 2nd ed. McGraw-Hill. New York.

Storck S, Bretinger H and Maier WF (1998) Characterization of micro- and mesoporous solids by physisorption methods and pore size analysis. *Appl. Catal. A: General*, **174**: 137–146.

Ternan M (1994) Large pore aluminas. *J. Catal.* **146**: 598–600.

Thomas JM and Thomas WJ (1997) *Principles and Practice of Heterogeneous Catalysis.* VCH, Weinheim.

Valenzuela DP and Myers AL (1989) *Adsorption Equilibrium Data Handbook.* Prentice Hall NJ.

Wartnaby CE, Stuck A, Yeo YY and King DA (1996) Microcalorimetric heats of adsorption of CO, NO and oxygen on Pt{110}. *J. Phys. Chem.* **100**: 12483–12488.

Weisz, PB (1982) The science of the possible. *Chemtech* July 424–425.

Yeo YY, Vattuone L and King DA (1997) Calorimetric heats for CO and oxygen adsorption and for the catalytic CO oxidation on Pt{111}. *J. Chem. Phys.* **106**: 392–401.

Zabetakis MG (1965) *Flammability Characteriztics of Combustible Gases and Vapors.* US Dept of the Interior, Bureau of Mines, US Government Printing Office, Washington, DC 20402.

# 3

# General Principles of Heterogeneous Catalytic Oxidation


3.1 Generic Features of Catalytic Oxidation Reactions 66
3.1.1 Kinetic Isolation of the Reaction Products in Selective Oxidation 66
3.2 The Mode of Activation of the Hydrocarbon and Discriminating Capacity of Active Sites in Oxidation Catalysts 72
3.2.1 Activation of the Hydrocarbon 72
3.2.2 Discriminating Mechanism of the Active Site 78
3.3 Types of Active Oxygen Species Involved in Oxidation Catalysis 81
3.3.1 Reactivity of Gas Phase Oxygen Species 84
3.3.2 Nature and Reactivity of Adsorbed Oxygen Species 85
3.3.3 Lattice Oxygen Species 86
3.4 Structural Features of Oxidation Catalysts 90
3.4.1 Macroscopic Features of Oxidation Catalysts 90
3.4.2 A Description of Active Sites on Selective Oxidation Catalysts 96
References to Chapter 3 99



Summary

Catalytic oxidation reactions can be conveniently divided into selective and total oxidations. Catalysts used in the former tend to be unsupported metal oxides (or chlorides in the case of oxychlorination), whereas supported precious metals are most often used in total oxidation reactions. An exception is the use of supported silver catalysts for ethylene epoxidation. In spite of some differences, it is nevertheless possible to identify many features of oxidation catalysis that are common to both types of reaction, such as the mode of activation of oxygen and hydrocarbons on catalyst surfaces, and the basic structure of oxidation catalysts. It should be noted that many materials are capable of acting as selective or total oxidation catalysts, but the emphasis throughout this work will be on those catalysts that operate at or close to industrial performance specifications.

Four generic features of oxidation catalysts are identified in this chapter. They are

- Kinetic isolation of the reaction products in selective oxidations,
- The mode of activation of the hydrocarbon and discriminating capacity of active sites in oxidation catalysts,
- Types of active oxygen species involved in oxidation catalysts,
- The general composition and structure of oxidation catalysts.

## 3.1 GENERIC FEATURES OF CATALYTIC OXIDATION REACTIONS

### 3.1.1 Kinetic isolation of the reaction products in selective oxidation

The basic thermodynamics of a number of the reactions used for the commercial production of bulk organic chemicals by catalytic oxidation, ammoxidation and oxychlorination are presented in Table 3.1 along with the data for the corresponding total oxidation reactions (Cox and Pilcher, 1970). A number of general features emerge. All the reactions are exothermic and the Gibbs free energies for the listed reactions closely follow the standard enthalpies of reaction, so all are thermodynamically favoured. All selective oxidation reactions must compete with the more thermodynamically favoured and more exothermic total oxidation reactions. Even over the most selective of catalysts undesirable reactions occur. Selective and total oxidation reactions have some mechanistic features in common and the major challenge in catalyst design is to favour the formation of the selective oxidation route over the total oxidation route. A technological factor to be considered is that the strong exothermicity of the selective and total oxidation reactions makes temperature control within large commercial scale reactors very difficult and hot spots, as shown in Figure 3.1, are usually reported.

A second feature to be pointed out is that most of the catalysts listed in Table 3.1 are base metal oxides of differing complexities and for the most part the catalytic material is unsupported. Epoxidation of ethylene over silver catalysts is exceptional in that it is the only process involving the selective oxidation of a hydrocarbon which uses a noble metal catalyst and in this case, also, a support is used.

In addition to the list of catalysts in Table 3.1, platinum and palladium play a predominant role in catalytic combustion processes. In general these materials serve as excellent catalysts for all the total oxidation reactions considered in Table 3.1.

All of the catalysts listed in Table 3.1 readily operate within the so called

**Table 3.1** Thermodynamics of selective oxidation reactions (Cox and Pilcher, 1970)

| *Reaction* | $\Delta H^\circ_{298}$/kJ mol$^{-1}$ [a] *Selective oxidation* | $\Delta H^\circ_{298}$/kJ mol$^{-1}$ [a] *Total oxidation* | *Catalyst* |
|---|---|---|---|
| 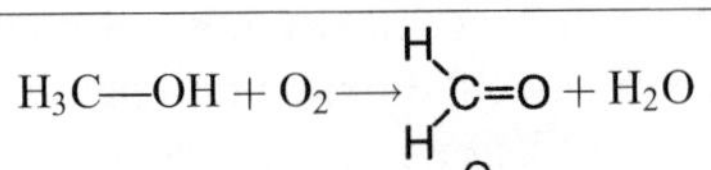 | −311 | −675 | $Fe_2(MO_4)_3$ |
| 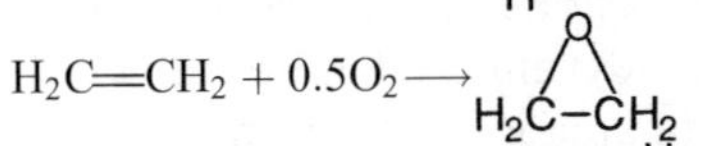 | −438 | −1323 | $Ag/\alpha\text{-}Al_2O_3$ |
| 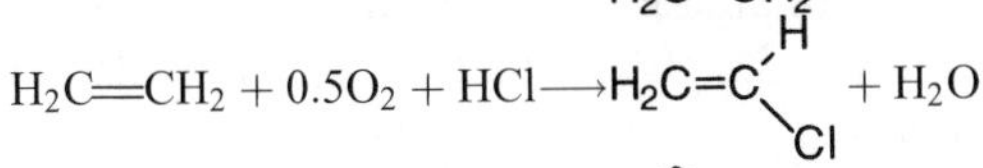 | −333 | −520 | $CuCl_2/Al_2O_3$ |
| 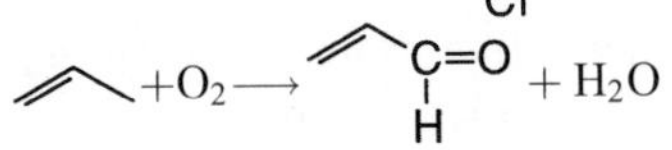 | −365 | −1959 | $FeCoBiMoO_x$ |
| 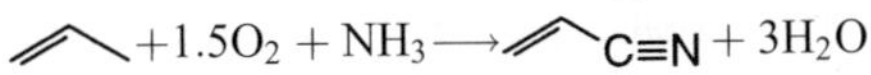 | −547 | −961 | $USb_3O_{10}$ |
| 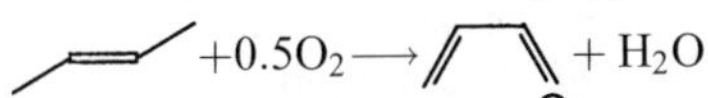 | −242 | −2522 | V/MgO |
| 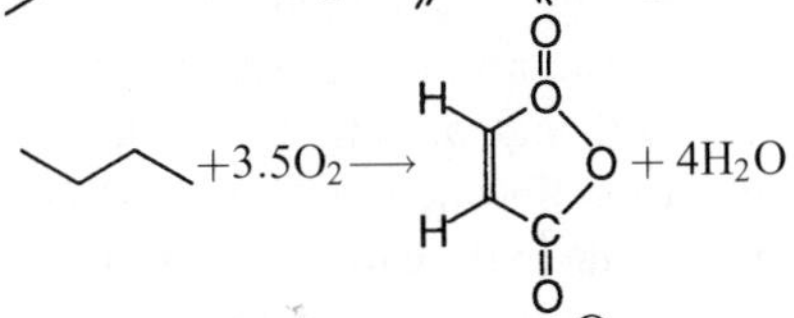 | −1237 | −2657 | $(VO)_2P_2O_7$ |
| 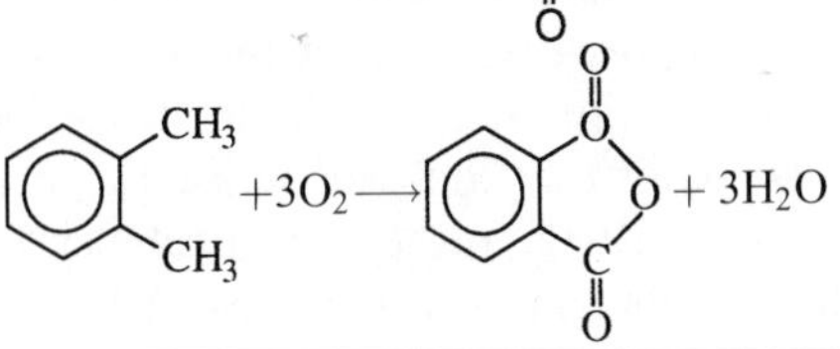 | −1115 | −4377 | $V_2O_5/TiO_2$ |

[a]kJ mol$^{-1}$ of the first reactant cited in column 1.

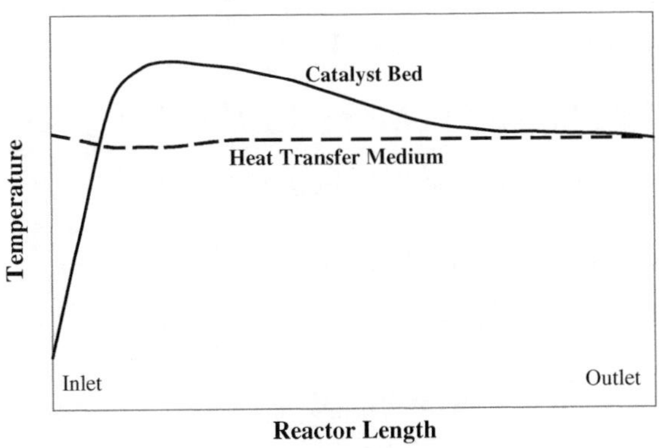


**Figure 3.1** Temperature profile in a tubular fixed bed reactor and in the surrounding salt bath heat transfer medium during a typical oxidation reaction.

Weisz window, whereby the reaction rate falls within the limits $10^{-6}$–$10^{-5}$ mol (cm catalyst)$^{-3}$ s$^{-1}$ identified by Weisz for the operation of many industrial processes (Weisz, 1982) (see Section 2.3) and a large variety of kinetic rate laws have been reported for these reactions, some of them demonstrating great complexity (Mezaki and Inoue, 1991).

Bearing in mind the difficulties associated with temperature control of oxidation reactions and the complexity of the kinetic rate laws encountered, we can still develop a useful model of selective oxidation reactions based on the simplest of kinetic networks. A key factor in process design in selective oxidation is the avoidance of the thermodynamic equilibrium, which in all cases favours total oxidation reactions. In practice, this means that, if allowed to remain in contact with the catalyst for sufficient time, total oxidation reaction products only will be produced.

This gives rise to the need to kinetically isolate the reaction products before the thermodynamically stable total oxidation products can become established. The extent to which kinetic isolation of the reactions products is successful depends on the relative rates for the formation of a specific selective oxidation product and its rate of decomposition. The situation is represented in Scheme 3.1, where $k_1$ is the rate coefficient for the formation of the selective oxidation product (SOP) and $k_2$ and $k_3$ are the rate coefficients for the formation of total oxidation products ($CO_x$) via decomposition of the SOP and via a direct oxidation of the hydrocarbon, respectively.

This reaction network, further simplified to neglect the $k_3$ route, can be represented as a simple consecutive reaction. In these circumstances equations (3.1)–(3.6) for a consecutive first order reaction can be used to predict the relative amounts of reactant, SOP and $CO_x$ present in a tubular fixed bed catalytic reactor for a given value of contact time ($t$). In these calculations it is assumed that the temperature gradient along the reactor is zero, rather than the real situation, such as shown in Figure 3.1.

$$[\mathrm{HC}]_t = [\mathrm{HC}]_0 \exp(-k_1 t) \tag{3.1}$$

$$[\mathrm{SOP}]_t = [\mathrm{HC}]_0 (k_1/k_2 - k_1)[\exp(-k_1 t) - \exp(-k_2 t)] \tag{3.2}$$

$$[\mathrm{CO}_x] = 1 - [\mathrm{HC}]_t - [\mathrm{SOP}]_t \tag{3.3}$$

$$\%\ \mathrm{Conversion} = 100\{[\mathrm{HC}]_0 - [\mathrm{HC}]_t\} \tag{3.4}$$

$$\%\ \mathrm{Selectivity} = 100\{[\mathrm{SOP}]_t/[\mathrm{HC}]_0 - [\mathrm{HC}]_t\} \tag{3.5}$$

$$\%\ \mathrm{Yield} = 100\{[\mathrm{SOP}]_t/[\mathrm{HC}]_0\} \tag{3.6}$$

The data in Figures 3.2 and 3.3 were generated using equations (3.1)–(3.6) and serve to illustrate the relationship between selective oxidation catalysts

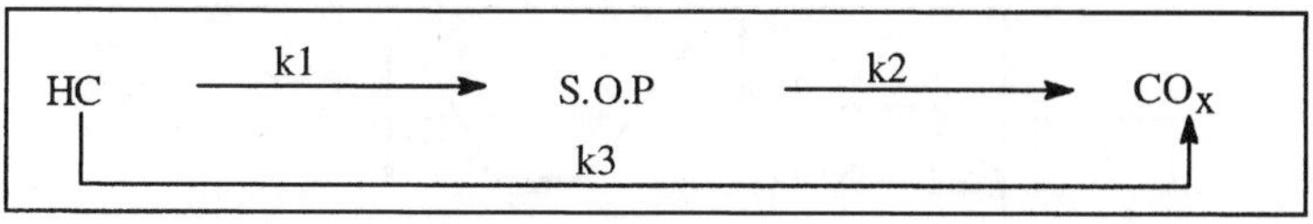


**Scheme 3.1** Reaction network in selective oxidation.

and reactor design. Without the appropriate catalyst and reactor design, high selectivities cannot be realized.

Figure 3.2(a) presents an energy profile for the selective oxidation of a hydrocarbon to SOP and further to $CO_x$. The catalyst alone determines the relative values of $Ea_1$ and $Ea_2$, but does not influence the thermodynamics. This figure simulates the situation for a catalyst which gives an activation energy for the $k_1$ step of 90 kJ mol$^{-1}$ and 100 kJ mol$^{-1}$ for the $k_2$ step. For these activation energies the Arhenius expression can be used to calculate the ratio of $k_1 : k_2$. At 673 K this ratio is 6 : 1 if the pre-exponential factors are taken to be equal. This implies that the reactant is quickly converted into the SOP, but more slowly into $CO_x$. The concentration profile of reactant, SOP and $CO_x$ along the reactor bed is presented in Figure 3.2(b). The contact time can be adjusted to coincide with the optimum SOP concentration. This optimum contact time corresponds to a reactant conversion of about 90 %. These data are also presented as a selectivity–yield–conversion plot in Figure 3.2(c) and confirm that the optimum yield is predicted at a conversion of about 90 %.

The sensitivity of the system to the catalyst is demonstrated by comparison with Figure 3.3. Conditions are identical to those in Figure 3.2 except that the situation now simulates a catalysts with an activation energy for the $k_1$ step of 100 kJ mol$^{-1}$ and for the $k_2$ step the value is 90 kJ mol$^{-1}$. Now the ratio of $k_1 : k_2$ is 1 : 6. Figure 3.3(b) and (c) show that there is never an appreciable concentration of SOP in the reactor irrespective of the contact time and the selectivity and yield are poor for all conversions.

Figure 3.4 shows a typical multiple selectivity–conversion plot for the oxidation of propane to acrolein (Cassidy, 1999) and demonstrates the additional complexity involved when the $k_3$ route of Scheme 3.1 becomes a significant factor. This plot was constructed from literature selectivity–conversion values irrespective of the experimental conditions or the nature of the catalysts used. This is a reaction for which a catalyst has not yet been found which exhibits a very high selectivity to acrolein, even at low propane conversions. It is clear from the data that the optimum yield occurs at less than 100 % conversion. A second feature of Figure 3.4 is that when the selectivity–conversion plot is extrapolated to close to zero conversion the selectivity to acrolein is not more than 60 %. This is an indication that even at very low conversions, about 40 % of the propane is converted directly into

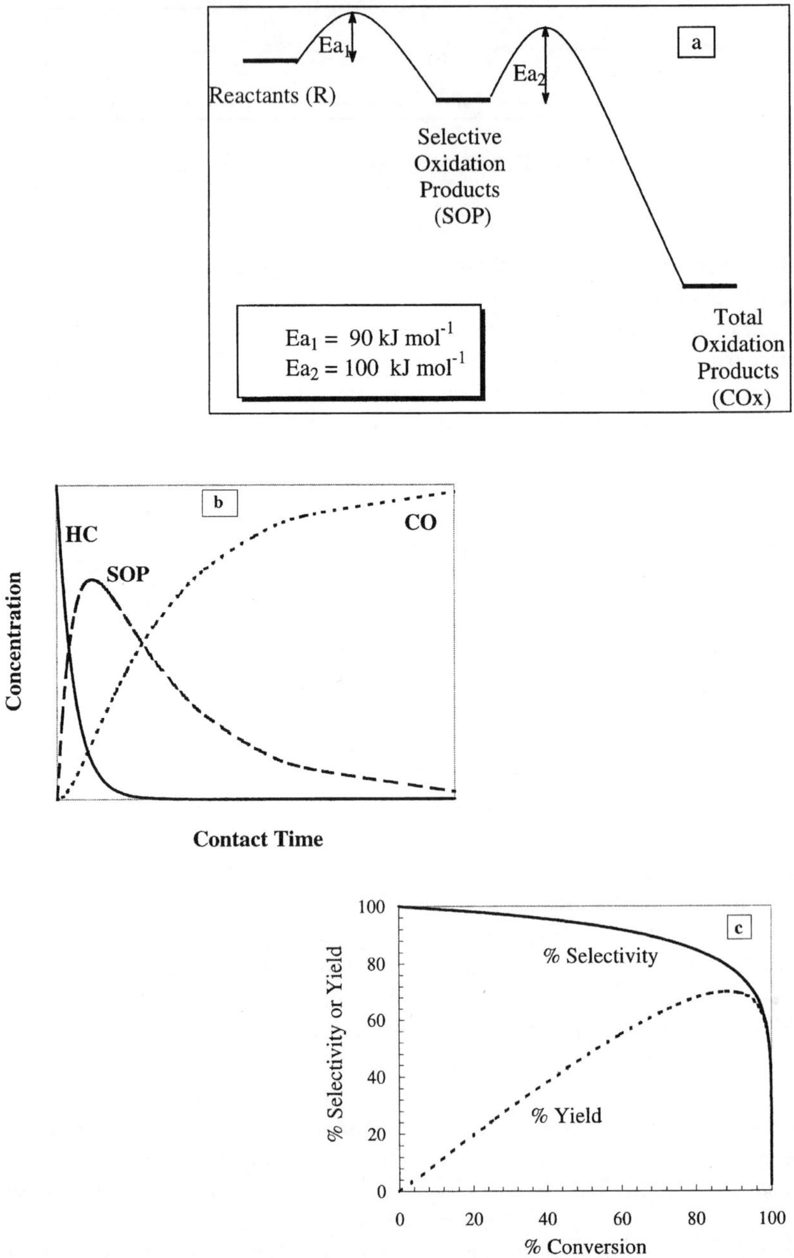


**Figure 3.2** (a) Schematic diagram of reaction profile for selective and total oxidation reactions, (b) concentration profile along a fixed bed reactor and (c) the corresponding selectivity–conversion and yield–conversion plots for the a pseudo-first order consecutive reaction with a ratio of $k_1 : k_2$ in Scheme 3.1 of 6 : 1.

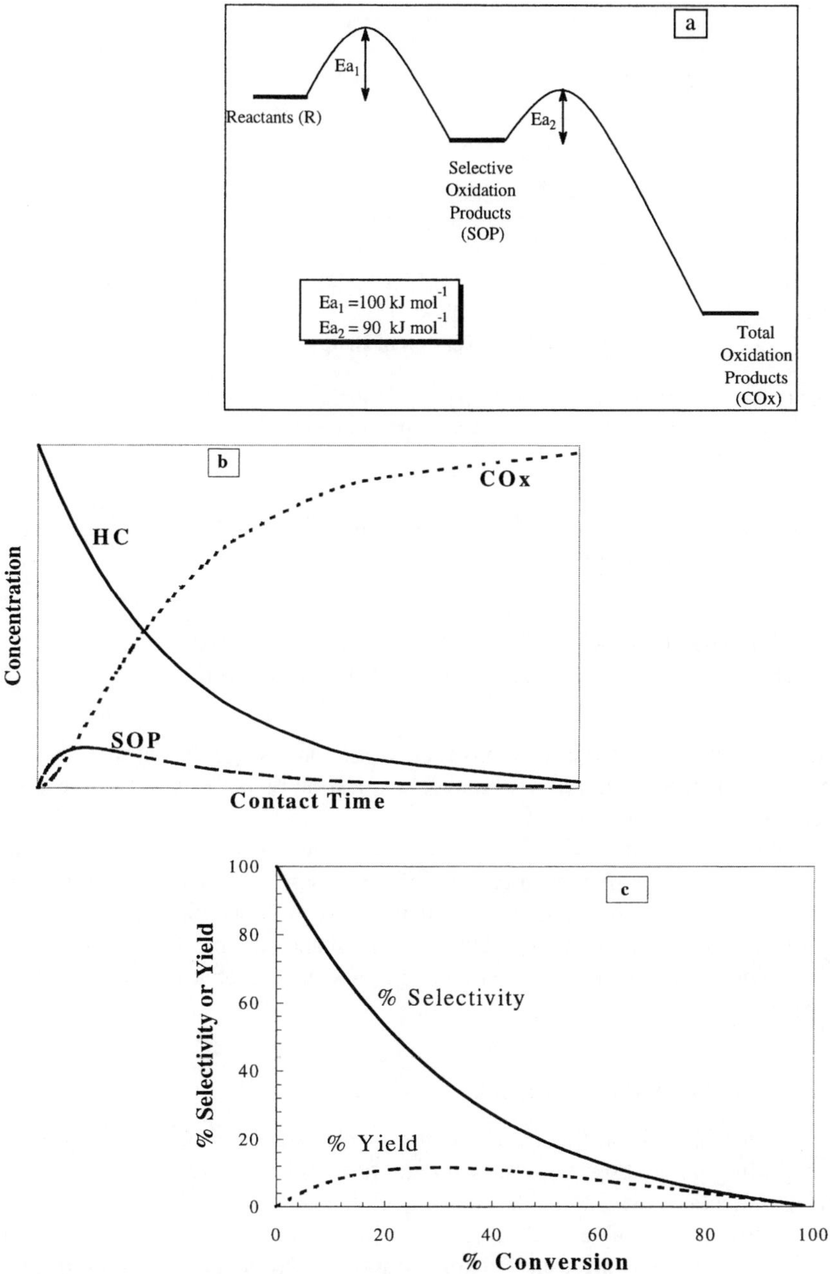


**Figure 3.3** (a) Schematic diagram of reaction profile for selective and total oxidation reactions, (b) concentration profile along a fixed bed reactor and (c) the corresponding selectivity–conversion and yield–conversion plots for the a pseudo-first order consecutive reaction with a ratio of $k_1 : k_2$ in Scheme 3.1 of 1 : 6.

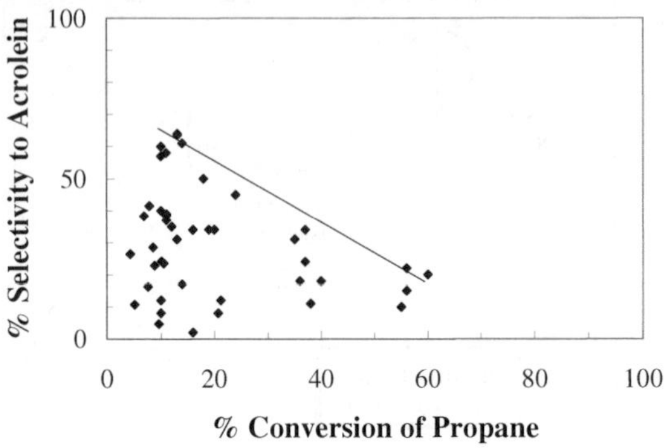


**Figure 3.4** Multiple selectivity–conversion plot for the ammoxidation of propane into acrylonitrile (Cassidy, 1999) (reproduced by permission of Elsevier Science).

non-selective products, or in terms of Scheme 3.1, the relative ratio of $k_1 : k_3$ is approximately 60 : 40.

## 3.2 THE MODE OF ACTIVATION OF THE HYDROCARBON AND DISCRIMINATING CAPACITY OF ACTIVE SITES IN OXIDATION CATALYSTS

Many selective oxidation reactions require, at some stage during the reaction mechanism, rupture of a C—H bond. The question then arises: is this the slow step during hydrocarbon activation?

In this section the mode of activation of the hydrocarbon by the active site is examined. An active site may be defined as an area of the catalyst at or around which the reaction occurs, hence it is the location at which the hydrocarbon is activated. In addition, the extent and mechanism whereby active sites discriminate between desirable and undesirable reactions will be explored.

### 3.2.1 Activation of hydrocarbon

If the slow step in a chemical reaction is cleavage of a C—H bond, then substituting the hydrogen atom for a deuterium atom will result in a lowering of the reaction rate, typically by a factor of about 2 for the range of temperatures normally encountered in selective oxidation reactions. The result of such experiments are usually expressed as $k_H/k_D$, namely the ratio of rate coefficients for the normal and deuterated forms of the reactant (substrate). This phenomenon is called the kinetic isotope effect (KIE) and has been used to very good effect in heterogeneous catalysis as a tool for the investigation of reaction mechanisms by substituting deuterium for hydrogen into a specific

C—H bond in a substrate. The origin of the KIE is summarized below. An essential feature needed to observe the full KIE is that the C—H bond in question must be fully stretched in the transition state. If a C—H bond other than that involved in the slow step is substituted with deuterium then no effect is observed. The basis for the phenomenon is the difference in activation energy that can arise when a C—H and C—D bond are involved in an otherwise identical transition state.

Although chemically identical, replacing a hydrogen atom by a deuterium atom in a C—H bond leads to a significant change in the reduced mass of the bond. The reduced mass $\mu_m$ for the R—H bond is defined as:

$$\mu_{mRH} = M_R M_H/(M_R + M_H) \tag{3.7}$$

where $M_R$ and $M_H$ are the molecular masses of the R and H fragments respectively. For many organic molecules in the size range of interest in selective oxidation the ratio of $\mu_m$ for RH and RD bond is about 2. The vibrational frequency, as measured by infrared spectroscopy, is related to the reduced mass by the expression:

$$\nu = (1/2\pi)(f/\mu_m)^{1/2} \tag{3.8}$$

and

$$\nu_{RH}/\nu_{RD} = (\mu_{mRD}/\mu_{mRH})^{1/2} = (M_H/M_D)^{1/2} \tag{3.9}$$

where $\nu$ is the vibrational frequency, $f$ the force constant for the bond in question and $\mu_m$ the reduced mass. Typically the $CH_2$ antistretching frequency for in $CH_3CH_2CH_3$ occurs at 2910 $cm^{-1}$ and it occurs at 2182 $cm^{-1}$ for $CH_3CD_2CH_3$ (Brazdil *et al.*, 1996). The ground vibrational level is at $\frac{1}{2}h\nu$ above the zero point of energy. This means that the C—H bond in its ground vibrational state has more vibrational energy than the C—D bond. Typically for C—H bonds this energy difference is about 5 kJ $mol^{-1}$. The ground vibrational levels of these bonds are superimposed on the free energy and activation energy profile for a reaction in Scheme 3.2. A difference in activation energy arises if the deuterated bond is fully stretched (broken) in the transition state.

If the C—H bond is fully stretched or ruptured in the transition state then the vibrational energy of this bond will not be a factor in the point of zero energy in the transition state, so that the activation energy difference will depend only on the zero point of energy difference for the C—H and C—D bonds in the reactants.

Values of $k_h/k_d$ may be calculated on the basis of equations (3.10) or (3.11). The first equation takes account of the stretching vibrations of the

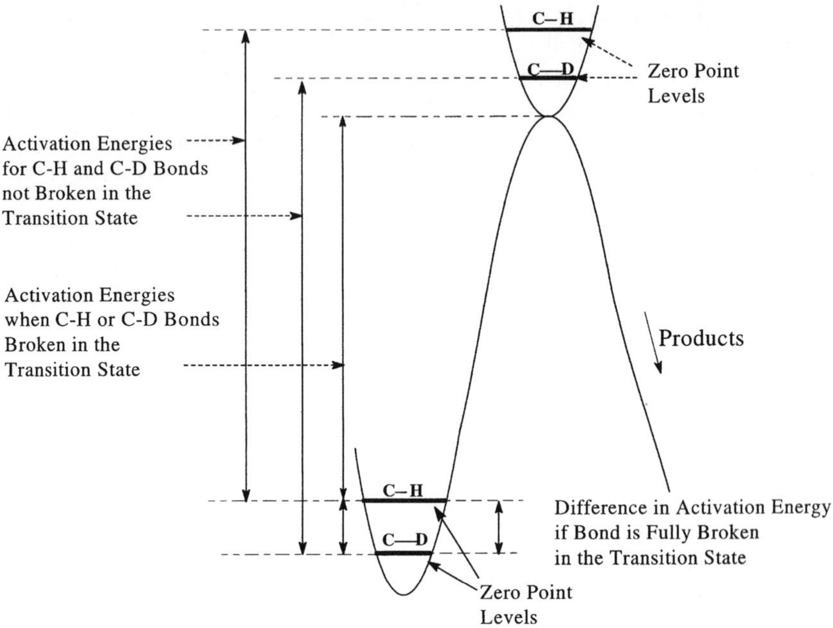


**Scheme 3.2** Origin of the KIE showing zero points of energy in the starting materials and in the transition state. A difference in activation energy arises if the C—H bond is fully broken in the transition state.

R—H (D) bond only whereas second equation allows for bending vibrations also (Melander, 1960):

$$\frac{k_H}{k_D} = \exp\left[\frac{hC}{2kT}(\nu_{R-D} - \nu_{R-D})\right] \quad (3.10)$$

$$\frac{k_H}{k_D} = \frac{M_H}{M_D}\exp\left[\frac{hC}{2kT}\sum_1^3(\nu_{R-H} - \nu_{R-D})\right] \quad (3.11)$$

where $h$ is Planck's constant, $c$ the speed of light and $h$ the Boltzmann constant. In equation (3.10) the C—H stretching vibration at an average value of 3000 cm$^{-1}$ only was used. In equation (3.11) the C—H stretching vibration at an average value of 3000 cm$^{-1}$ and two bending vibrations at 1000 cm$^{-1}$ each were used. Values of $k_H/k_D$ for a range of temperatures, calculated according to equations (3.10) and (3.11) are presented in Table 3.2.

Table 3.3 summarizes a number of KIE studies for oxidation and ammoxidation reactions. The reactions presented include methanol oxidation to formaldehyde, propene oxidation to acrolein, $n$-butane oxidation to maleic

**Table 3.2** Kinetic isotope effects for C—H bonds

| *Temp/K* | $k_H/k_D$ *[equation (3.10)]* | $k_H/k_D$ *[equation (3.11)]* |
|---|---|---|
| 273 | 10.1 | 23.7 |
| 373 | 5.4 | 8.4 |
| 473 | 3.8 | 4.6 |
| 573 | 3.0 | 3.1 |
| 673 | 2.6 | 2.4 |
| 773 | 2.3 | 2.0 |
| 873 | 2.1 | 1.7 |
| 973 | 1.9 | 1.5 |

anhydride, methane oxidation to ethane and propane ammoxidation to acrylinitrile. For each reaction the experimentally measured $k_H/k_D$ ratio for a particular isotope is compared with the theoretical value for rupture of each of the different types of C—H bonds within the substrate. In each case a KIE is observed, very close in value to that expected for the full stretching or rupture of a specific type of C—H bond in the transition state. For example, with *n*-butane there are two types of C—H bond, methylene (—$CH_2$—) and methyl (—$CH_3$). A KIE is observed only when the methylene (—$CH_2$—) C—H bond is deuterated. This implies that the slow step in selective oxidation can be associated with rupture of this particular type of bond in the reactant and that selective catalysts can target this bond when similar C—H bonds are available in the structure.

Moreover comparison of the bond dissociation enthalpies for specific C—H bonds indicates that KIEs are observed only when the weakest C—H bond in a particular structure is deuterated. In column 2 of Table 3.3 the bond dissociation enthalpy of each of the different types of C—H bonds is annotated next to the substrate structure. For substrates which present more than one type of C—H bond in the structure, the weakest bond is preferentially activated. For *n*-butane oxidation and propane ammoxidation this is the methylene (—$CH_2$—) bond, whereas propene is preferentially activated at the methyl (—$CH_3$) bond.

Complementary evidence relating to the activation of propene over $Bi_2O_3$, a component in commercial propene oxidation catalysts, has been presented by Driscoll *et al.* (1987). Their work relates to the detection of free radical species at the exit of a catalytic reactor, often by matrix isolation followed by electron spin resonance (ESR) analysis. They studied the formation of radicals derived from $C_3H_6$ and $C_3D_6$ in the temperature range 638–748 K and compared the predicted and experimentally measured values for the KIE. The results are summarized in Table 3.4. Over the entire temperature range the predicted and experimental values are entirely consistent and point to the hypothesis that over certain catalytic surfaces the primary activation step is

**Table 3.3** Kinetic isotope effects in selective oxidation and ammoxidation reactions (Cassidy and Hodnett, 1998)

| *Reaction, catalyst and temperature* | *C—H bond energies/ kJ mol$^{-1}$* | *Isotopic form of the substrate* | *Experimental* $k_H/k_D$ | *Theoretical* $k_H/k_D$[a] | *Theoretical* $k_H/k_D$[a] | *Theoretical* $k_H/k_D$[a] | *Reference* |
|---|---|---|---|---|---|---|---|
| | | | | —$CH_2$— | —$CH_3$— | | |
| $n$-$C_4H_{10} + 3.5O_2 \longrightarrow C_4H_2O_3 + 4H_2O$<br>Catalyst $(VO)_2P_2O_7$ Temp 673 K | 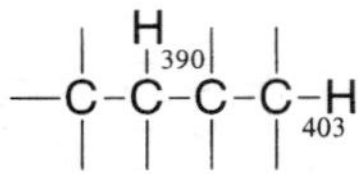 | $CH_3CD_2CD_2CH_3$<br>$CD_3CH_2CH_2CD_3$ | 2.18<br>1.05 | 2.3<br>1.0 | 1.0<br>2.3 | | Pepera *et al.* 1985 |
| | | | | $CH_2$═ | ═CH— | —$CH_3$ | |
| $C_3H_6 + O_2 \longrightarrow C_3H_4O + H_2O$<br>Catalyst BiMoCeFeOx, Temp 723 K | 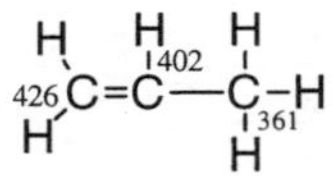 | $CH_2$═CH—$CH_2D$<br>CHD═CH—$CH_3$<br>$CH_2$═CD—$CD_3$ | 2.04<br>1.02<br>1.78 | 1.00<br>1.82<br>1.00 | 1.00<br>1.00<br>1.75 | 1.82<br>1.00<br>1.75 | Adams and Jennings, 1963, 1964. Keulks and Krenzke, 1976 |
| | | | | —$CH_2$— | —$CH_3$ | | |
| $CH_3CH_2CH_3 + 2O_2 + NH_3 \longrightarrow$ $CH_2$═CH—CN + $4H_2O$<br>Catalyst 50 %$VSb_{3.5}P_{0.5}WO_x$ –50 %$Al_2O_3$ Temp 743 K | 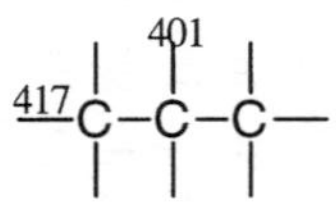 | $CH_3CD_2CH_3$<br>$CD_3CH_2CD_3$<br>$CD_3CD_2CD_3$ | 1.7<br>1.1<br>2.0 | 2.02<br>0.98<br>2.03 | 0.99<br>2.04<br>2.04 | | Bradzil *et al.*, 1996 |
| | | | | $CH_3$— | —OH | | |
| $CH_3OH + 0.5O_2 \longrightarrow HCHO + H_2O$<br>Catalyst $Fe_2(MoO_4)_3$ Temp 423 K | 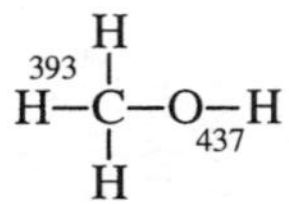 | $CH_3OD$<br>$CD_3OD$ | 1.2<br>3.6 | 1.0<br>4.0 | 4.0<br>4.0 | | Machiels, and Sleight, 1982 |
| | | | | $CH_4$ | | | |
| $2CH_4 + 0.5O_2 \longrightarrow C_2H_6 + H_2O$<br>Catalyst NaOH/CaO, Temp 973 K | 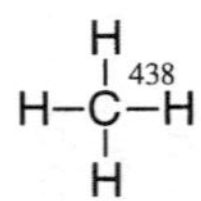 | $CD_4$ | 2.18 | 1.96 | | | Lehmann and Baerns, 1992 |

[a]Predicted value of $k_H/k_D$ if rupture of the indicated C—H bond in the corresponding substrate is the slow step in the reaction (reproduced by permission of Baltzer Scientific Publishers).

**Table 3.4** Kinetic isotope effect ($k_H/k_D$) for radical formation from $C_3H_6$ and $C_3D_6$ over $Bi_2O_3$ (Driscoll *et al.*, 1987)

| *Temperature/K* | *Experimental* | *Predicted* |
|---|---|---|
| 638 | $2.3 \pm 0.4$ | 2.6 |
| 658 | $2.2 \pm 0.4$ | 2.5 |
| 698 | $2.2 \pm 0.4$ | 2.2 |
| 723 | $2.1 \pm 0.4$ | 2.1 |
| 749 | $1.8 \pm 0.4$ | 2.0 |



the generation of free radical species, whose further reaction will lead to the eventual stable oxidation product.

In general, the chemistry involved in the majority of selective oxidation reactions occurring in the gas phase over heterogeneous oxidation catalyst is consistent with activation of the hydrocarbon via the weakest C—H bond in the reacting substrate. However, there appears to be an exception to this general rule and it concerns the reactivity of the so called $\alpha$-oxygen species generated when nitrous oxide is exposed to an iron–silicalite catalyst (Panov *et al.*, 1998). The resulting species reacts with benzene to form phenol and this is now the basis for the commercial production of phenol. Since this is a zeolite based catalyst, in which diffusion limitations can normally be expected, KIE studies are not useful. However, the $\alpha$-oxygen species does give a very unusual reactivity trend when exposed to a hydrocarbon with more than one type of C—H bond. In a study of the oxidation of isopropyl benzene to yield a mixture of alcohols, the $\alpha$-oxygen species showed a preference for attacking the strongest available C—H bond. This is demonstrated by the product distribution shown for the reaction of $\alpha$-oxygen with isopropylbenzene in Scheme 3.3, which shows a preference for attack at the strongest C—H bonds, namely those attached to the aromatic ring.

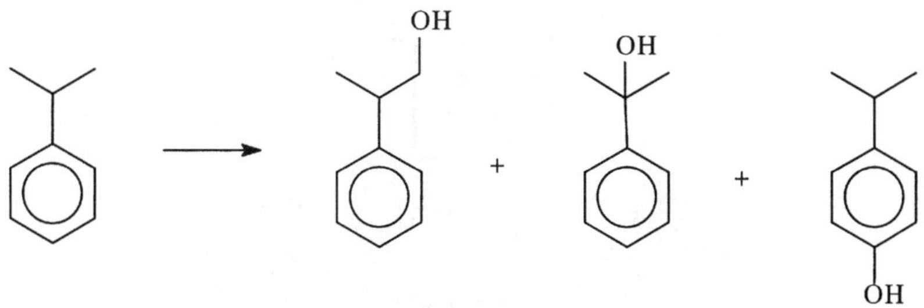


**Product Distribution** **2** **1** **12**

**Scheme 3.3** Reactivity of $\alpha$-oxygen towards isopropylbenzene (Panov *et al.*, 1998).

### 3.2.2 Discriminating mechanisms of the active site

If activation of the weakest C—H bond in the reactant directs the reaction towards selective oxidation, cleavage of any bond in the selective oxidation product will result in a lowering of selectivity. The question then arises as to how the active site in a selective oxidation catalyst distinguishes between C—H bonds in a reactant and similar C—H bonds in the selective oxidation product. This situation is represented in Scheme 3.4 for a metal oxide surface on which an active site for C—H bond rupture is located. The nature of this site is not elucidated, but a number of possible adsorbed and lattice oxygen structures are shown as well as an oxygen anion vacancy. It is assumed that C—H bonds can present themselves to the active site by collision from the gas phase or through adsorption. If the active site distinguishes between C—H bonds in the reactant on the basis of bond strength, it is reasonable to propose that it distinguishes in a similar fashion between all C—H bonds that present themselves to the active site on a bond strength basis irrespective of whether these bonds appear in the reactant or selective oxidation product. Scheme 3.4 represents an active site sensing the presence of the C—H bond, but the interaction with the substituent groups (R) may be secondary in importance to the properties of the bond being activated. Clearly steric effects can be important here. Scheme 3.4 presents three types of C—H bond in propene and three in acrolein. On what basis does the active site distinguish between them?

It is possible through appropriate use of the chemical literature to determine the best performance that has been reported for a wide range of selective

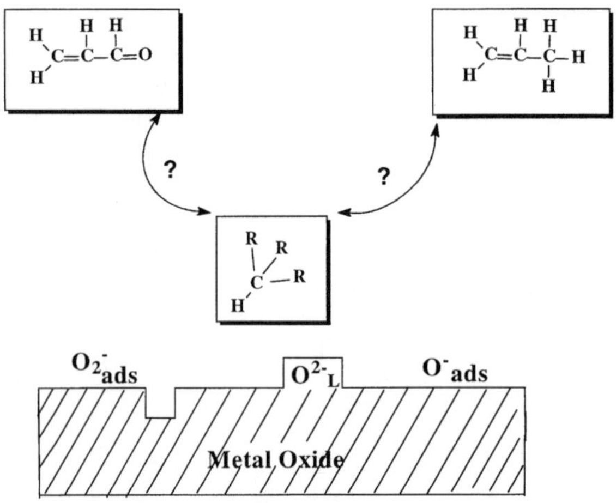


**Scheme 3.4** Schematic diagram of the interaction of a species representing a reactant (propene) or a selective oxidation product (acrolein) with an active site on a metal oxide surface.

oxidation reactions. This approach does not attempt to determine kinetic parameters or reaction networks, but simply sets out to use selectivity–conversion plots to determine if there is a limiting selectivity associated with individual reactions when conventional oxidation catalysts are used. Figures 3.5–3.8 present collations of selectivity–conversion data for four selective oxidation reactions. For each reaction the data were collected for a range of catalysts types and the plots were constructed without regard to the experimental conditions used. These plots demonstrate that for each selective oxidation reaction there is an upper limit in terms of selectivity–conversion beyond which a particular selective oxidation reaction has not yet proceeded. It should be stressed that these limits are determined by the available selective oxidation catalysts and there is no reason why these limits cannot be surpassed in future through further catalyst development work. For the reactions shown in Figures 3.5–3.8 the upper selectivity limit is very close to ideal behaviour for methanol oxidation, but falls short of the ideal for propene, *n*-butane and methane oxidation. A total of 24 oxidation reactions have been

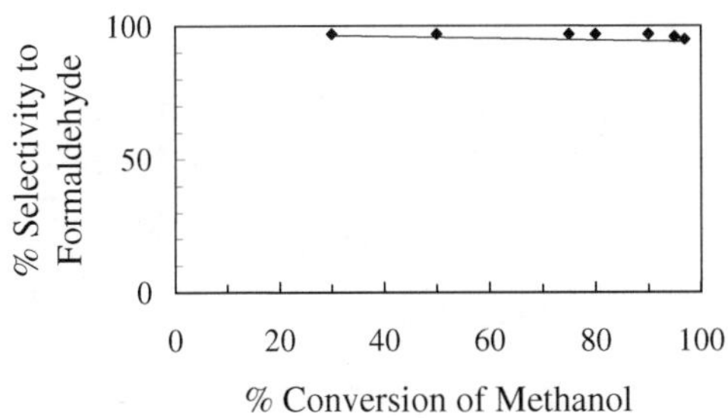


**Figure 3.5** Multiple selectivity–conversion plot for the oxidation of methanol to formaldehyde (Batiot and Hodnett, 1996) (reproduced by permission of Elsevier Science).

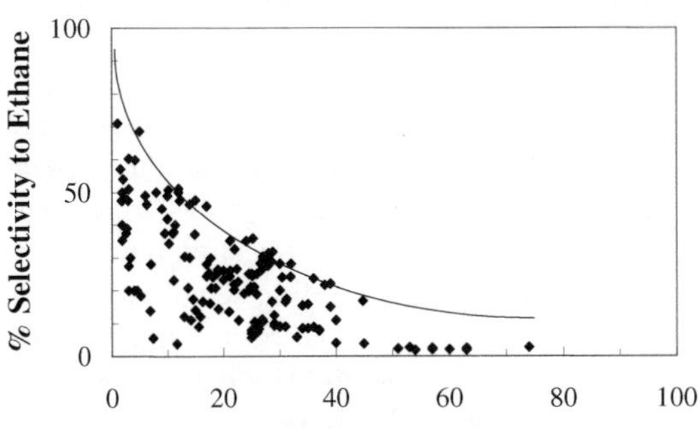


**Figure 3.6** Multiple selectivity–conversion plot for the oxidation of methane to ethane (Batiot and Hodnett, 1996) (reproduced by permission of Elsevier Science).

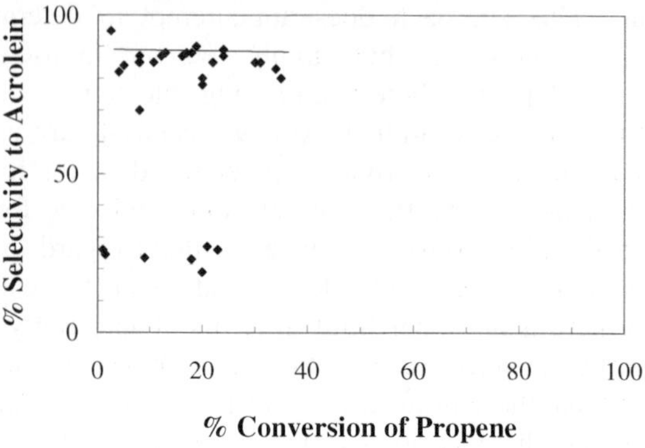


**Figure 3.7** Multiple selectivity–conversion plot for the oxidation of propene to acrolein (Batiot and Hodnett, 1996) (reproduced by permission of Elsevier Science).

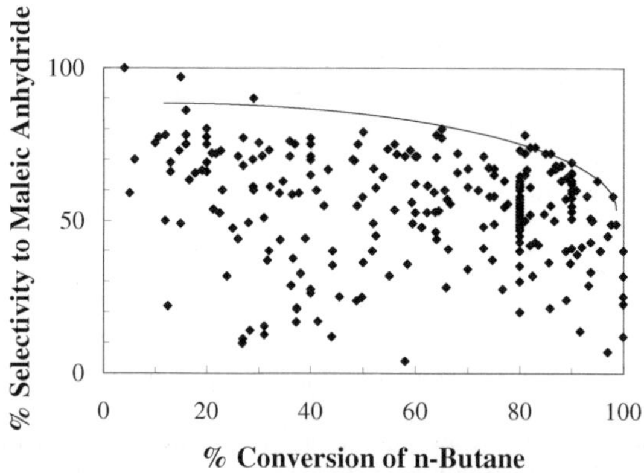


**Figure 3.8** Multiple selectivity–conversion plot for the oxidation of *n*-butane to maleic anhydride (Batiot and Hodnett, 1996) (reproduced by permission of Elsevier Science).

examined in this way and each reaction has its own characteristic upper limit. To arrive at this upper limit for each reaction one must operate with a good reactor configuration and with an optimized catalyst composition, as emphasized in Section 3.1. However, the primary question now is whether there was a generic factor which determines the limiting selectivity behaviour for which multiple selectivity–conversion plots have been complied (Batiot and Hodnett, 1996; Batiot *et al.*, 1997; Cassidy and Hodnett, 1998).

A general hypothesis is that selectivity in oxidation catalysis involves activation of the reactant through rupture of specific a C—H bond (the $k_1$ route in Scheme 3.1), whereas diminishing selectivity is associated with rupture of any bond in the selective oxidation product (the $k_2$ route in

**Table 3.5** Bond dissociation enthalpies for reactants and products in four selective oxidation reactions (Batiot and Hodnett, 1996)

| *Reaction* | *Weakest C—H in the reactant/kJ mol$^{-1}$* | *Weakest C—H or C—C in the product/kJ mol$^{-1}$* | $D^\circ H_{C—H\ (reactant)} - D^\circ H_{C—H\ or\ C—C\ (product)}$/kJ mol$^{-1}$ |
|---|---|---|---|
| $CH_3OH + 0.5O_2 \longrightarrow HCHO + H_2O$ | 393<br>[C—H$_3$]—OH | 364<br>[C—H$_2$]=O | 29 |
| $2CH_4 + 0.5O_2 \longrightarrow C_2H_6 + H_2O$ | 438<br>[C—H$_4$] | 376<br>[H$_3$C—CH$_3$] | 62 |
| $C_3H_6 + O_2 \longrightarrow C_3H_4O + H_2O$ | 361<br>CH$_2$=CH—[C—H$_3$] | 347<br>CH$_2$=[HC—CH]=O | 14 |
| $n$-$C_4H_{10} + 3.5O_2 \longrightarrow C_4H_2O_3 + 4H_2O$ | 391<br>CH$_3$—[C—H$_2$]—CH$_2$—CH$_3$ | 412<br>COOH—[C—H]=CH—COOH | −21 |

Scheme 3.1). As a means of validating this hypothesis the function (Batiot and Hodnett, 1996):

$$D^\circ H_{\text{C—H (reactant)}} - D^\circ H_{\text{C—H or C—C (product)}} \tag{3.12}$$

where $D^\circ H_{\text{C—H (reactant)}}$ is the bond dissociation enthalpy of the weakest C—H bond in the substrate and $D^\circ H_{\text{C—H or C—C (product)}}$ is the bond dissociation enthalpy of the weakest bond in the selective oxidation product was evaluated in 24 selective oxidation reactions. The values for this function are tabulated in Table 3.5 for the four reactions for which selectivity–conversion data are presented in Figures 3.5–3.8. This table also indicates the weakest bonds in reactants and selective oxidation products used to evaluate equation (3.12).

The multiple selectivity–conversion plots can be used to evaluate the optimum selectivity at a conversion of 30 % that has been reported for each of the 24 reactions studied. It is important to work at a fixed conversion when comparing selectivities because selectivity in many of these types of reaction depends strongly on conversion (see Figures 3.2 and 3.3). The observed correlation (Figure 3.9) shows that there is a clear relationship between limiting selectivities and weakest bonds in the reactants and products.

For some reactions, such as the oxidation of *n*-butane to maleic anhydride, there is no bond in the product weaker than the weakest C—H bond in the reactant (see Table 3.5). This combination always gives a good selectivity, because the selective oxidation product can survive the reaction environment without an excessive amount of decomposition. Conversely, formaldehyde is much less stable than methane and there does not appear to be any catalyst which is capable of selectively activating the strong C—H bonds in methane rather than the similar, but much weaker, C—H bonds in formaldehyde. In a general sense, Figure 3.9 shows that active sites on oxide catalysts are capable of activating target bonds in the reactant that are up to 30–40 kJ $mol^{-1}$ weaker than similar bonds in the selective oxidation product. When bigger differences in bond energies arise drastic reductions in selectivities result because the discriminating capacity of the active site has been exceeded (Batiot and Hodnett, 1996; Batiot *et al.*, 1997; Cassidy and Hodnett, 1998).

## 3.3 TYPES OF ACTIVE OXYGEN SPECIES INVOLVED IN OXIDATION CATALYSIS

A central question in oxidation catalysis is the identity of the reactive oxygen species. At least three distinct types of oxygen are available in the reaction medium, namely gas phase, adsorbed and lattice oxygen. It is important to understand the relationship between these types of oxygen species and the role of each in the oxidation process.

Figure 3.10 shows the enthalpy of formation of a number of neutral and negatively charged gas phase oxygen species (Bielanski and Haber, 1990). No

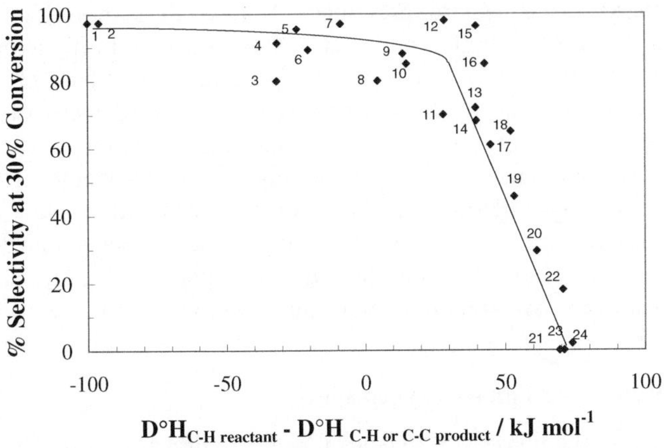


**Figure 3.9** Selectivity at 30% conversion for the reactions indicated as a function of $D°H_{C—H\ (reactant)} - D°H_{C—H\ or\ C—C\ (product)}$ (Cassidy and Hodnett, 1999) (reproduced by permission of Baltzer Scientific Publishers): 1, ethylbenzene to styrene; 2, 1-butene to 1,3-butadiene; 3, toluene to benzoic acid; 4, acrolein to acrylic acid; 5, ethane to ethylene; 6, *n*-butane to maleic anhydride; 7, benzene to phenol; 8, toluene to benzaldehyde; 9, propene to acrolein; 10, 1-butene to 2-butanone; 11, isobutane to isobutene; 12, methanol to formaldehyde; 13, methacrolein to methacylic acid; 14, propane to propene; 15, ethanol to acetaldehyde; 16, isobutene to methacrolein; 17, *n*-butane to butene; 18, benzene to maleic anhydride; 19, propane to acrolein; 20, methane to ethane; 21, ethane to acetaldehyde; 22, isobutane to methacrylic acid; 23, methane to formaldehyde; 24, isobutane to methacrolein.

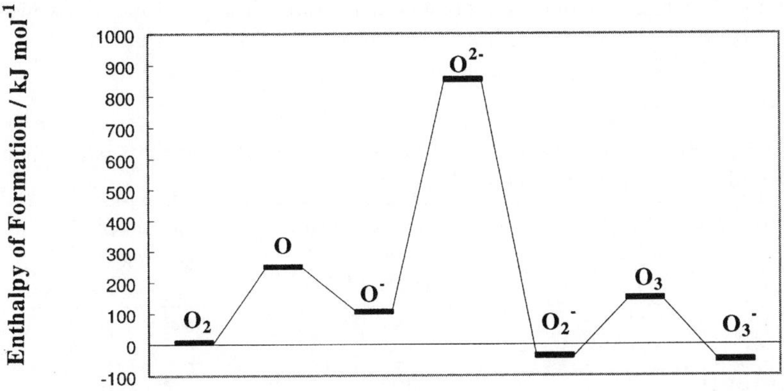


**Figure 3.10** Enthalpy of formation of neutral and negatively charged gas phase oxygen species (Bielanski and Haber, 1990).

positively charged species are presented because all show enthalpies of formation in excess of +1000 kJ mol$^{-1}$. The species shown in Figure 3.10 are important in oxidation catalysis because a significant number ($O^{\bullet -}$, $O_2^-$, $O_3^-$) have been observed by ESR spectroscopy on the surfaces of a wide variety of oxides, and $O^{2-}$ is the predominant anion in oxides (Lunsford, 1973). The $O_2^-$ and $O_3^-$ species are thermodynamically stable in the gas phase with respect to $O_2$. However, adsorption or incorporation into a crystalline lattice has a major influence on the stability of these species. Indeed, there are good chemical reasons to think that adsorbed species $O^{\bullet -}$, $O_2^-$ and $O_3^-$ play roles in selective and total oxidation catalysis.

### 3.3.1 Reactivity of gas phase oxygen species

The reactivity of gas phase $O^{\bullet -}$ species towards hydrocarbons has been well documented. The predominant gas phase reaction of methane with the oxygen anion radical is (Lee and Grobowski, 1992):

$$O^{\bullet -} + CH_4 \longrightarrow OH^- + CH_3^{\bullet} \tag{3.13}$$

In general, the predominant reaction of gas phase generated $O^{\bullet -}$ with alkanes is the abstraction of a hydrogen atom and the generation of the corresponding radical species, as shown in equation (3.13) for the reaction of $O^{\bullet -}$ with methane. The link with catalytic oxidation is that the methyl radical has been identified as a key intermediate in the oxidation of methane to ethane over oxide catalysts. In addition, there is evidence to suggest that hydrocarbon derived radicals are important in a wide range of selective and total oxidation reactions (Driscoll *et al.*, 1987).

Another factor is that the reactivity of $O^{\bullet -}$ towards saturated hydrocarbons shows some correlation with C—H bond dissociation energy of the hydrocarbon. The relevant data are presented in Table 3.6 (Lee and Grobowski,

**Table 3.6** Room temperature rate coefficients for the reaction of $O^{\bullet -}$ with saturated hydrocarbons (Lee and Grobowski, 1992)

| *Hydrocarbon* | *Bond dissociation enthalpy of the weakest C—H bond/kJ mol$^{-1}$* | *Rate coefficient/ cm$^{-3}$ molec$^{-1}$ s$^{-1}$* |
|---|---|---|
| $CH_4$ | 438 | $1 \times 10^{-10}$ |
| $CH_3CH_3$ | 420 | $7.3 \times 10^{-10}$ |
| $CH_3CH_2CH_3$ | 401 | $9.3 \times 10^{-10}$ |
| $CH_3CH_2CH_2CH_3$ | 390 | $12 \times 10^{-10}$ |

1992). For the sequence $C_1$–$C_4$, in which the C—H bond strength decreases as the chain length increases, the reactivity towards $O^{\bullet -}$ increases. This is reminiscent of the reactivity of hydrocarbons at oxide surfaces (see Section 3.2). Work described in more detail in Chapter 6 demonstrates a similar effect to that described in Table 3.6, namely a correlation between the rate of total oxidation of a range of linear alkanes over a $Cu_2O$ catalyst and the bond dissociation enthalpy of the weakest C—H bond in the alkane (Sokolovskii, 1990).

The reactions of $O^{\bullet -}$ with alkenes also closely resemble some features observed when the same alkenes are mixed with oxygen and passed over a selective oxidation catalyst (Lee and Grobowski, 1992). Hence ethylene oxide is the predominant product of the reaction of $O^{\bullet -}$ with ethylene, just as it is when a mixture of ethylene and oxygen is passed over a silver catalyst:

$$\begin{aligned} O^{\bullet -} + CH_2{=}CH_2 &\longrightarrow C_2H_4O + 1e && 68\,\% \\ &\longrightarrow CH_2{=}C^{\bullet -} + H_2O && 29\,\% \\ &\longrightarrow CH_2{=}CHO^- + H^{\bullet} && 32\,\% \end{aligned} \qquad (3.14)$$

The predominant product of the reaction of $O^{\bullet -}$ with propene is the allyl radical:

$$\begin{aligned} O^{\bullet -} + CH_2{=}CH{-}CH_3 &\longrightarrow C_3H_5{}^{\bullet} + OH^- && 89\,\% \\ &\longrightarrow C_3H_4{}^{\bullet -} + H_2 && 11\,\% \end{aligned} \qquad (3.15)$$

We will see below that this is an important intermediate in the oxidation of propene to acrolein (Chapter 4) and in its ammoxidation to acrylonitrile (Chapter 8).

Hence, we can conclude that reaction of gas phase $O^{\bullet -}$ species with hydrocarbons lead to many of the reaction intermediates and reactivity trends commonly observed in oxidation catalysts.

### 3.3.2 Nature and reactivity of adsorbed oxygen species

Adsorbed $O^{\bullet -}$ can be generated on an oxide surface by exposure to $O_2$ or $N_2O$. Generally this species can be observed at and below room temperature using the ESR technique (Lunsford, 1973). The adsorption process stabilizes the $O^{\bullet -}$ relative to the situation which pertains in the gas phase by an amount corresponding to the heat of adsorption. The heat of adsorption is always negative (see Section 2.5.2) so the adsorbed species can be regarded as occupying a thermodynamic sink along the reaction pathway. The characteristic $O^{\bullet -}$ ESR signal disappears when the sample is exposed to hydrocarbons, even at room temperature. However, it is often necessary to increase the sample temperature above 423 K to observe reaction products.

Generally $O^{\bullet -}$ seems to be the more reactive of the negatively charged adsorbed oxygen species observed by the ESR technique. $O^{\bullet -}$ on MgO reacts with a 1 : 1 stoichiometry with $C_1$–$C_4$ alkanes. At low desorption temperatures the main products are the corresponding alkenes and at higher temperatures $CO_2$ and $CH_4$ form. No alkyl radicals were observed but it is felt that the overall reaction is similar to those outlined for the corresponding gas phase reaction. Alkyl radicals may react with surface $O^{2-}$ anions to give surface alkoxides or they may undergo a second hydrogen abstraction to give small amount of alkenes (Che and Tench, 1982).

The $O_2^-$ species is much less reactive than the $O^{\bullet -}$ species. The ESR signal disappears slowly when contacted with alkenes or alkanes at room temperature. It appears that propene reacts with $O_2^-$ to give an allyl radical in a similar fashion to the analagous gas phase reaction (Che and Tench, 1983).

### 3.3.3 Lattice oxygen species

The relationship between gas phase, adsorbed and lattice oxygen can also be probed using isotopic exchange of oxygen. In this technique $^{18}O_2$ or a mixture of $^{18}O_2$ and $^{16}O_2$ is contacted with the surface of an oxide and the reaction is monitored for the appearance of $^{16}O^{18}O$. Three types of exchange mechanisms are possible and these are outlined in Scheme 3.5. In the first of these (the $R_0$ mechanism) the exchange occurs exclusively between adsorbed oxygen species; lattice oxygen does not become incorporated into the reaction products. In the second and third mechanisms, one ($R_1$) or two ($R_2$) lattice $^{16}O$ anions are extracted from the oxide and replaced by $^{18}O$. Table 3.7 lists a number of simple oxide systems that have been investigated using the oxygen isotope exchange method and the type of exchange observed ($R_0$, $R_1$ or $R_2$) observed for each (Bielanski and Haber, 1990). In general, exchange of lattice and gas phase oxygen is possible in the conditions typically encountered in oxidation catalysis.

A large number of studies have now appeared in the literature in which a mixture containing a hydrocarbon and $^{18}O_2$ is reacted over a selective oxidation catalyst. The catalyst is prepared in the normal fashion, so that all the lattice oxygen present is as isotope 16. A consistent finding of these studies is that the oxygen which becomes incorporated into the selective oxidation product is $^{16}O$. In many cases the total oxidation products are also formed with $^{16}O$ derived from the lattice. The results of one such study are presented in Figure 3.11. Here a mixture of propene and $^{18}O_2$ was passed over a bismuth molybdate catalyst and the reaction products analysed in terms of the percentage oxygen-18 incorporated into the product acrolein. Because of the expense associated with the use of $^{18}O_2$, this type of reaction was originally carried out in closed loop type systems and then, as techniques

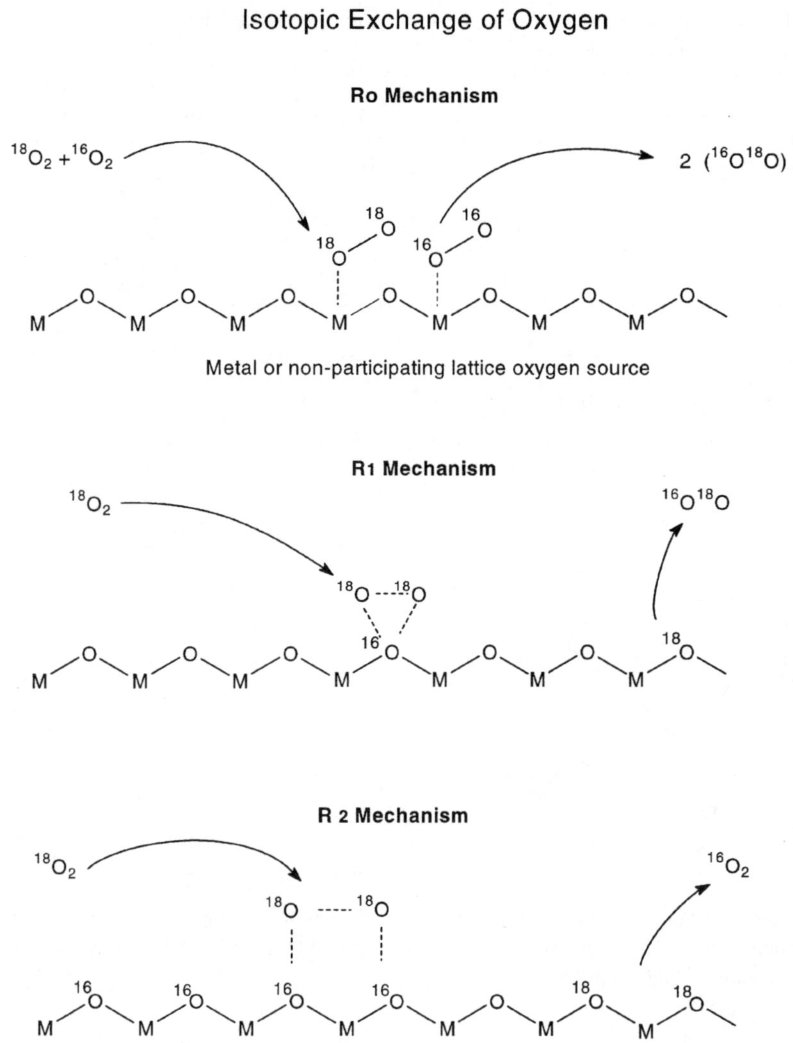


**Scheme 3.5** Schematic diagram of the $R_0$, $R_1$ and $R_2$ mechanisms of oxygen isotope exchange.

improved, in continuous flow systems. A feature of these studies was that $^{16}O$ derived from the lattice became replenished with $^{18}O$ derived from $^{18}O_2$ as the reaction proceeded, so that the percentage of $^{16}O$ in the selective oxidation product became less as the reaction proceeded. However, when the data in Figure 3.11 are extrapolated to close to zero reaction time it appears that all the acrolein product is formed from oxygen derived from the lattice.

**Table 3.7** Mechanism of oxygen isotope exchange on selected oxides (Bielanski and Haber, 1990)

| *Oxide* | *Temperature range/K* | *Mechanism* |
|---|---|---|
| CuO | 500–600 | $R_2$ |
| $V_2O_5$ | 750–810 | $R_1$ |
| $MoO_3$ | 873–933 | $R_1$ |
| | 933–963 | $R_2$ |
| $La_2O_3$ | 623–723 | $R_0$, $R_1$ |
| $Al_2O_3$ | 773–873 | $R_1$ |

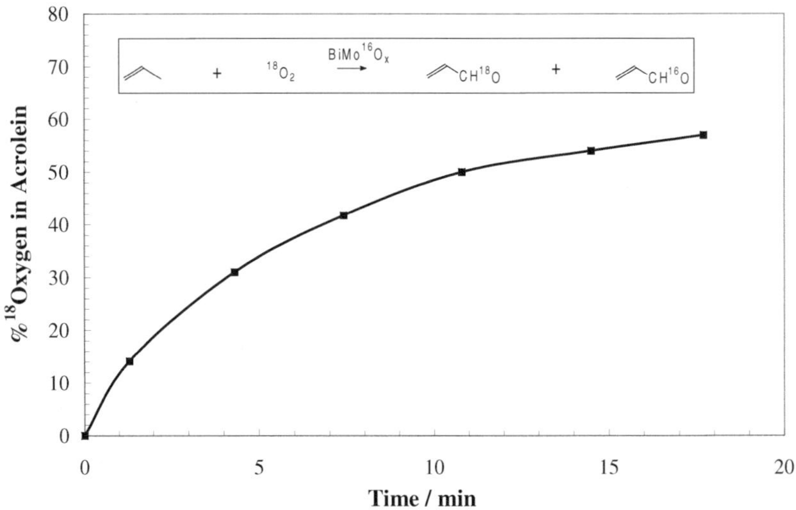


**Figure 3.11** Percentage of $^{18}O$ in acrolein formed as a function of reaction time from propene and $^{18}O_2$ over $Bi_{0.85}V_{0.55}Mo_{0.45}{}^{16}O_4$ (Ueda *et al.*, 1986) (reproduced by permission of Academic Press).

The more recent development of the TAP apparatus (see Section 2.3.2) has allowed rapid and very sensitive analysis of the submonolayer equivalents, namely it allows the detection by mass spectral analysis of the reaction products at a point in the reaction where most active sites on the catalyst surface have not yet been involved in one catalytic cycle.

The results of one such study are presented in Scheme 3.6 for the oxidation of methane to formaldehyde over a vanadium oxide catalyst supported on silica. In this study a mixture of $^{18}O_2$ and $CH_4$ was pulsed over the catalyst sample held in a TAP reactor. The major finding was that the $HCH^{16}O$ was the only selective oxidation product observed (Kartheuser *et al.*, 1993). The mechanism presented in Scheme 3.6 is consistent with Mars and van Krevelen

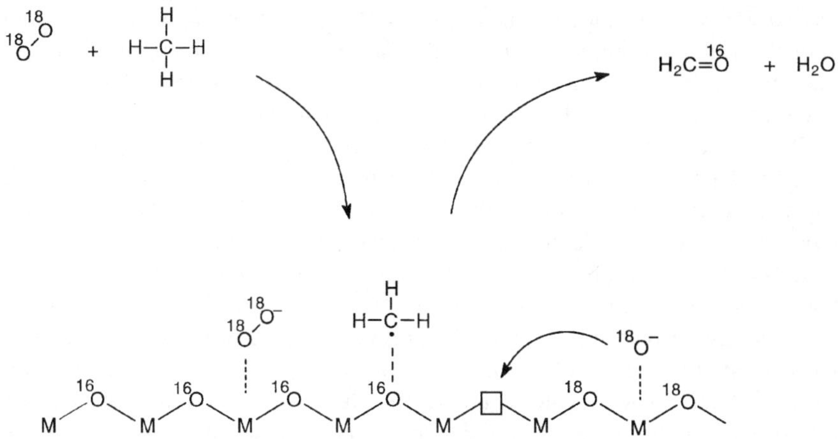


**Scheme 3.6** Schematic diagram of the reaction of $^{18}O_2$ with $CH_4$ over a $^{16}O$ oxide catalyst (Kartheuser *et al.*, 1993).

type kinetics wherein the hydrocarbon, or a species derived from it, reacts and extracts a lattice oxygen anion from the surface layer of the metal oxides (Mars and van Krevelen, 1954). This reaction brings about a reduction of the metal oxide surface, but the original surface oxidation state is restored by reaction of the reduced site with an adsorbed oxygen species.

In Scheme 3.6 two types of adsorbed oxygen species, $O_2^-$ and $O^-$, are shown. There is no evidence that the reacting methane ever chemisorbs on the catalyst surface. It probably reacts directly from the gas phase, or through a very weakly physisorbed state. Other evidence from this same study indicates that over vanadium oxide supported on silica, methane does not react in the absence of adsorbed oxygen species. Accordingly, the detailed mechanism, in this case, involves the initial activation of the methane by an adsorbed oxygen species with the generation of a methyl radical. There is ample evidence from the work of Lunsford (1989) that methyl radicals are generated in these conditions and moreover they can react with lattice oxygen. Formaldehyde is then generated by reaction of the methyl radical with lattice oxygen. Details are as follows (Kartheuser *et al.*, 1993):

$$CH_4 + O_n^- \longrightarrow CH_3^\bullet + O_nH^- \tag{3.16}$$

$$CH_3^\bullet + O_L^{2-} \longrightarrow CH_3O^\bullet + 2e \tag{3.17}$$

$$CH_3O^\bullet \longrightarrow HCHO + H^\bullet \tag{3.18}$$

where $n = 1$ or 2. A complication encountered in this type of work is that oxygen isotope exchange via the $R_0$, $R_1$ or $R_2$ mechanisms may be more rapid than the rate of oxidation of the hydrocarbon, and there may be additional possibilities for exchange between the reaction products and $^{18}O_2$. In such circumstances care must be taken in interpreting the results.

The respective roles of lattice and adsorbed oxygen species in selective and total oxidation reactions has not yet been fully elucidated. An alternative approach is that proposed by Haber and co-workers (Haber, 1997). Here two types of oxygen are identified, namely electrophilic species, including $O^{\bullet -}$ and $O_2^-$, and nucleophilic species, namely $O^{2-}$. The central idea here is that the $O^{2-}$ species is fully reduced, hence does not attack electron rich sites in hydrocarbons. However, $O^{\bullet -}$ and $O_2^-$, prevalent as adsorbed species, may be considered as electrophilic reagents, which, especially in reactions with olefins, attack the molecule in the region of highest electron density. Such electrophilic additions result in the formation of peroxy and epoxy structures, which serve as intermediates for the degradation of the carbon skeleton and lead to total oxidation.

Lattice oxygen ions on the other hand are considered to be nucleophilic reagents, with no oxidizing properties. The hydrocarbon molecule is activated by interaction with the metal cation of the oxide catalyst. The metal cation is reduced in this process. Oxygen anions are inserted into the activated hydrocarbon molecules by nucleophilic addition. When the oxygenated product desorbs it leaves an oxygen anion surface vacancy. This vacancy is replenished from the pool of gas phase oxygen and this action also restores the original oxidation state of the metal cation (Haber, 1997).

## 3.4 STRUCTURAL FEATURES OF OXIDATION CATALYSTS

In general oxidation catalysts are used to treat large gas volumes, so ease of flow of gases through catalyst beds is essential. Thus the need to design suitable catalyst pellets, extrudates or honeycomb structures for fixed bed operation and fluidizible particles for fluidized bed operation. In the case of pellets, sizes range from 1 to 10 mm, but the primary particles which are compacted to make up pellets are typically in the size range 20–1000 nm. A good deal of information is available on the size of primary particles (macroscopic scale) that constitute oxidation catalysts and on the nature of active sites (microscopic scale) where the reactions of interest occurs. Structural features of oxidation catalysts are evaluated on the macroscopic scale, in terms of the size of the primary particles which make up oxidation catalysts and on the microscopic scale in terms of the structure of active sites.

### 3.4.1 Macroscopic features of oxidation catalysts

A small subset of oxidation catalysts is presented in Table 3.8 with a minimum amount of characterization data. It is intended to illustrate the simple finding that, in general, good oxidation catalysts are not composed of highly dispersed materials. This is amply illustrated by reference to the standard Ag/$\alpha$-$Al_2O_3$ catalyst used in the epoxidation of ethylene. This catalyst, which will be explored extensively in Chapter 7, has a surface area of less than 1 $m^2\ g^{-1}$ and features very poorly dispersed silver particles on a non-porous alumina support.

A similar observation can be made concerning the vanadium phosphorus oxide and bismuth molybdate multicomponent catalysts used in *n*-butane and propene oxidation and ammoxidation reactions. In neither case is a support used in spite of the economic advantages associated with dispersing the rather expensive active phases. (In ammoxidation reactions, which operate in fluidized beds, silica is added to the basic catalyst to give it the necessary attrition resistant characteristics.)

The $V_2O_5/TiO_2$ catalyst used for the oxidation of *o*-xylene into phthalic anhydride appears, at first sight, to be an exception in this regard, because it features a support onto which the vanadium oxide phase is dispersed. However, electron microscopy evidence on the nature of these catalysts indicates that the vanadium oxide phase appears to fully cover the $TiO_2$ phase. The overall result, presented in Figure 3.12 appears to be consistent with the formation of several layers of vanadium oxide over a titanium oxide support (Sanati *et al.*, 1991). Similar structures have also been proposed for the structure of the multicomponent catalysts used in propene oxidation (Wolfs and Batist, 1974) and this was recently illustrated by electron microscopy (He *et al.*, 1992) (see Chapter 4 for full details).

Total oxidation catalysts also fall into this pattern. Platinum or palladium is normally dispersed to the ultimate extent in many applications. However, the situation is somewhat different in oxidation catalysis. In this case there is ample evidence that larger precious metal particles are superior to smaller particles in total oxidation reactions. The reasons for this will be entered into in detail in Chapter 6. Platinum particle sizes in the range 1–30 nm are commonly encountered in oxidation catalysts, a long way from atomic dispersion. In Chapter 6 combustion catalysts will be discussed in detail and the relationship between metal particle size and heat of adsorption of oxygen will be developed.

The recently developed titanium silicalite family of catalysts appear to feature a genuinely atomically dispersed form of the active material (Ti) inside a silicalite support. In this case the catalyst is made up of a solid solution of titanium in silicalite, whereby a small fraction of the silicon atoms in the silicalite (no more than 2.5 mol %) are replaced by titanium.

In view of the structural nature of many oxidation catalysts the question of whether oxidation reactions on heterogeneous catalysts are structure sensitive has often been addressed. The term 'structure sensitive' (Boudart, 1969) is applied to those reactions which are characterized by a strong dependence of turnover rate on the structure of the catalytic surface. The structure of the

**Table 3.8** Textural features oxidation catalysts

| *Reaction* | *Catalyst* | *Phase composition* | *Surface area/ $m^2\ g^{-1}$* | *Texture* | *Reference* |
|---|---|---|---|---|---|
| 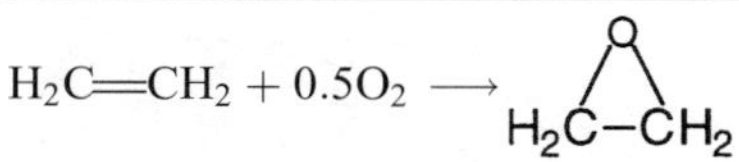 | Ag/$\alpha$-$Al_2O_3$ | Silver metal and $\alpha$-alumina | <1 | Large silver particles >100 nm covering the alumina support | De Jong, 1998 |
| 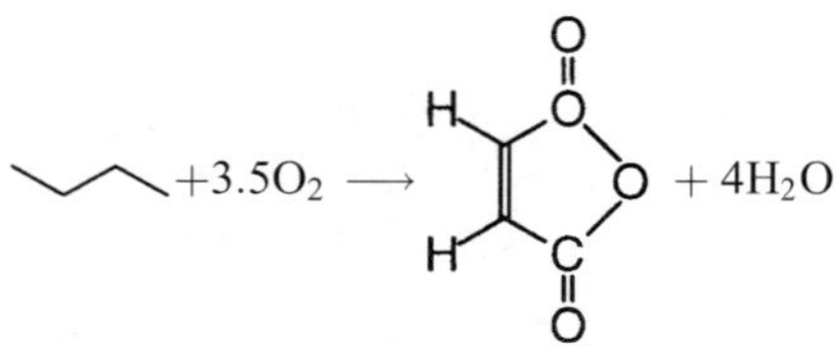 | Vanadium phosphorus oxide | $(VO)_2P_2O_7$ | <30 | Crystalline platelet morphology Dimensions (70 × 70 × 10 nm) | Horowitz *et al.*, 1988 |
| 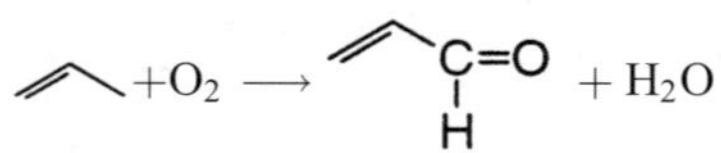 | Bismuth molybdates promoted with cobalt and iron | Mixture of cobalt and iron molybdates covered by a bismuth molybdate layer | <15 | Surface layer of bismuth molybdate 0.2–1 nm in depth. Particle diameter ca 300 nm | He *et al.*, 1992; Moro-oka and Ueda, 1994 |

| | | | | | |
|---|---|---|---|---|---|
|  $+5O_2 \longrightarrow 3CO_2 + 4H_2O$ | $Pt/Al_2O_3$ | Pt particles on carrier | | 1–30 nm | Yao, 1980 |
| 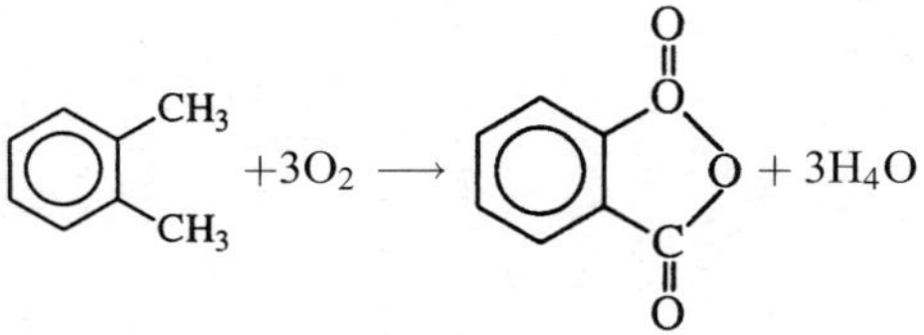 | $V_2O_5/TiO_2$ | Vanadium oxide layer spread on titanium dioxide | <30 | Surface layer thickness 1–5 nm. Support diameter ca 100 nm | Sanati *et al.*, 1991 |
| 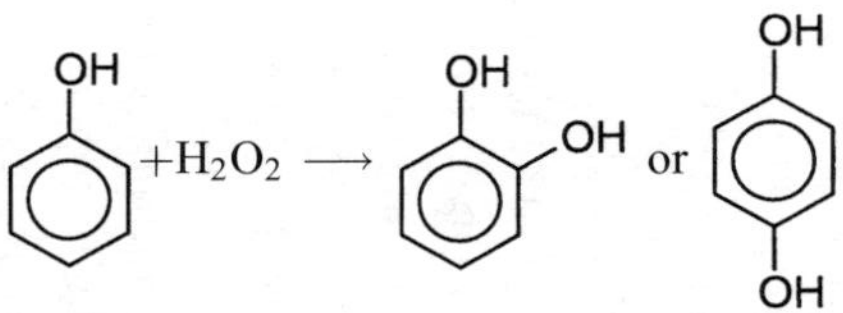 | Titanium silicalite | Solid solution of titanium in silicalite | micro-porous | Isolated titanium sites in TS-1, to a maximum of 2.5 mol% | Millini *et al.*, 1992 |

(a)
(b)

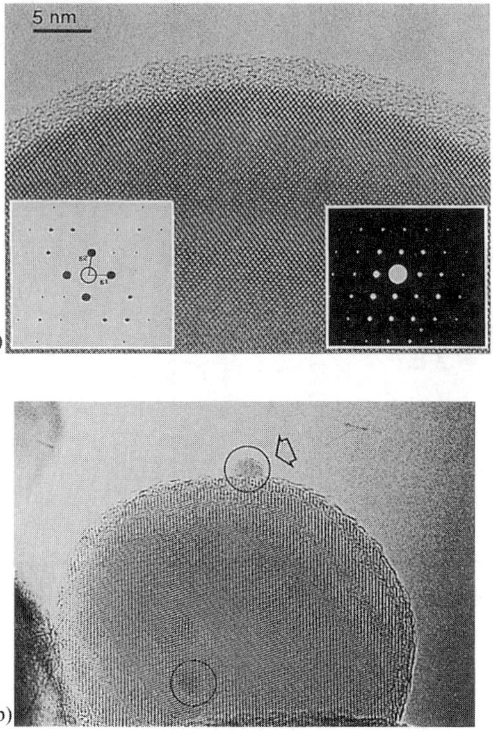


**Figure 3.12** High resolution electron microscopy of vanadium oxide overlayers on (a) $TiO_2$ anatase and (b) the emergence of vanadium oxide crystallites (Sanati *et al.*, 1991) (reproduced by permission of Academic Press).

catalytic surface is determined in large part by the crystallographic face of which it forms part. The effectiveness of a particular catalytic site is evaluated in terms of a turnover rate of frequency, which is usually expressed as molecules of reactant converted (or of product formed) per active site per second (molec site$^{-1}$ s$^{-1}$). This is normally expressed in terms of s$^{-1}$. In circumstances where the active site density or size cannot be determined, areal rates are used, such as molec cm$^{-2}$ s$^{-1}$. The term 'structure sensitive' was first developed for reactions catalysed by supported metals and for which a strong relationship was observed between metal particle size and reaction rate. More recently this concept has been applied to reactions catalysed on oxide surfaces, although as we will see below this development may not be altogether appropriate.

There are two approaches to determining if a reaction is structure sensitive. The first of these, and the one originally applied, was to vary the metal particle size below about 20 nm, but more preferably below 5 nm (Sajkowski and

Boudart, 1987). Variations in this size range can generate large changes in the proportion of individual crystal planes that are exposed to the reactants.

The second approach uses modern surface science techniques which often make it possible to evaluate the catalytic activity of less than 1 $cm^2$ of an exposed crystal face. The reaction of interest is normally carried out at pressures and temperatures close to those used in standard testing, but the single crystal can be examined by a range of UHV techniques before and after testing.

For the surface science approach rates are often reported as molec $cm^{-2}\,s^{-1}$, whereas in the particle size approach a turnover frequency is quoted. The latter implies a knowledge of the number of the active sites, but in reality with supported metals this is taken as the number of exposed metal atoms determined by a technique such as hydrogen or carbon monoxide chemisorption (see Section 2.7.3).

The classic example of a structure sensitive reaction is the nitrogen fixation reaction on iron single crystals or on supported iron catalysts. Both the approaches described above were applied to this reaction, and the results are summarized in Table 3.9.

These data illustrate a change of 35 in the turnover frequency for ammonia formation for the Fe/MgO catalysts studied, with the highest frequencies associated with the larger iron particles. In the surface science approach, a difference in areal rates of over 400 was recorded between Fe(1 1 1) and Fe(1 1 0).

To prove a structure sensitivity effect a large difference in turnover frequencies or areal rates must be demonstrated. This point is illustrated in Table 3.10 for a structure insensitive reaction, namely the hydrogenolysis of cyclopropane on platinum catalysts. Here the turnover frequency of a platinum single crystal surface is compared with that of a number of supported catalysts.

Although the rates or turnover frequencies differ by a factor of 4, these values are regarded as identical within experimental error. Variations in areal

**Table 3.9** Evidence for structure sensitivity for the reaction $2N_2 + 3H_2 \longrightarrow 2NH_3$ from single crystal and supported iron catalysts

| *Surface science approach (Spencer et al., 1982)* | | *Particle size approach (Boudart and Djega-Mariadassou, 1984)* | | |
|---|---|---|---|---|
| *Surface* | *Areal rate Molec $cm^{-2}\,s^{-1}$* | *Catalyst* | *Fe particle diameter d/nm* | *Turnover frequency $s^{-1}$* |
| Fe(1 1 1) | $2.8 \times 10^{16}$ | 1 %Fe/MgO | 1.5 | 1.0 |
| Fe(1 0 0) | $1.7 \times 10^{15}$ | 5 %Fe/MgO | 4 | 9.0 |
| Fe(1 1 0) | $6.6 \times 10^{13}$ | 40 %Fe/MgO | 30 | 35.0 |

**Table 3.10** Rates of cyclopropane hydrogenolysis on platinum single crystals and on a $Pt/Al_2O_3$ catalyst
(Somorjai and Carrazza, 1986)

| *Catalyst* | *Areal rate molec* $cm^{-2}\ s^{-1}$ | *Turnover frequency* $s^{-1}$ |
|---|---|---|
| Pt (1 1 1) | $2.0 \times 10^{16}$ | 13.5 |
| 0.04 wt% $Pt/\eta$-$Al_2O_3$ | $7.7 \times 10^{15}$ | 6.8 |
| 0.3 wt% $Pt/\eta$-$Al_2O_3$ | $8.9 \times 10^{16}$ | 8.0 |
| 0.6 wt% $Pt/\gamma$-$Al_2O_3$ | $2.5 \times 10^{16}$ | 22.3 |

rates of the size shown in Table 3.10 can easily be brought about by poisoning effects and by other factors, so extreme care must be taken in describing a reaction as structure sensitive.

There are a number of instances in the literature where attempts have been made to associate activity of a particular catalyst with active sites located on a particular crystal plane. One example is in the epoxidation of ethylene to ethylene oxide over supported silver catalysts. Because the silver particle size in supported catalysts often follows the particle size of the alumina support, particle diameters are always greater than 50 nm. In this particle size range it is difficult to see a structure sensitivity effect because all the predominant crystalline faces are present on the silver particle and changes in particle size above 5 nm do not usually change the proportion of individual crystal faces exposed (Verykios *et al.*, 1980; Sajkowski and Boudart, 1987). Differences have been measured in terms of areal rates for different preparations, which can often be associated with the presence of impurities, such as chlorine in the preparation step, which strongly inhibits the silver system (Seyedmonir *et al.*, 1990). However, silver catalysts are unique among selective oxidation catalysts because single crystal surfaces can be readily prepared and tested. When turnover frequencies were measured using single crystals of silver with the (1 1 0) or (1 1 1) faces exposed in conditions close to those in which supported catalysts are normally tested, the differences were less than a factor of 2 and as such too small to assign to a structure sensitive effect.

### 3.4.2 A description of active sites on selective oxidation catalysts

The concept of active sites on heterogeneous catalysts was first introduced by HS Taylor and has been widely used since, although not always fully defined (Thomas and Thomas, 1997). Here a working definition will be adopted whereby the active site is defined as that part of the catalyst surface over which reactants are activated, surface reaction occurs and from which products desorb.

The concept of active site has been widely applied in enzyme chemistry and a useful comparison of active sites in enzymes and on oxide catalysts can be carried out within certain limitations. Figure 3.13 presents this comparison on the basis of four criteria common to both. These are (i) operating temperature, (ii) molecular mass of the active site, (iii) turnover frequency and (iv) selectivity. The normal definitions of terms (i), (iii) and (iv) have been used for Figure 3.13 and the molecular mass of the active site for the enzyme was estimated assuming just one active site per enzyme molecule (Voet and Voet, 1995). The lower limit of active site size for heterogeneous catalysts has been estimated at the size of a single metal atom on a support whereas the upper limit is calculated from the concentration of acid sites in high silica to alumina ratio zeolites or in titanium silicalite. This measure is intended to determine the number of active sites per unit volume that could be fitted into a reactor so that the contribution by the support or the underlying bulk is taken into account in making the calculation. Realistically, on this basis, the active site size range in oxide catalysts lies in the range $10^2$–$10^4$ amu, the lower value referring to unsupported catalysts and the upper value to supported or diluted systems (Batio *et al.* 1977).

The general overview presented in Figure 3.13 is sufficient to indicate that in the majority of cases active sites on heterogeneous catalysts can be bigger than conventionally represented and overlap to some extent the size of corresponding sites in enzymes. In oxidation catalysis this finding is not surprising in view of the size of the primary particles that constitute oxidation catalysts, hence the rather low surface to mass ratios encountered. The normal operating temperature range for enzymes is restricted by comparison with heterogeneous catalysts but the turnover frequencies that can be achieved on

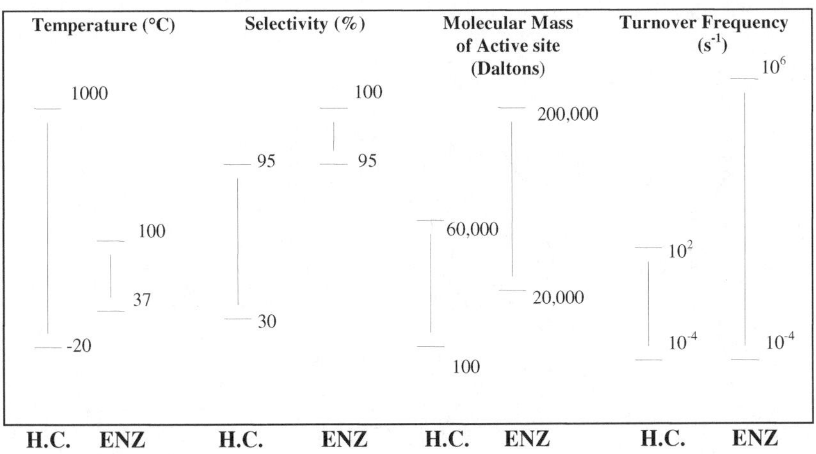


**Figure 3.13** Comparison of active sites in heterogeneous and enzymatic catalysis. (H.C. = heterogeneous catalysis; ENZ = enzymatic catalysis) (Batiot *et al.*, 1997) (reproduced by permission of Elsevier Science).

enzymes far outperform those achieved with heterogeneous catalysts. In addition, the selectivity achieved is normally very close to 100 % for enzymes and the transformations that can be achieved are chemically much more complex than those over heterogeneous catalysts.

Literature description of active sites on oxide catalysts often describe a picture of the surface site by extrapolation of the bulk structure. This is the approach taken by Grasselli to explain a proposed active site on the surface of a $USb_3O_{10}$ propene ammoxidation catalysts (Grasselli and Suresh, 1972; Grasselli, 1983). This compound was identified as the active phase in propene ammoxidation from a series of catalysts prepared with varying U : Sb ratios. Optimum catalytic performance, as indicated by the yield of acrylonitrile, was observed at an Sb : U ratio of 3 : 1. This was also the composition where the largest amount of $USb_3O_{10}$ was observed in the catalyst by X-ray diffraction and by infrared spectroscopy. The catalytic performance and characterization data are collected into Figure 3.14 and the proposed structure of the active surface for the catalyst is presented in Figure 3.15.

Total oxidation or ammoxidation reactions require more oxygen than the corresponding selective oxidation or ammoxidation reactions. For example, in the case of propene ammoxidation the selective reaction has the following stoichiometry:

$$CH_2{=}CH{-}CH_3 + 1.5O_2 + NH_3 \longrightarrow CH_2{=}CH{-}CN + 3H_2O \qquad (3.19)$$

The non-selective reaction has a greater oxygen requirement:

$$CH_2{=}CH{-}CH_3 + 4O_2 + NH_3 \longrightarrow HCN + 2CO_2 + 4H_2O \qquad (3.20)$$

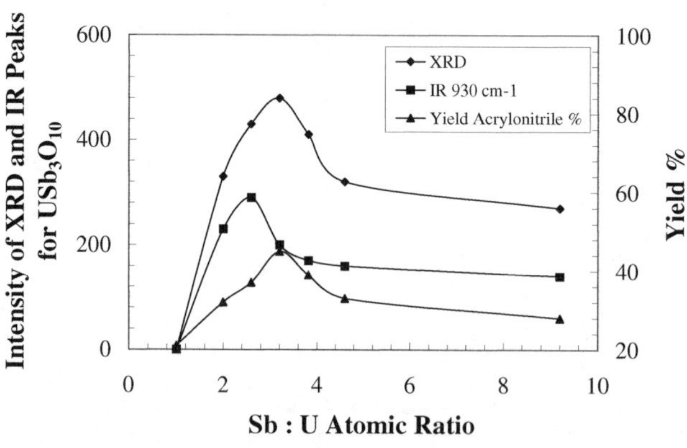


**Figure 3.14** Correlation of the catalytic performance and physico-chemical characterization for a series of $USb_3O_{10}$ catalysts with varying Sb : U ratios (Grasselli, 1983) (reproduced by permission of Kluwer Academic Publishers).

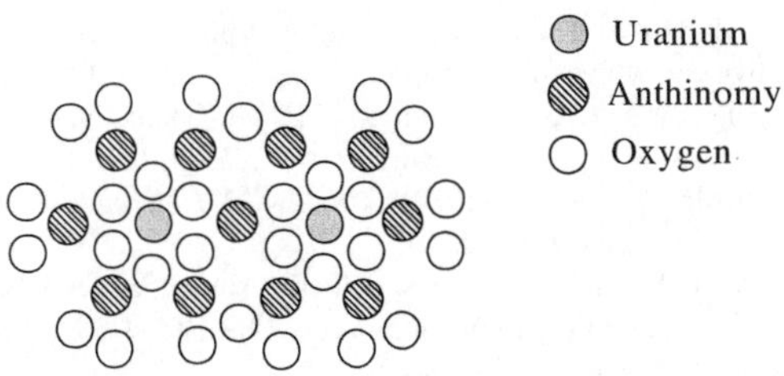


**Figure 3.15** Schematic diagram of the active site geometry, with dilution of the active uranium oxide component by antimony oxide (Grasselli, 1983) (reproduced by permission of Academic Press).

In general, the oxygen requirement of the selective reactions is less demanding than that of the non-selective reactions. It follows that restricting the availability of lattice oxygen at and around the active site will favour the selective oxidation reactions over the total oxidation reactions.

A point to note in the data presented in Figure 3.14 is that when the Sb : U ratio in the catalyst is 1 : 1 the total ammoxidation reaction predominates. This implies the over-abundance of surface oxygen species when the concentration of uranium in the catalyst is too high. Adding Sb to the optimum extent dilutes the concentration of reactive uranium oxide, and according to the stoichiometric requirements of selective and total oxidation reactions, this diluting effect should favour the selective oxidation reaction.

The surface structure proposed for $USb_3O_{10}$ is presented in Figure 3.15 as a cross-section of the (0 0 *l*) planes. An important feature is the dilution of the ions within the active site capable of redox behaviour (U) brought about by the presence of metal ions (Sb) which do not readily change oxidation states. Change in the oxidation state of uranium is a necessary condition for the release of lattice oxygen during the reduction–reoxidation cycle. The diluting effect is a consequence of the crystallographic cleavage plane on which the active site is located. For $USb_3O_{10}$ this diluting effect will be observed irrespective of where along the *c*-axis of the unit cell the cleavage plane occurs. As shown in Figure 3.15 there are no U—O—U structures on any of the (0 0 *l*) planes of this compound. In the absence of this diluent the propene will be overoxidized, lessening the selectivity.

## REFERENCES TO CHAPTER 3

Adams CR and Jennings TJ (1963) Investigation of the mechanism of catalytic oxidation of propylene to acrolein and acrylonitrile. *J. Catal.* **2**: 63–68.

Adams CR and Jennings TJ (1964) Mechanistic studies of the catalytic oxidation of propylene. *J. Catal.* **3**: 549–558.

Batiot C, Cassidy FE, Doyle AM and Hodnett BK (1997) Selectivity of active sites on oxide catalysts. *Stud. Surf. Sci. Catal.* **110**: 1097–1107.

Batiot C and Hodnett BK (1996) The role of reactant and product bond energies in determining limitations to selective catalytic oxidations. *Appl. Catal. A General* **137**: 179–191.

Bielanski A and Haber J (1990) *Oxygen in Catalysis*. Marcel Dekker, New York.

Boudart M. (1969) Catalysis by supported metals. *Adv. Catal.* **20**: 153–166.

Boudart M and Djega-Mariadassou G (1984) *Kinetics of Heterogenous Catalytic Reactions*. Princeton University Press, Princeton, NJ.

Brazdil LA, Ebner AM and Brazdil JF (1996) Kinetic isotope effect in the ammoxidation of propane over alumina supported vanadium antimony oxide catalyst. *J. Catal.* **163**: 117–121.

Cassidy F (1999) Thesis, University of Limerick.

Cassidy F and Hodnett BK (1998) Selective oxidation catalysts: An evaluation of the discriminating mechanism of active sites on oxide catalysts with molecular oxygen as oxidant. *CATTECH* **2**(2): 173–180.

Centi G (1996) Nature of active layer in vanadium oxide supported on titanium oxide and control of its reactivity in the selective oxidation and ammoxidation of alkylaromatics. *Appl. Catal. A General* **149**: 267–298.

Che M and Tench AJ (1982) Characterization and reactivity of mononuclear oxygen species on oxide surfaces. *Adv. Catal.* **31**: 77–133.

Che M and Tench AJ (1983) Characterization and reactivity of molecular oxygen species on oxide surfaces. *Adv. Catal.* **32**: 1–148.

Cox JD and Pilcher D (1970) *Thermochemistry of Organic and Organometallic Compounds*. Academic Press, New York, pp. 140–492.

De Jong KP (1998) Assembly of solid catalysts. *CATTECH* **2**: 87–95.

Dricoll DJ and Lunsford JH (1983) Kinetic isotope effect in the partial oxidation of propylene over $Bi_2O_3$. *J. Phys. Chem.* **87**: 301–303.

Driscoll DJ, Campbell KD and Lunsford JH (1987) Surface-generated gas-phase radicals: Formation, detection and role in catalysis. *Adv. Catal.* **35**: 139–186.

Grasselli RK (1983) Selective oxidation and ammoxidation catalysis: History of catalyst design. In *Surface Properties and Catalysis by Non-Metals*, Bonnelle JP *et al.* (Eds). Reidel, Dordrecht, pp. 273–288.

Grasselli RK and Suresh DV (1972) Aspects of structure and activity in U–Sb–Oxide acrylonitrile catalysts. *J. Catal.* **25**: 273–291.

Haber J. (1997) Molecular mechanism of heterogeneous oxidation – organic and solid state chemists' view. In *Proceedings of the Third World Congress on Oxidation Catalysis*, Grasselli RK, Oyama ST, Gaffney AM and Lyons JE (Eds). Elsevier Science, Amsterdam.

He D-HH, Ueda W and Moro-oka Y (1992) Promotion effect of molybdate support on $Bi_2Mo_3O_{12}$ catalyst in the selective oxidation of propylene. *Catal. Lett.* **12**: 35–44.

Horowitz HS, Blackstone CM, Sleight AW and Teufer G (1988) V–P–O catalysts for oxidation of butane to maleic anhydride. Influence of $(VO)_2H_4P_2O_9$ precursor morphology on catalytic properties. *Appl. Catal.* **38**: 193–210.

Kartheuser B, Hodnett BK, Zanthoff H and Baerns M (1993) Transient experiments on the selective oxidation of methane to formaldehyde over $V_2O_5/SiO_2$ studied in the temporal-analysis-of-products reactor. *Catal. Lett.* **21**: 209–214.

Keulks GW and Krenzke LD (1976) The role of lattice oxygen in selective oxidation of propylene over binary oxide catalysts. In *Proceedings of the Sixth International Congress on Catalysis*. The Chemical Society, London, pp. 806–818.

Lee J and Grobowski JJ (1992) Reaction of the atomic radical anion and the synthesis of organic reactive intermediates. *Chem. Rev.* **92**: 1611–1647.

Lehmann L and Baerns M (1992$) Mechanistic aspects of the oxidative coupling of methane over a NaOH/CaO catalyst. *Catal. Today.* **13**: 265–272.

Lunsford JH (1973) ESR of adsorbed oxygen species. *Catal. Rev.* **8**: 135–157.

Lunsford JH (1989) The role of surface-generated gas-phase radicals in catalysis. *Langmuir* **5**: 12–16.

Machiels CJ and Sleight AW (1982) Kinetic isotope effect in the selective oxidation of methanol to formaldehyde over some molybdate catalysts. *J. Catal.* **76**: 238–239.

Mars P and van Krevelen DW (1954) Oxidations carried out by means of vanadium oxide catalysts. *Chem. Eng. Sci. Suppl.* **3**: 41–59.

Melander L (1960) *Isotope Effects on Reaction Rates*. The Ronald Press, New York.

Mezaki R and Inoue H (1991) *Rate Equations of Solid-Catalysed Reactions*. University of Tokyo Press, Tokyo.

Millini R, Previde Massara E, Perego G and Bellussi G (1992) Framework composition of titanium silicalite—1. *J. Catal.* **137**: 497–503.

Moro-oka Y and Ueda W (1994) Multicomponent bismuth molybdate catalyst: A highly functionalized catalyst system for the selective oxidation of olefin. *Adv. Catal.* **40**: 233–273.

Panov GI, Uriarte AK, Rodkin MI and Sobolev VI (1998) Generation of active oxygen species on solid surfaces. Opportunity for novel oxidation technologies over zeolites. *Catal. Today* **41**: 365–385.

Pepera MA, Callahan JL, Desmond MJ, Milberger EC, Blum PR and Bremer NJ (1985) Fundamental study of the oxidation of butane over vanadyl pyrophosphate. *J. Am. Chem. Soc.* **107**(17): 4883–4892.

Sajkowski DJ and Boudart M (1987) Structure sensitivity of the catalytic oxidation of ethene by silver. *Catal. Rev.-Sci. Eng.* **29**(4): 325–360.

Sanati M, Wallenberg LR, Andersson A, Jansen S and Tu Y (1991) Vanadia catalysts on anatase, rutile and $TiO_2$(B) for the ammoxidation of toluene: An ESR and high-resolution electron microscopy characterization. *J. Catal.* **132**: 128–144.

Seyedmonir SR, Plischke JK, Vannice MA and Young HW (1990) Ethylene oxidation over small silver crystallites. *J. Catal.* **123**: 534–549.

Sokolovskii VD (1990) Principles of oxidative catalysis on solid oxides. *Catal. Rev.-Sci. Eng.* **32**: 1–49.

Somorjai GA and Carrazza J (1986) Structure sensitivity of catalytic reactions. *Ind. Eng. Chem. Fundam.* **25**: 63–69.

Spencer ND, Schoonmaker RC and Somorjai GA (1982) Iron single crystals as ammonia synthesis catalysts: Effect of surface structure on catalytic activity. *J. Catal.* **74**: 129–135.

Thomas JM and Thomas WJ (1997) *Principles and Practice of Heterogeneous Catalysis*. VCH, Weinheim.

Ueda W, Asakawa K, Chen C-L, Moro-oka Y and Ikawa T (1986) Catalytic properties of tricomponent metal oxides having the scheelite structure. I. Role of bulk diffusion of lattice oxide ions in the oxidation of propylene. *J. Catal.* **101**: 360–368.

Verykios XE, Stein FP and Coughlin RW (1980) Influence of metal crystalline size and morphology on selectivity and activity of ethylene oxidation catalysed by supported silver. *J. Catal.* **66**: 368–382.

Voet D and Voet JG (1995) *Biochemistry*. J. Wiley & Sons, London.

Weisz PB (1982) The science of the possible. *Chemtech*, July, 424–425.

Wolfs MWJ and Batist PhA (1974) The selective oxidation of 1-butene over multicomponent molybdate catalysts: Influence of various elements on structure and activity. *J. Catal.* **32**: 25–36.

Yao Y-FY (1980) Oxidation of alkanes over noble metal catalysts. *Ind. Eng. Chem. Prod. Res. Dev.* **19**: 293–298.

# 4

# Propene Oxidation to Acrolein: Allylic Oxidation over Catalysts with Highly Mobile Oxygen Anions

4.1 Introduction 103
4.2 The Composition of Multicomponent Bismuth Molybdate Catalysts for Propene Oxidation 105
4.2.1 Bulk Composition Multicomponent Bismuth Molybdate Catalysts for Propene Oxidation 105
4.2.2 Additives, Promoters, Dopants, Solid Solutions, Mixed Oxides and Complex Oxides: What's the Difference? 107
4.2.3 Surface Composition of Multicomponent Bismuth Molybate Catalysts 112
4.3 The Surface Allyl Species 116
4.4 Source of Oxygen Inserted into Propene to Form Acrolein 119
4.5 Oxygen Transport within Bismuth Molybdate Catalysts 124
4.6 The Overall Reaction Mechanism 126
References to Chapter 4 129

Summary

Propene oxidation to acrolein is a key step in the production of acrylic acid for use in the polymer industry. The reaction is carried out on a range of bismuth molybdate catalysts to which at least 3–4 additional metal oxides are added to boost the activity. The final catalysts are mixtures of binary and ternary oxides and some solid solutions. A feature of these materials is the ability of lattice oxygen to transfer readily at the reaction temperature between the multiple phases that exist within the catalyst and to the reacting propene. Another key feature is that the initial point of activation of the propene is one of the methyl ($CH_3$) C—H bonds with the production of a surface allyl intermediate, hence the term allylic oxidation

This chapter describes

- The composition of multicomponent bismuth molybdate catalysts,
- The surface allyl species, an important intermediate in the reaction,
- Source of oxygen inserted into propene to form acrolein,
- Oxygen transport within bismuth molybdate catalysts,
- The overall reaction mechanism.

## 4.1 INTRODUCTION

Acrolein is the simplest unsaturated aldehyde and is highly reactive owing to conjugation of the carbonyl group with the vinyl group. It is produced as refined acrolein (approximately 120 ktonne per year) and as crude acrolein (approximately 1400 ktonne per year). Crude acrolein forms the feedstock for about 85 % of worldwide acrylic acid production through the catalytic oxidation of acrolein on molybdenum and vanadium catalysts according to equation (4.1):

$$CH_2{=}CH{-}CHO + 0.5O_2 \rightarrow CH_2{=}CH{-}COOH \qquad (4.1)$$

The major use for refined acrolein is in the synthesis of methionine, an important animal food supplement, via

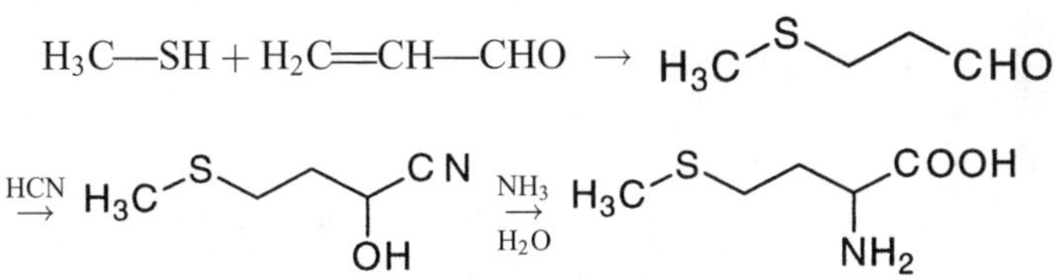


(4.2)

Table 4.1 presents pertinent summary data for the reaction of propene to acrolein, including the basic thermodynamics of the reaction and an indication of the catalyst used and operating conditions for normal commercial operation. Literature reports of the performance of bismuth molybdate catalysts are summarized as a multiple selectivity–conversion plot in Figure 4.1. As before, these data were plotted irrespective of the catalyst used and without taking into account the reaction conditions. The purpose of the plot is simply to show the best selectivity that can be attained over a range of conversions.

The vapour phase catalytic oxidation of propene to acrolein is usually carried out in multi-tubular fixed bed reactors operating at atmosphere pressure. Air is the most commonly used oxidant and steam is added as a diluent to suppress the formation of flammable gas mixtures. The reaction is

**Table 4.1** Summary data for acrolein production
(Etzkorn *et al.*, 1991; Zabetakis, 1965)

Annual production $1.5 \times 10^6$ tonne
comprising $1.4 \times 10^6$ tonne of crude acrolein for acrylic acid production and $0.1 \times 10^6$ tonne of refined acrolein

Selective oxidation (4 electron)
$C_3H_6 + O_2 \rightarrow C_3H_4O + H_2O$ $\Delta G_{298\,K} = -397\ kJ\,mol^{-1}$

Total oxidation (10 electron)
$C_3H_6 + 4.5O_2 \rightarrow 3CO_2 + 3H_2O$ $\Delta G_{298\,K} = -1973\ kJ\,mol^{-1}$

| Catalyst | Support | Additives | Temp/K | Pressure/bar |
|---|---|---|---|---|
| $BiMoO_x$ | none | Fe,Co | 560–620 | 1 |

| | LFL$_{298}$ [a] | UFL$_{298}$ [a] | AIT [a]/K |
|---|---|---|---|
| $C_3H_6$ | 2.4 | 11 | 733 |
| $C_3H_4O$ | 2.8 | 31 | 508 |
| CO | 12.5 | 74 | |

[a] Lower and upper flammability limits at 298 K and autoignition temperature.

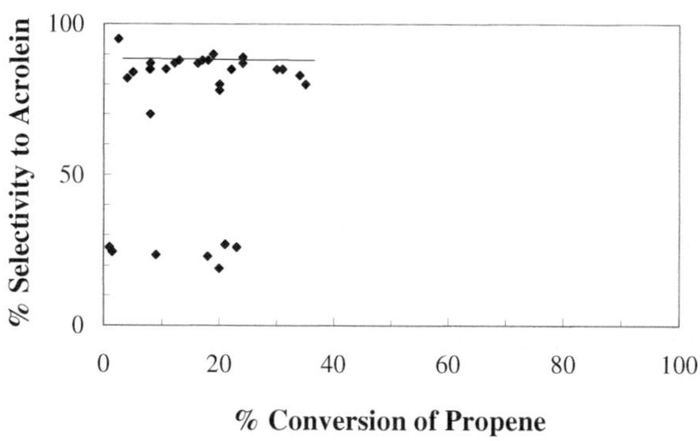


**Figure 4.1** Multiple selectivity–conversion plot for propene oxidation to acrolein (Cassidy, 1999) (reproduced by permission of Elsevier Science).

strongly exothermic, as expected, and the main side products are acrylic acid, acetaldehyde, acetic acid and carbon oxides.

A typical process flowsheet for acrolein production is presented in Figure 4.2. The reactor effluent gases are cooled in two steps. In the first step acrylic acid is condensed in a water stream. In a second condensation step this stream is further cooled, acrolein is condensed and absorbed in a second water stream. Low boiling components, such as unreacted propene, air and carbon oxides are separated overhead. Part of this stream can be recycled to the reactor. Further distillation steps are necessary to separate acrolein from water

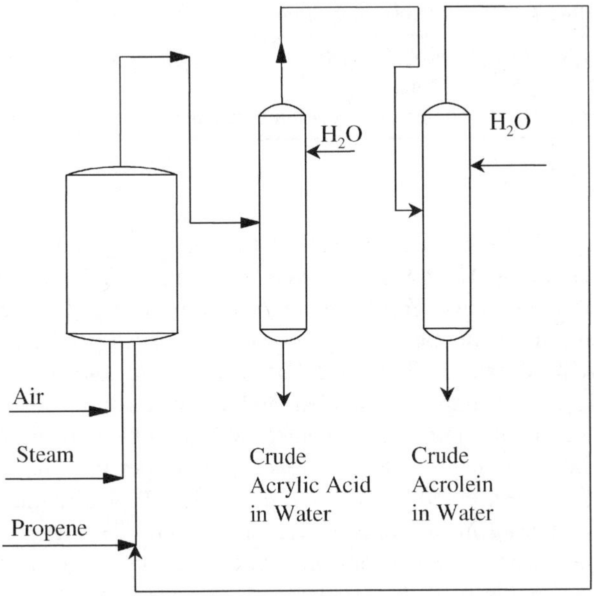


**Figure 4.2** Flow sheet for the production of acrolein from propene.

and acetaldehyde, and a refined acrolein product is obtained from a final distillation (Etzkorn *et al.*, 1991).

## 4.2 THE COMPOSITION OF MULTICOMPONENT BISMUTH MOLYBDATE CATALYSTS FOR PROPENE OXIDATION

### 4.2.1 Bulk composition multicomponent bismuth molybdate catalysts

The allylic oxidation of propene to acrolein over a $Cu_2O$ catalyst was first reported in 1948 by workers from the Shell Development Company (Hearne and Adams, 1948):

$$CH_2{=}CH{-}CH_3 + O_2 \rightarrow CH_2{=}CH{-}CHO + H_2O \qquad (4.3)$$

Ten years late the bismuth molybdate catalysts were revealed by the SOHIO company, which formed the basis of the first commercial operation (Idol, 1959). It is now recognized that there are three binary oxides of bismuth and molybdenum, with distinguishing Bi : Mo atomic ratios, that are important in propene oxidation (Moro-oka and Ueda, 1994). These are presented in Table 4.2

Subsequent research has led to the development of the so called multi-

**Table 4.2** Binary oxides of bismuth and molydenum

| | |
|---|---|
| $Bi_2Mo_3O_{12}$ | $\alpha$-phase distorted Scheelite structure |
| $Bi_2Mo_2O_9$ | $\beta$-phase |
| $Bi_2MoO_6$ | $\gamma$-phase Koechlinite structure |

component bismuth molybdate catalysts. Knapsak (1962) found that when half of the bismuth in $Bi_9P_1Mo_{12}O_{52}$ was replaced with Fe to give $Fe_{4.5}Bi_{4.5}P_1Mo_{12}O_x$ an extensive improvement in activity resulted. Since then a large number of multicomponent catalysts have been developed based on each of the three structure types in Table 4.2, but in which the bismuth component is largely replaced by a mixture of divalent and trivalent cations.

The advantages of the multicomponent catalysts include superior activity (sometimes exhibiting 20 times greater reaction rates per unit mass than the binary oxides), selectivity and mechanical properties. Typically multicomponent catalysts operating in the temperature range 563–623 K gave yields of 90–95 % acrolein whereas the simple binary oxides operating in the range 683–733 K gave yields of 40 % at identical contact times (Moro-oka and Ueda, 1994). A typical preparation method for a multicomponent bismuth molybdate catalyst is:

Typical preparation method for a multi-component bismuth molybdate catalyst (Grasselli *et al.*, 1972)

A specific catalyst was made as follows: 36.4 g bismuth nitrate, $Bi(NO_3)_3{\cdot}5H_2O$, were dissolved in 700 ml dilute nitric acid in a large beaker. 7.6 g potassium nitrate, $KNO_3$, 120 g cobaltous nitrate, $Co(NO_3)_2{\cdot}6H_2O$, and 182 g ferric nitrate, $Fe(NO_3)_3{\cdot}9H_2O$ were added to the solution and dissolved. 8.65 g phosphoric acid (85 %) were added to the solution while it was continuously stirred. 159 g ammonium molybdate, $(NH_4)_6Mo_7O_{24}{\cdot}4H_2O$, were dissolved in a small quantity of water and slowly added to the foregoing acidic solution. Precipitation was instantaneous. The solution was constantly stirred and heated over a hot plate until the entire mass gelled. The gel was spooned out into trays and dried in a conventional convection oven at 403 K for 16 h. The dried mass was calcined in air at 700 K for 16 h. A final heat-treatment in air at 810 K for 3 h was applied prior to loading into the reactor.

The resulting catalyst had composition $KFe_6BiMo_{12}Co_{5.5}O_x$

A very wide variety of multicomponent bismuth molybdate catalysts have been prepared for propene oxidation. Compositional ranges and examples of elemental and phase compositions are presented in Table 4.3. In addition to the binary oxides listed in Table 4.3 some terniary oxides and solid solutions, such as $Co_{11/12}Fe_{1/12}MoO_x$ have also been observed in these catalysts.

**Table 4.3** Compositional range for multicomponent bismuth molybdate catalysts
(Moro-Oka and Ueda, 1994)

| *Component* | *Range/atom %* |
|---|---|
| $Mo^{6+}$ | 50–55 |
| $Bi^{3+}$ | 3–7 |
| $Me^{2+}$ (Co, Ni, Fe, Mg, Mn) | 30–35 |
| $Me^{3+}$ (Fe, Cr, Al) | 8–15 |

*Examples*

| *Elemental composition* | *Phase composition by XRD* |
|---|---|
| $Mo_{12}Bi_1Co_{11}O_X$ | $\beta$-$CoMoO_4$, $\alpha$-$Bi_2(MoO_4)_3$ |
| $Mo_{12}Bi_1Co_8Fe_3O_X$ | $\beta$-$CoMoO_4$, $\alpha$-$Bi_2(MoO_4)_3$, $Fe_2(MoO_4)_3$, $FeMoO_4$ |
| $Mo_{12}Bi_1Mg_8Fe_3O_X$ | $\beta$-$CoMoO_4$, $\alpha$-$Bi_2(MoO_4)_3$, $Fe_2(MoO_4)_3$, $FeMoO_4$ |
| $Mo_{12}Bi_1Co_8Al_3O_X$ | $\beta$-$CoMoO_4$, $\alpha$-$Bi_2(MoO_4)_3$, $Al_3(MoO_4)_3$ |

Reproduced by permission of Academic Press.

### 4.2.2 Additives, promoters, dopants, solid solutions, mixed oxides and complex oxides: what's the dfference?

A newcomer to the field of oxidation catalysis is often confronted by a bewildering array of terms used to describe material added to a basic catalyst to improve its performance. In many cases the reason for this additional material is not explained, especially in the patent literature, and it should also be pointed out that catalysts are often not pure compounds. They can be non-stoichiometric, are sometimes amorphous and additives are often present in very small quantities. These factors often makes classification into a single category difficult. The following offers a simple definition for each of the terms used above and the minimum amount of experimental evidence needed to justify the use of each one.

*Solid solutions*

A solid solution is a crystalline compound of variable composition. These materials are commonly encountered in oxidation catalysis, for example in the bismuth molybdate catalysts with the Scheelite structure. The Scheelite structure is that of the mineral $CaWO_4$. It crystallizes into a tetragonal unit cell with four molecules per unit cell. Starting from $PbMoO_4$ a series of Scheelite type phases of composition $Pb_{1-3x}{}^{2+}Bi_{2x}{}^{3+}MoO_4$ for $0 < x < 0.15$.

This series of compounds shows all the characteristics of a solid solution. Over the entire range of compositions the X-ray diffraction patterns of all samples in the range are almost identical with just small changes occurring in unit cell dimensions as the value of $x$ changes. The changes in unit cell parameter for this tetragonal lattice tend to be equally distributed over the $a$ and $c$ axes. Taken together the unit cell volume shrinks about 3 % in a regular fashion as $x$ varies from 0 to 0.15 (see Figure 4.3) because the $Bi^{3+}$ ion is smaller than the $Pb^{2+}$ ion it replaces. With the addition of each two bismuth ions a cation vacancy is introduced to maintain electrical neutrality. In fact, the compound is often written as $Pb_{1-3x}{}^{2+}Bi_{2x}{}^{3+}\phi_xMoO_4$ where $\phi$ represents a cation vacancy. When this series of solid solutions was tested in propene oxidation to acrolein the rate of acrolein formation increased directly in line with the defect concentration (Sleight, 1997; Sleight *et al.*, 1975).

*Complex oxide*

This term is used to describe an oxide with more than one cation. Examples would include the persovskite $BaTiO_3$, the spinel $MgAl_2O_4$ or the Schellite structure $CaWO_4$. Each is a single phase compound, crystallizes into a particular unit cell and exhibits a unique X-ray diffraction pattern. Binary, ternary and quaternary complex oxides are known and are frequently studied in oxidation catalysis. It would appear that the best catalysts for the oxidation of $n$-butane to maleic anhydride are about 95 % vanadyl pyrophosphate,

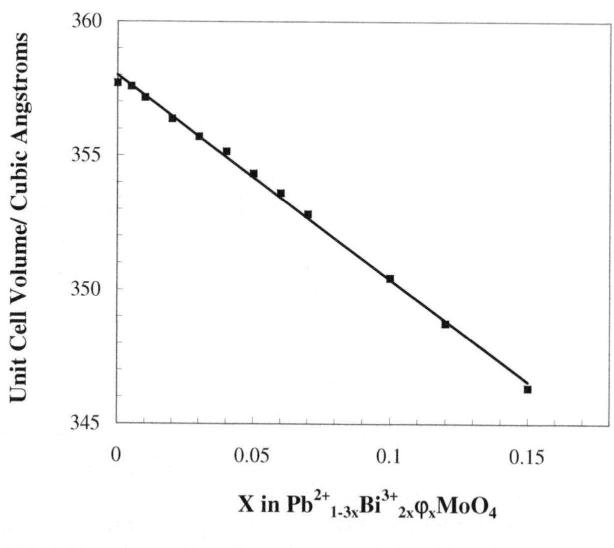


**Figure 4.3** Variation in unit cell volume with added bismuth for the system $Pb_{1-3x}{}^{2+}Bi_{2x}{}^{3+}\phi_xMoO_4$ (Sleight *et al.*, 1975).

$(VO)_2P_2O_7$, which is known to be a binary oxide of vanadium and phosphorus. When examined by X-ray diffraction this catalyst exhibits the pattern of a single crystallographic phase without the changes in unit cell parameters characteristic of solid solutions (Horowitz *et al.*, 1988).

### *Mixed oxides*

The term mixed oxide refers to a mixture of distinct oxides and can include mixtures of simple and complex oxides and solid solutions. The X-ray diffraction pattern of the mixture should indicate the presence of more than one phase. One such system is the mixed oxides of $MoO_3$ and $\alpha$-$Sb_2O_4$ which have been used for the selective oxidation of isobutene to methacrolein (Gaigneaux *et al.*, 1996). A synergy in terms of catalytic activity has been observed between the two phases, but there is no contamination of one phase by the other and no complex oxide formation. Such a mixture can be prepared in any composition required and the mixture would exhibit the characteristic X-ray diffraction peaks of the individual oxides. The term mixed oxide should be used therefore when the evidence from X-ray diffraction points to any mixture of simple or complex oxides, in other words when a multiphase system is observed. The multicomponent bismuth molybdate catalysts can also be described as mixed oxides. The catalyst $Mo_{12}Bi_1Co_8Fe_3O_X$ is in fact a mixture of at least four distinct binary oxides, $\beta$-$CoMoO_4$, $\alpha$-$Bi_2(MoO_4)_3$, $Fe_2(MoO_4)_3$, $FeMoO_4$ and a solid solution $\beta$-$Co_{11/12}Fe_{1/12}MoO_4$ (Moro-oka and Ueda, 1994).

Supported oxides such as $V_2O_5$ supported on $SiO_2$ may also be classified as a mixed oxide, but usually are not.

### *Dopant*

This term describes dissolution of a small amount of a foreign material in a catalyst to modify the rate of a particular reaction. Dopants can substitute into regular lattice or interstitial positions. This term differs from a solid solution mainly in terms of the amount of material that is incorporated into the host. Usually the amounts of dopants is so small that changes in lattice parameter are not seen. Instead the presence of dopants is often confirmed by changes in electrical or ionic conductivities of the catalysts. In reality, it is difficult to find definitive evidence in the literature dealing with heterogeneous catalysis for doping. This term is often incorrectly used even when there is no evidence that doping has been achieved.

Doped catalysts have been widely applied in the oxidative coupling of methane to ethane and ethylene. Incorporation of lower valent cations, such as $Li^+$ or $Zn^{2+}$, or higher valent cations such as $Ta^{5+}$, $Sb^{5+}$ or $W^{6+}$ into the $TiO_2$ host lattice results in changes to its electronic structure and oxygen anion mobility (Efstathiou *et al.*, 1994).

Catalysts for fluorene oxidation to fluorenone have compositions such as $V_1Fe_{0.74}Cs_{0.06}O_x$. Energy dispersive X-ray analysis (EDXA) points to the presence of three separate phases, which can be distinguished on the basis of electron microscopy and EDXA (see Figure 4.4). The phases are $FeVO_4$, $Fe_2V_4O_{13}$ and an amorphous phase which contains the dopant. Because this part of the catalyst is amorphous it is very difficult to determine with any certainty the position of the dopant within this phase. This system can be regarded as a mixed oxide containing $FeVO_4$, $Fe_2V_4O_{13}$ and an amorphous phase which incorporates the caesium (Bruckner and Baerns, 1997).

*Promoter*

A promoter a substance of little or no catalytic activity added to a catalyst in a relatively small quantity to increase the rate of reaction, selectivity or the useful working life of the catalyst. Promoters can be present as dopants, solid solutions or can form mixed oxides. One of the best known examples is the cobalt promoted molybdenum sulfide catalysts used for hydrodesulfurization processes. Here the basic catalyst is $MoS_2$, but small amounts of added cobalt boost the catalytic activity (Topsoe *et al.*, 1984).

*Additive*

This is the least specific of the terms considered in this glossary. This terms can be used to describe any material which is added to a basic catalyst system and does not imply that the added material carries any beneficial effects. Use of this term does not imply where in the catalyst the added material may be located (it may be present as a dopant, solid solution or a mixed oxide but the material has not been characterized sufficiently). For example, when an

**Figure 4.4** Electron micrograph of $V_1Fe_{0.74}Cs_{0.06}O_x$ and EDXA of the indicated regions (Bruckner and Baerns, 1997) (reproduced by permission of Elsevier Science).

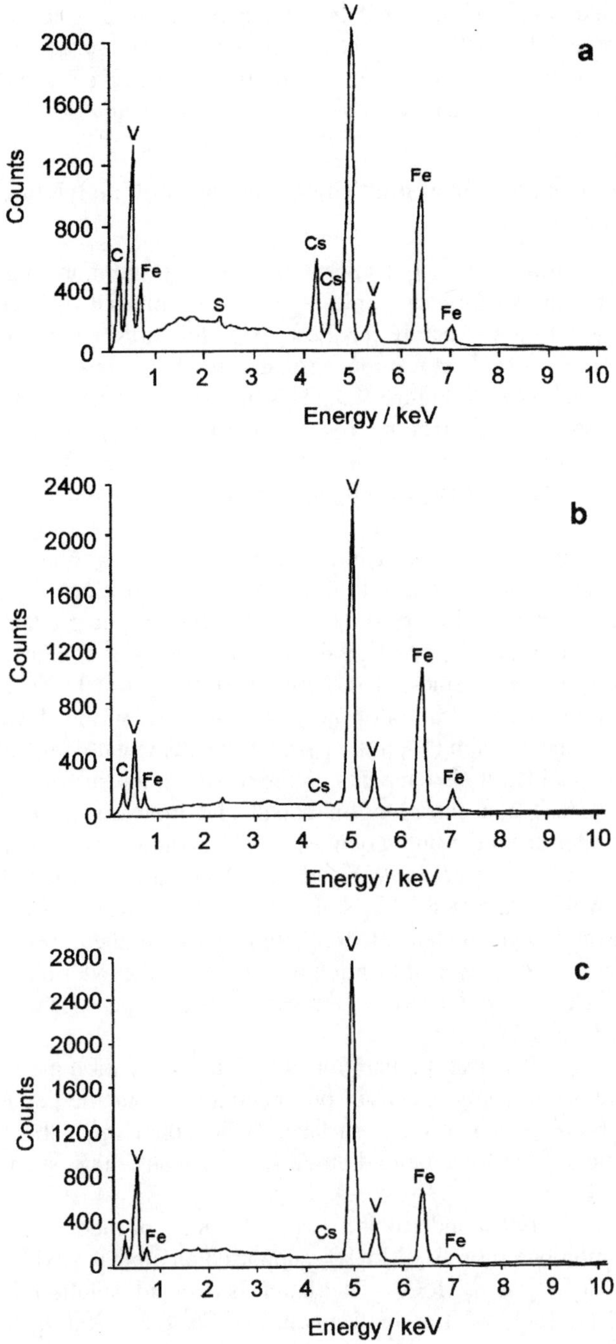


**Figure 4.4** (*continued*)

additive is added to a supported catalyst it may attach itself to the support or to the supported phase. In many cases, especially with poorly crystalline materials or where small amounts of an additive are involved, it may not be possible to determine its location.

### 4.2.3 Surface composition of multicomponent bismuth molybdate catalysts

A feature of multicomponent catalysts is that they are not single phase systems, but mixtures of binary oxides and solid solutions. Wolfs and Batist (1974) have proposed a structure for these oxide mixtures comprising a core of $Me^{2+}MoO_4$ and $Me_2{}^{3+}(MoO)_4$ encapsulated inside a thin shell of bismuth molybdates, as shown in Figure 4.5. This model has been confirmed by combined transmission electron microscopy and EDXA. Figure 4.5 shows an electron micrograph of a cross-section of a multicomponent bismuth molybdate catalyst as a layer of $Bi_2Mo_3O_{12}$ supported on $Co_{11/12}Fe_{1/12}MoO_x$ (He *et al.*, 1992).

This type of structure has been further confirmed by X-ray photoelectron spectroscopy, which has been used to illustrate the composition of the surface layers of these catalysts. Figure 4.6 presents a typical set of data for the surface composition for a range of catalysts with varying Bi contents, namely $Bi_yCo_8Fe_3Mo_{12}O_X$ (for values of $y$ between 0 and 1.0). These materials always show a high concentration of molybdenum in the surface layers, more than would be expected simply on the basis of the elemental composition but when bismuth is added it also appears preferentially in the surface layers. By contrast the surface concentrations of cobalt and iron are much less than would be expected from the bulk composition. This finding is consistent with the TEM work of He *et al.* (1992) and the much earlier structural model proposed by Wolfs and Batist (Wolfs and Batist, 1974) which envisages the cobalt and iron molybdates located predominantly in the interior of the catalyst particles, with a layer of bismuth molybdate on the exterior. In a sense the cobalt and iron molybdates may be regarded as a support for the bismuth molybdate surface layer.

Another feature of catalyst preparation is that the preparation methods used for multi-component catalysts do not preferentially favour the deposition of any particular component on the surface. Hence there must be a strong thermodynamic driving force behind the formation of structures such as those shown in Figure 4.5.

Moro-oka *et al.* (1991) undertook a series of studies of model catalysts for propene oxidation based on $Bi_2(MoO_4)_3$ supported on a binary oxide, namely $CoMoO_4$ or $Co_{11/12}Fe_{1/12}MoO_4$. The latter is a solid solution based on $CoMoO_4$ and $FeMoO_4$ with a small amount of $Fe^{3+}$ substituting for $Co^{2+}$. The bare supports are poor propene selective oxidation catalysts and give carbon oxides as the principal oxidation products. As the $Bi_2(MoO_4)_3$ loading

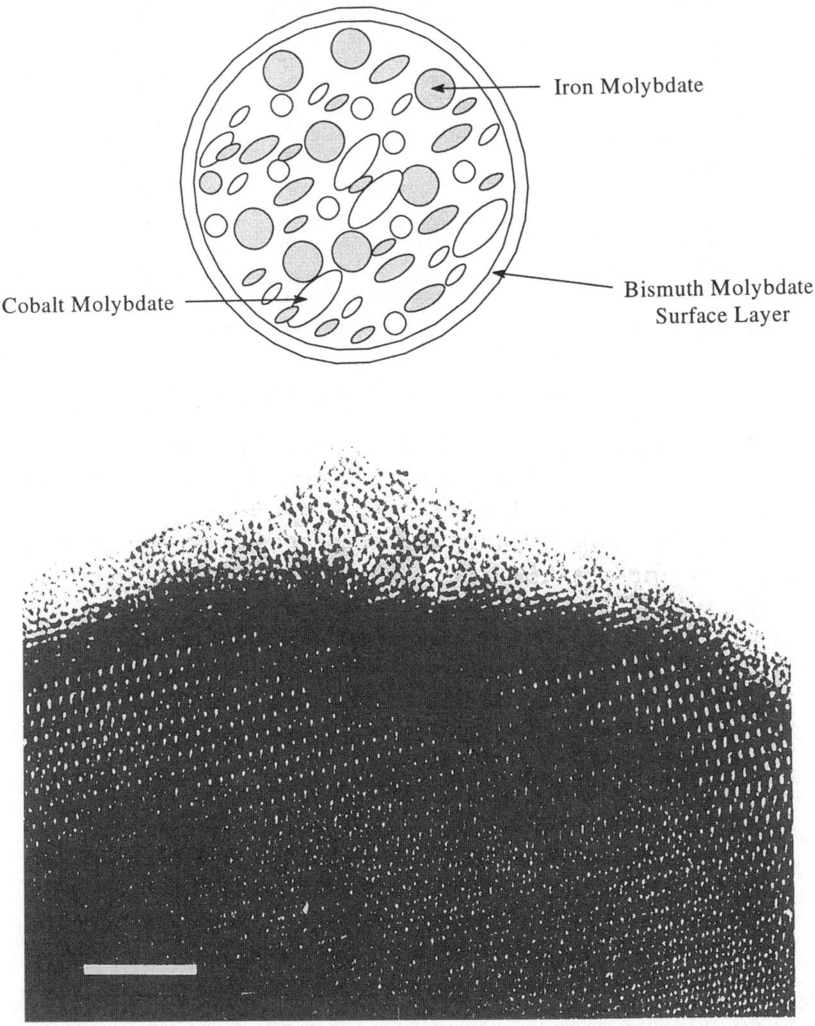


**Figure 4.5** (a) Schematic diagram of the structure of a multicomponent bismuth molybdate catalyst (Wolfs and Batist, 1974). (b) Transmission electron micrograph of $Bi_2Mo_3O_{12}/Co_{11/12}Fe_{1/12}MoO_x$ (He *et al.*, 1992) (reproduced by permission of Baltzer Scientific Publishers).

on $Co_{11/12}Fe_{1/12}MoO_4$ was increased, the rate of acrolein formation rapidly reached the level exhibited by the unsupported $Bi_2(MoO_4)_3$ and then surpassed this level (Figure 4.7). The amount of $Bi_2(MoO_4)_3$ needed to give a rate of acrolein formation equal to that of the unsupported $Bi_2(MoO_4)_3$ corresponds to about one monolayer coverage of the solid solution. The change in rate is less pronounced for the $CoMoO_4$ support, but this serves to

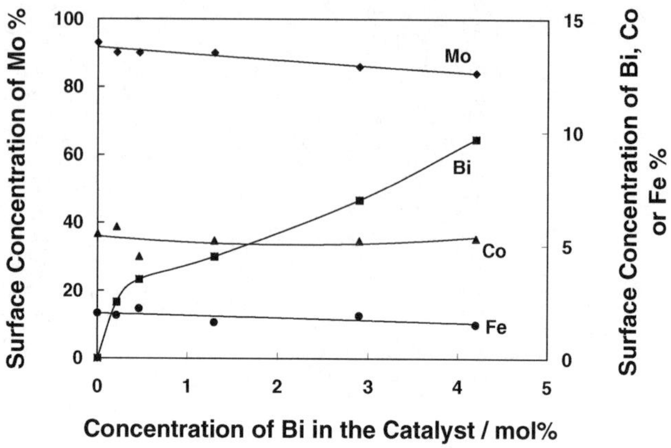


**Figure 4.6** The surface concentration of each metal element in $Mo_{12}Bi_{0-1}Co_8Fe_3O_X$. by XPS (Moro-oka and Ueda, 1994) (reproduced by permission of Academic Press).

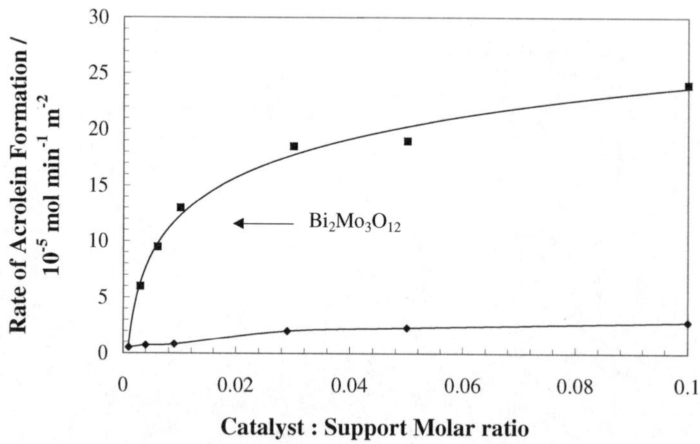


**Figure 4.7** Rate of acrolein formation per unit of surface area at 723 K over (◆)$Bi_2Mo_3O_{12}/CoMoO_4$ and (■) $Bi_2Mo_3O_{12}/Co_{11/12}Fe_{1/12}MoO_4$ (Moro-oka *et al.*, 1991) (reprinted by permission of Elsevier Science).

illustrate the desirability of promoting the support through the addition of a small amount of iron. Striking changes in selectivity are also observed as the loading of $Bi_2(MoO_4)_3$ increases (Figure 4.8). Since the rate data shown in Figure 4.7 are specific reaction rates expressed per unit of surface area, the observed effects cannot be associated with differences in surface area of $Bi_2(MoO_4)_3$ and the supported form of this material. Instead these data point to a strong interaction between the support and $Bi_2(MoO_4)_3$ and illustrate the importance of the interactions between the surface layer of bismuth molybdate and the underlying solid solution (Moro-oka *et al.*, 1991).

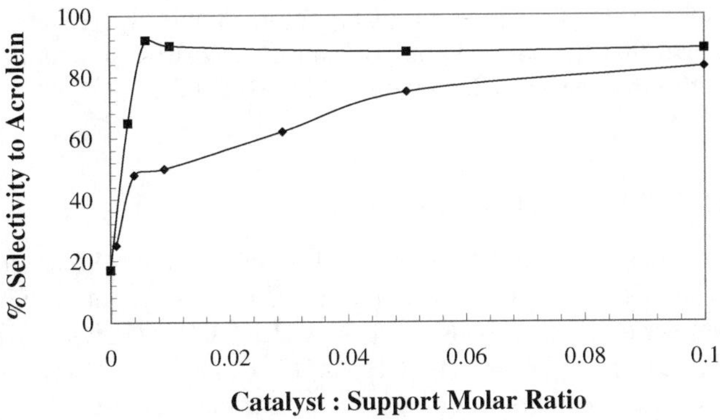


**Figure 4.8** Selectivity to acrolein at 723 K over (◆) $Bi_2Mo_3O_{12}/CoMoO_4$ and (■) $Bi_2Mo_3O_{12}/Co_{11/12}Fe_{1/12}MoO_4$ (Moro-oka *et al.*, 1991) (reprinted by permission of Elsevier Science).

A feature of the bismuth molybdate catalyst prepared by the standard preparation methods which points to their robustness has been pointed out by Matsuura *et al.* (1980). These workers have shown that when the bulk Bi : Mo ratio is changed in the range 0.8 : 1–1.8 : 1 the surface composition as measured by X-ray photoelecron spectroscopy does not change. The data are presented in Figure 4.9 and these workers also showed that the conversion of

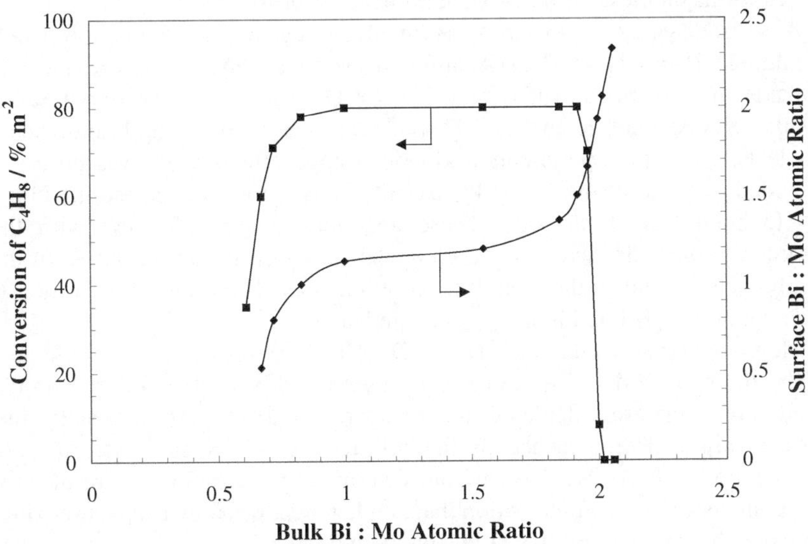


**Figure 4.9** Conversion of 1-butene at 753 K per unit of surface area and the surface composition as a function of the bulk composition for bismuth molybdate catalysts (Matsuura *et al.*, 1980) (reproduced by permission of Academic Press).

1-butene also does not change in this composition range. These results effectively mean that the surface composition is insensitive to changes in the bulk composition for the specified range of bulk Bi : Mo ratios.

## 4.3 THE SURFACE ALLYL SPECIES

A number of classical studies were performed in the 1960s in which isotopically labelled propenes were fed to $Cu_2O$ or bismuth molybdate catalysts. One series of studies used $CH_2{=}CH{-}^{13}CH_3$, $^{13}CH_2{=}CH{-}CH_3$ (Voge *et al.*, 1963), $^{14}CH_2{=}CH{-}CH_3$ and $CH_2{=}CH{-}^{14}CH_3$ (Sachtler, 1963). The two key features to emerge were that the isotopic label is never found in the central carbon atom of the acrolein product, but the label is equally distributed between the terminal carbons as 50 % $^{13}CH_2{=}CH{-}CHO$ and 50 % $CH_2{=}CH^{13}CHO$.

The first finding means that a cyclopropane type intermediate is not involved in the reaction mechanism and the second finding points to a linear symmetrical intermediate. The only plausible linear symmetrical intermediate is a $\pi$-allyl species:

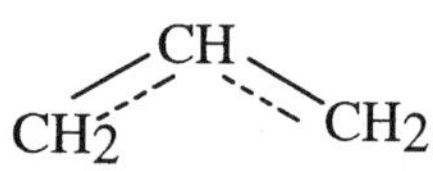


which forms on these surfaces when propene adsorbs.

A second series of experiments involved feeding deuterated propenes, including $CH_2{=}CH{-}CH_2D$ (Adams and Jennings, 1963), $CHD{=}CH{-}CH_3$ (Adams and Jennings, 1964) and $CH_2{=}CD{-}CD_3$ (Krenzke and Keulks, 1980a; Keulks and Krenzke, 1976) over a series of bismuth molybdate catalysts. In these experiments a kinetic isotope effect (KIE) was observed provided that one of the C—H bonds in the $CH_3$ group was substituted by a C—D bond (see Table 4.4). These experiments are consistent with the abstraction of a $CH_3$ hydrogen atom as the slow step in the formation of the $\pi$-allyl species and in the overall reaction kinetics. The point of activation is the weakest C—H bond in the propene molecule.

More detailed results for $CH_2{=}CD{-}CD_3$ are presented in Table 4.5, where the predicted $k_H/k_D$ values for a number of possible slow steps in the overall reaction are calculated and compared with the experimentally observed values. The presence of the full KIE at 723 K indicates that the abstraction of the $\alpha$-hydrogen is rate determining. At 623 K the results are less conclusive, with an indication that another step, possibly catalyst reoxidation, may be rate determining at this temperature.

The expected isotope distribution in the product acrolein can also be calculated on the basis of these results and the results are consistent with abstraction of an $\alpha$-hydrogen being the slow step. An example of this type of

**Table 4.4** Theoretical and experimental KIE ($k_H/k_D$) for the oxidation of propene over bismuth molybdate catalysts at 723 K

| *Isotope* | *Catalyst* | *Expt.* $k_H/k_D$ | *Theory*[a] $k_H/k_D$ | *Reference* |
|---|---|---|---|---|
| $CH_2{=}CH{-}CH_2D$ | Bismuth molybdate | 2.04 | 1.82 | Adams and Jennings, 1963 |
| $CHD{=}CH{-}CH_3$ | Bismuth molybdate | 1.02 | 1.00 | Adams and Jennings, 1964 |
| $CH_2{=}CD{-}CD_3$ | $\alpha$-$Bi_2Mo_3O_{12}$ | 1.78 | 1.75 | Krenzke and Keulks, 1980a |
| $CH_2{=}CD{-}CD_3$ | $\gamma$-$Bi_2MoO_6$ | 1.69 | 1.75 | Krenzke and Keulks, 1980a |

[a]The predicted KIE if the slow step is cleavage of a methyl C—H bond.

**Table 4.5** Theoretical and experimental KIE ($k_H/k_D$) for the oxidation of $CH_2{=}CD{-}CD_3$ over $\alpha$-$Bi_2Mo_3O_{12}$ and $\gamma$-$Bi_2MoO_6$ at 623 and 723 K
(Krenzke and Keulks, 1980a; Snyder and Hill Jr, 1989)

| | $k_H/k_D$ | |
|---|---|---|
| | *623 K* | *723 K* |
| *Theory Slow Step* | | |
| First hydrogen abstraction | 2.12 | 1.75 |
| Second hydrogen abstraction | 1.37 | 1.27 |
| Formation of a C—O bond | 1.0 | 1.0 |
| Catalyst reoxidation | 1.0 | 1.0 |
| *Experimental* | | |
| $\alpha$-$Bi_2Mo_3O_{12}$ | 1.69 | 1.79 |
| $\gamma$-$Bi_2MoO_6$ | 1.49 | 1.69 |



calculation is shown in Scheme 4.1 for $CH_2{=}CH{-}CH_2D$ where the rate constants should be regarded as probabilities. In this scheme the first hydrogen abstraction can only occur on the $CH_2D$ group and the predicted proportions of $d_0$ and $d_1$ acrolein are shown using a $k_H/k_D$ ratio of 2.12. The experimental results are shown in the first two entries in Table 4.6 and compared with the theoretical predictions of Scheme 4.1. An implicit assumption in this calculation is that isotropic discrimination may occur at each abstraction and will be independent of the kinetically slow step.

Some additional experiments point to the formation of a second allyl species, namely an oxygen–$\sigma$-allyl species ($-O{-}CH_2{-}CH{-}CH_2$) prior to the extraction of the second hydrogen from the evolving acrolein species.

**Scheme 4.1**
**Isotopic Distribution in Acrolein from $CH_2$=CH-$CH_2$D**
( Adams and Jennings, 1963)

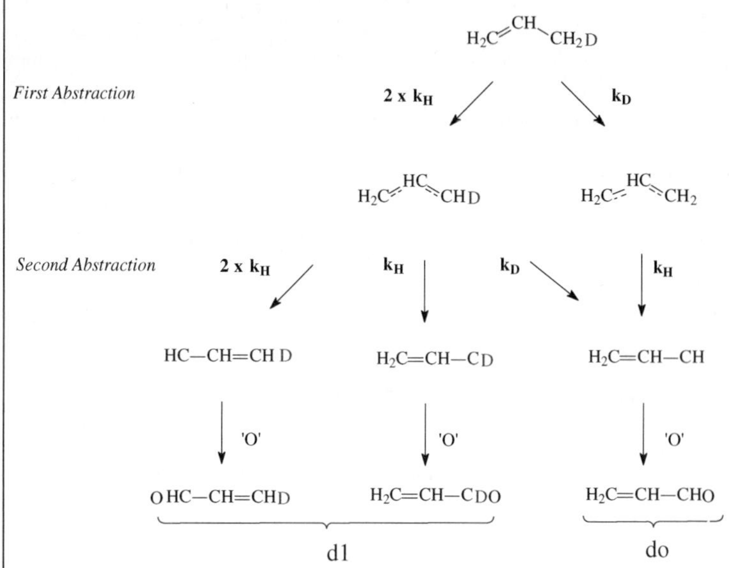


$$d_o = \{(k_D / (3k_H + k_D) \bullet 2k_H / (2k_H + k_D)\} + \{ k_D / (2k_H + k_D)\} = 0.30$$

$$d_1 = \{ 3k_H / (3k_H + k_D) \bullet 2k_H / (2k_H + k_D)\} = 0.70$$

**Scheme 4.1** Isotopic distribution in acrolein from $CH_2$═CH—$CH_2$D (Snyder and Hill Jr, 1989; Adams and Jennings, 1963) (reproduced by permission of Academic Press).

**Table 4.6** Deuterium content of acrolein obtained from isotopically labelled propenes over bismuth molybdate catalysts at 723 K
(Adams and Jennings, 1963, 1964; Snyder and Hill Jr, 1989)

| *Reactant* | *Conversion/%* | *Observed* | | *Theoretical* | |
|---|---|---|---|---|---|
| | | $d_0$/% | $d_1$/% | $d_0$/% | $d_1$/% |
| $CH_2$═CH—$CH_2$D | 39 | 32.3 | 67.7 | 33.7 | 66.3 |
| $CH_2$═CH—$CH_2$D | 81 | 29.4 | 70.6 | 33.7 | 66.3 |
| CHD═CH—$CH_3$ | 63 | 19.6 | 80.4 | 15.4 | 84.5 |
| CHD═CH—$CH_3$ | 77 | 20.4 | 79.6 | 15.4 | 84.5 |

Reproduced by permission of Marcel Dekker.

Based on the relative stability of allyl ($CH_2$—CH—$CH_2$) and carbene ($CH_2$═CH—CH) species, it is probable that sequential removal of hydrogens (as shown in Scheme 4.1) would require more energy for the removal of the second hydrogen species (Callaghan *et al.*, 1970). This is not consistent with the experimental results, which clearly show that extraction of the first hydrogen is rate determining and as such should require more energy for its removal. Burrington *et al.* (1980) fed allyl alcohol to bismuth molybdate catalyst and observed rapid exchange of the type:

$$CD_2{=}CH{-}CH_2OH \rightleftharpoons CH_2{=}CH{-}CD_2OH \qquad (4.4)$$

Based on this result they postulated that either end of the $\pi$-allyl species, which forms initially from propene, has an equal probability of collapsing on a lattice oxygen, and the exchange observed with allyl alcohol indicates that this process is rapid and reversible. The same isotopic product distribution is expected as for the sequential hydrogen removal mechanism shown in Scheme 4.1 and the presence of a $\sigma$ C—O bond in —O—$CH_2$—CH—$CH_2$ could reduce the activation energy for removal of the second hydrogen.

These arguments point to a mechanism in which the rate determining step is abstraction of the first $\alpha$ hydrogen to form a $\pi$-allyl surface species. The next step is the reversible formation of a $\sigma$ bond between a catalyst oxygen atom and either of the terminal carbon allyl atoms. The second hydrogen abstraction, which is facilitated by the presence of the C—O bond, follows with formation of the carbonyl bond and desorption of the product acrolein. The next question to be addressed is the source of oxygen which appears in the product acrolein.

## 4.4 SOURCE OF OXYGEN INSERTED INTO PROPENE TO FORM ACROLEIN

There is now a considerable body of evidence to indicate that propene oxidation to acrolein occurs via a Mars and van Krevelen mechanism whereby the reacting hydrocarbon, or a species derived from it, extracts lattice oxygen from the catalyst and in a separate step the lattice oxygen is replenished from the gas phase. This process is shown in Scheme 4.2 for the extraction of lattice oxygen from a surface layer of bismuth molybdate.

The first indications that the lattice oxygen might be involved in this way with propene oxidation over bismuth molybdate catalysts was the finding that the reaction could proceed for a time in the absence of gas phase oxygen until the supply of lattice oxygen became exhausted (Wragg *et al.*, 1973; Bielanski and Haber, 1979). A second indicator is that on $\beta$-$Bi_2Mo_2O_9$, at reaction temperatures above 673 K, the reaction is zero order in oxygen partial pressure, as shown in Table 4.7. At lower temperatures the reaction was non-

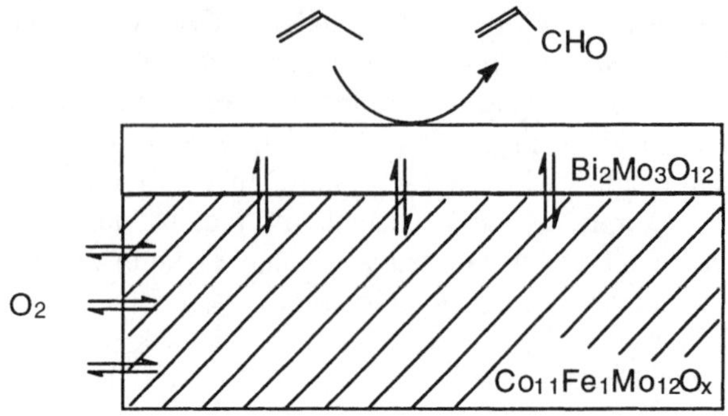


**Scheme 4.2** Illustration of Mars and van Krevelen mechanism with incorporation of gas phase oxygen into one phase and transfer to a separate surface phase for reaction with propene (Moro-oka and Ueda, 1994) (reproduced by permission of Academic Press).

**Table 4.7** Reaction orders in propene and oxygen over $\beta$-$Bi_2Mo_2O_9$ (Monnier and Keulks, 1981)

| *Temperature/K* | *Pressure* $C_3H_6$*/atm* | *Pressure* $O_2$*/ atm* | *Reaction order in* $C_3H_6$ | *Reaction order in* $O_2$ |
|---|---|---|---|---|
| 673 | 0.2–0.3 | 0.1–0.3 | 0.60 | 0.2 |
| | 0.04–0.2 | 0.1–0.3 | 1.0 | 0.2 |
| 698 | 0.00–0.3 | 0.1–0.3 | 1.0 | 0.0 |
| 723 | 0.00–0.3 | 0.1–0.3 | 1.0 | 0.0 |
| 748 | 0.00–0.3 | 0.0–0.3 | 1.0 | 0.0 |

Reproduced by permission of Academic Press.

zero order in oxygen and this was explained in terms of the slow step at these temperatures being reoxidation of the lattice by gas phase oxygen.

A number of experiments were carried out in which $^{18}O$ was used as the source of gas phase oxygen ($^{18}O_2$), or the source of lattice oxygen ($^{18}O_L$). These experiments were designed to probe the nature of the interaction between propene and the two sources of oxygen available in the reactor, namely gas phase and lattice oxygen. Preliminary experiments (Wragg *et al.*, 1973) confirmed that at 748 K the rate of exchange of gas phase oxygen for lattice oxygen was slow over bismuth molybdate:

$$^{18}O_2 + {}^{16}O_L \rightleftharpoons {}^{18}O^{16}O + {}^{18}O_L \tag{4.5}$$

as was the rate of exchange between the product acrolein and lattice oxygen:

$$C_3H_4{}^{16}O + {}^{18}O_L \rightleftharpoons C_3H_4{}^{18}O + {}^{16}O_L \tag{4.6}$$

Wragg *et al.* (1973) found that when propene and $^{18}O_2$ were reacted over a bismuth molybdate catalyst initally comprising only $^{16}O_L$ the fraction of $C_3H_4{}^{18}O$ in the reaction product was initially very small but increased as the reaction proceeded, consistent with the gradual replacement of $^{16}O_L$ by $^{18}O_L$ derived from $^{18}O_2$.

When the reaction was repeated with propene and $^{16}O_2$ in the gas phase over a bismuth molybdate catalyst comprising, initially, only $^{18}O_L$ the amount of $C_3H_4{}^{18}O$ in the reaction product was very high initially but decreased as the reaction proceeded, again consistent with the gradual replacement of $^{18}O_L$ by $^{16}O_L$ derived from $^{16}O_2$.

This work has been repeated by many others using a variety of batch, flow and, most recently, TAP reactors and with binary and multicomponent catalysts, and in all circumstances the essential result has been confirmed, namely that the oxygen which appears in the acrolein product is inserted via the lattice and not by reaction of the hydrocarbon directly with a gas phase or chemisorbed oxygen species.

An example of the more recent work with multicomponent catalysts is shown in Figure 3.11 in terms of the lapsed reaction time and in Figure 4.10 in terms of the number of monolayers of lattice oxygen which are stripped from the catalyst to arrive at a particular isotopic composition in acrolein.

The compilation of results in Table 4.8 is an indication of the surprisingly high fraction of the total lattice oxygen which can participate in the oxidation of propene over these catalysts. For example at 723 K all the lattice oxygen in

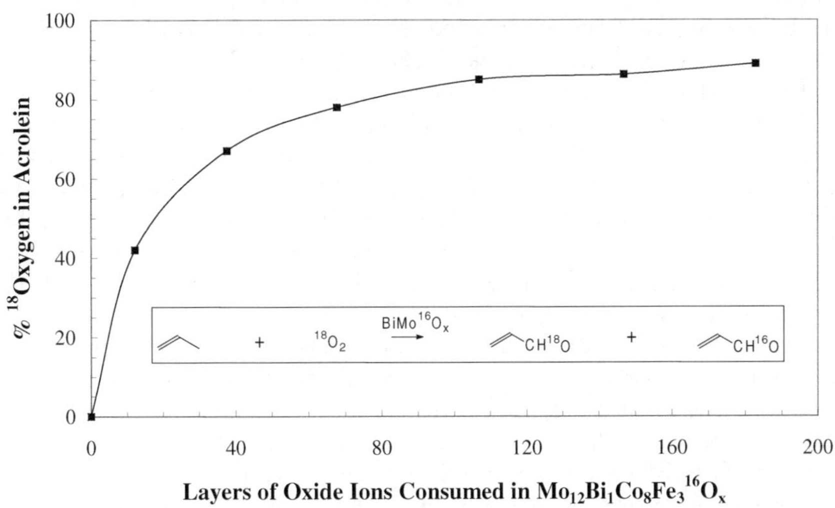


**Figure 4.10** Percentage of $^{18}O$ in acrolein formed from propene and $^{18}O_2$ over $Mo_{12}Bi_1Co_8Fe_3{}^{16}O_x$ as a function of the number of catalyst oxide layers consumed (Ueda *et al.*, 1986; Moro-oka and Ueda, 1994) (reproduced by permission of Elsevier Science).

**Table 4.8** Fraction of lattice oxygen participating in the oxidation of propene for the catalysts indicated
(Krenzke and Keulks, 1980a and b)

| *Catalyst* | *Temperature/K* | *Total lattice oxygen participating in the formation of acrolein, %* | *Number of oxide ion layers involved* |
|---|---|---|---|
| $Bi_2Mo_3O_{12}$ | 623 | 2.6 | 13 |
| | 673 | 4.4 | 22 |
| | 723 | 9.4 | 47 |
| $Bi_2MoO_6$ | 623 | 28 | 79 |
| | 673 | 45 | 126 |
| | 723 | 100 | 282 |
| $Bi_3FeMo_2O_{12}$ | 623 | 81 | 58 |
| | 673 | 89 | 64 |
| | 723 | 100 | 72 |



$Bi_2MoO_6$ is capable of being extracted from the lattice and becoming incorporated into the reaction products. The number of layers of oxide ions involved in this way has also been calculated in Table 4.8 on the basis of the BET surface area and assuming an oxide ion packing density of $1 \times 10^{19}$ ions $m^{-2}$ (Krenzke and Keulks, 1980b).

This level of involvement of lattice oxygen from deep within bismuth molybdate catalysts is confirmed from the results of Raman spectroscopy of catalysts which had been reduced for a short time with propene and reoxidized with $^{18}O_2$ in the gas phase (Grasselli, 1985). When $\gamma$-$Bi_2Mo^{16}O_6$ was subjected to a number of cycles of reduction by propene and reoxidation by $^{18}O_2$ the positions of the Mo—$^{16}$O stretching frequencies at 884, 803 and 725 $cm^{-1}$ shifted to lower values as outlined in Table 4.9, consistent with the gradual replacement of $^{16}O_L$ by $^{18}O_L$.

The surprisingly high level of participation of lattice oxygen in the oxidation of propene is made possible by the equally surprisingly low activation energy measured for the replenishment of lattice oxygen provided that the extent of reduction is not extensive. These experiments have been reviewed by Snyder and Hill Jr (1989). In all cases studied very low activation energies have been reported especially for temperatures in excess of 673 K, provided the extent of reduction is very small, as it is during catalyst operation. A typical set of data is presented in Table 4.10 (Bradzil *et al.*, 1980). Only data taken at very low degrees of reduction are important because several workers have reported that selectivity in acrolein formation is reduced

**Table 4.9** Raman band shifts in $\gamma$-$Bi_2MoO_6$ reduced in propene and reoxidized in $^{18}O_2$
(Grasselli, 1985)

| *Mo—$^{16}$O vibrational frequencies/cm$^{-1}$* | *Size of shift/cm$^{-1}$* | |
|---|---|---|
| | *After one cycle* | *After six cycles* |
| 844 | −9 | −14 |
| 803 | −11 | −18 |
| 725 | −10 | −17 |

Reprinted by permission of Elsevier Science.

**Table 4.10** Activation energies for the reoxidation of bismuth molybdate catalysts (Brazdil *et al.*, 1980)

| *Catalyst* | *Initial degree of reduction %* | *Temperature range for activation energy/K* | *Activation energy/kJ mol$^{-1}$* |
|---|---|---|---|
| $Bi_2Mo_3O_{12}$ | 0.04 | 593–733 | 5.4 |
| | 0.09 | 703–733 | 5.8 |
| | 0.09 | 593–673 | 102 |
| | 0.26 | 593–733 | 108 |
| $Bi_2MoO_6$ | 0.06 | 593–733 | 5.0 |
| | 0.16 | 703–733 | 2.9 |
| | 0.16 | 593–673 | 33.9 |
| | 0.42 | 593–733 | 33.0 |

Reproduced by permission of Academic Press.

by a factor of 10 or more when the degree of reduction of the catalyst exceeds 1 %. (Peacock *et al.*, 1969; Bielanski and Haber, 1979).

These experiments essentially indicate that the oxidation of propene over bismuth molybdate catalyst should be viewed as a two step reaction. In the first step propene reacts with lattice oxygen to yield the reaction products and a reduced catalyst:

$$C_3H_6 + 2O_L \rightarrow C_3H_4O + H_2O + 2\square_{O^=} \qquad (4.7)$$

In the second step lattice oxygen is replenished by gas phase oxygen:

$$2\square_{O^=} + O_{2(g)} \rightarrow 2O_L \qquad (4.8)$$

The mechanism of replenishment is not a simple reoxidation of the surface bismuth molybdate layer by gas phase oxygen, but involves the underlying phases as will be illustrated in the next section.

## 4.5 OXYGEN TRANSPORT WITHIN BISMUTH MOLYBDATE CATALYSTS

An interesting aspect of the multicomponent–multiphase bismuth molybdate catalysts is that the lattice oxygen which is incorporated into the oxidation product comes not just from the bismuth molybdate surface layer, but large amounts must also originate from the other phases inside the catalysts. Typical findings are summarized in Table 4.11. Between 5 % and 13 % of the lattice oxygen $^{16}O_L$ in a multicomponent catalyst is present in one of the bismuth molybdate phases. However, a much larger fraction of the total lattice oxygen of the catalyst is incorporated into the oxidation products. This finding is consistent with significant transfer of lattice oxygen from the bulk phases to the bismuth molybate phase located at the surface (see Figure 4.5) in these multi-phase catalysts (Moro-Oka and Ueda, 1994).

Figure 4.11 presents the results of an experiment in which propene was oxidized to acrolein over a catalyst composed of $Bi_2(Mo^{16}O_4)_3$ supported on $\beta$-$Co_{11/12}Fe_{1/12}Mo^{18}O_4$ (Moro-oka *et al.*, 1991). The source of gas phase oxygen used in this experiment was $^{16}O_2$. As seen in this figure, the percentage of $^{18}O$ in the acrolein product increased with reaction time and passed through a maximum, consistent with transfer of $^{18}O$ from the $\beta$-$Co_{11/12}Fe_{1/12}Mo^{18}O_4$ phase to the $Bi_2(Mo^{16}O_4)_3$ and hence to the acrolein, as shown in Scheme 4.2.

Clearly the mobility of lattice oxygen must be very rapid through these structures at the reaction temperature, and a number of studies have been undertaken to assess the influence of catalyst structure on the anion vacancy concentration and mobility. The Scheelite type bismuth molybdate phases have been very extensively studied in this regard. Sleight *et al.* (1975) prepared a series of these phases to which Pb was added to give a series of solid solutions of composition $Pb_{1-3x}Bi_{2x}\phi_x(MoO_4)$ (See Section 4.2.2 for further details). With two bismuth ions in the structure a cation vacancy ($\phi$) was generated. When this series of catalysts was tested a relationship was observed between vacancy concentration and the rate of propene oxidation, as shown in Figure 4.12.

**Table 4.11** Comparison of the amount of lattice oxygen incorporated into oxidation products during propene oxidation over multicomponent catalysts (Moro-Oka and Ueda, 1994)

| *Multicomponent catalyst composition* | *Percentage of the total catalyst $^{16}O_L$ present as the $Bi_2Mo_3O_{12}$ phase* | *Percentage of the total catalyst $^{16}O_L$ incorporated into the oxidation products* |
|---|---|---|
| $Mo_{12}Bi_1Co_{11}O_X$ | 12.5 | 13.0 |
| $Mo_{12}Bi_1Co_8Fe_3O_X$ | 5.5 | 19.1 |
| $Mo_{12}Bi_1Mg_8Fe_3O_X$ | 13.2 | 27.9 |
| $Mo_{12}Bi_1Co_8Al_3O_X$ | 12.5 | 18.4 |

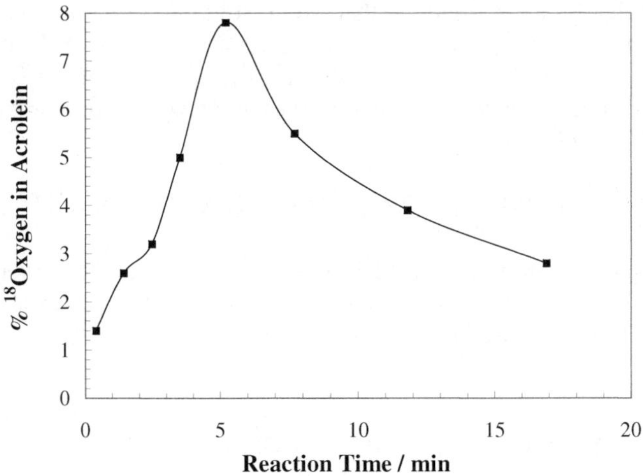


**Figure 4.11** Evolution of $^{18}O$ concentration of acrolein with reaction time during the oxidation of propene over $Bi_2(Mo^{16}O_4)_3$ supported on $\beta$-$Co_{11/12}Fe_{1/12}Mo^{18}O_4$ (Ueda *et al.*, 1986; Moro-oka *et al.*, 1991) (reproduced by permission of Elsevier Science).

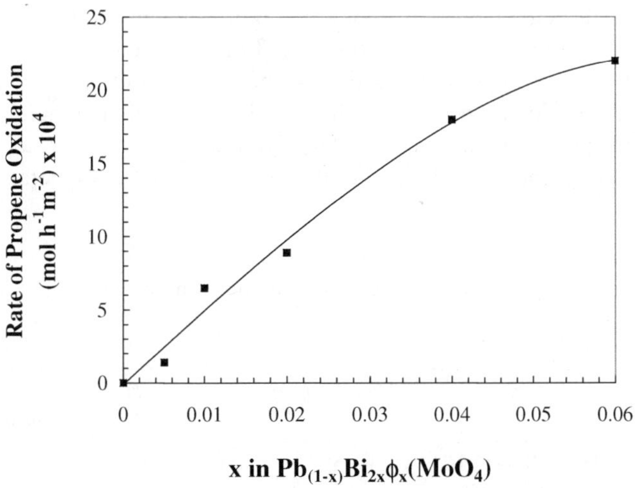


**Figure 4.12** Plot of propene oxidation at 673 K versus cation vacancy concentration for the solid solution $Pb_{1-x}Bi_{2x}\phi_xMoO_4$. (Sleight *et al.*, 1975; Sleight, 1977) (reproduced by permission of Academic Press).

What is clearly understood is that solid solution formation is a very important feature in the operation of these catalysts. For instance the presence of a small fraction of the iron component in $Co_{11/12}Fe_{1/12}MoO_4$ as $Fe^{3+}$ can give rise to cation vacancies according to the following equation proposed by Ponceblanc *et al.* (1992).

$$Me^{2+}MoO_4 + 2[Me^{3+}] \rightarrow M_{1-3x}{}^{2+}M_{2x}{}^{3+}\phi_x MoO_4 + 3[M^{2+}] \quad (4.9)$$

where $\phi$ is a cation vacancy. Similarly it is felt that $M^{3+}$ doping of $M_2{}^{2+}(MoO_4)_3$ could give rise to anion vacancies, although experimental evidence for this is less well documented.

## 4.6 THE OVERALL REACTION MECHANISM

Clearly from the preceding discussion there are three separate steps involved in acrolein formation from propene, namely, abstraction of two hydrogen atoms and insertion of an oxygen atom into the evolving acrolein product. Researchers have attempted, in so far as it is possible to do so, to associate the separate bismuth or molybdenum components of these catalysts with one or other of these steps (Otsubo *et al.*, 1975; Miura *et al.*, 1979).

Gryzbowska *et al.* (1977) reacted pulses of oxygen and allyl iodide or propene over bismuth oxide or molybdenum oxide and examined the resultant product distribution. The results are summarized in Table 4.12. A clear limitation of these experiments is that even the simplest bismuth molybdate catalysts contain neither bismuth oxide nor molybdenum oxide, but instead are made up of a binary oxide of bismuth and molybdenum, whose structure is different from that of bismuth oxide and molybdenum oxide. Gryzbowska *et al.* selected allyl iodide because of the very low bond dissociation enthalpy of the C—I bond, implying that a surface allyl species, would readily form from this starting material. In addition a lower reaction temperature was

**Table 4.12** Product distribution from the reactions of propene and allyl iodide with $Bi_2O_3$ and $MoO_3$
(Gryzbowska *et al.*, 1977)

| *Catalyst* | *Reactant* | *Temperature/K* | *Acrolein formed/%* | *Hexadiene formed/%* |
|---|---|---|---|---|
| $Bi_2O_3$ | $CH_2{=}CH{-}CH_2I$ | 593 | 5 | 70 |
| | $CH_2{=}CH{-}CH_3$ | 753 | 0 | 9 |
| $MoO_3$ | $CH_2{=}CH{-}CH_2I$ | 593 | 98 | 0 |
| | $CH_2{=}CH{-}CH_3$ | 753 | 0 | 0 |

required for allyl iodide than for propene reflecting the greater inherent reactivity of the former.

Over bismuth oxide no acrolein was formed from propene and just small amounts from allyl iodide. Instead the major reaction product was the dimer hexadiene, formed from the dimerization of two allyl species. Over molybdenum oxide no reaction occurred with propene but copious amounts of acrolein formed from allyl iodide.

These results led to the postulate that the bismuth component of bismuth molybdate catalysts is responsible for the abstraction of the first hydrogen from propene, but this oxide was not capable of inserting oxygen into the reaction intermediate which forms. Instead any allyl species which form from allyl iodide, and less easily from propene over $Bi_2O_3$, dimerize to hexadiene in the absence of a suitable source of lattice oxygen. By contrast, allyl species do not form from propene over molybdenum oxide. However, when this oxide is exposed to a facile source of allyl species, namely allyl iodide, acrolein forms readily, consistent with facile $\pi$-allyl formation from allyl iodide followed by oxygen insertion and the second hydrogen abstraction steps occurring over the molybdenum oxide component of the bismuth molybdate catalysts.

A similar approach has been used by Lunsford and co-workers (Lunsford, 1989; Driscoll *et al.*, 1987), where they monitored the presence of gas phase radical species at the exit of a reactor used for propene oxidation. The exit gases from the reactor were cooled and radical species stabilized using a matrix isolation technique. Table 4.13 presents the results. Allyl and peroxy

**Table 4.13** Radical species and product distribution obtained during the oxidation of propylene
(Lunsford, 1989)

| *Catalyst* | *Radical amount (nmol)* | | *Stable products (nmol)* | | |
|---|---|---|---|---|---|
| | *Allyl* | *Allyl peroxy* | *1,5-Hexadiene* | *Acrolein* | *3-Methylpentane* |
| $Bi_2O_3$ | 31.72 | 5.60 | 33.2 | 7.0 | 36.1 |
| $\gamma$-Bismuth molybdate | 3.72 | 0.65 | 0.0 | 2920 | 31.0 |
| $\alpha$-Bismuth molybdate | 0.82 | 0.15 | 0.0 | 311 | 24.3 |
| $MoO_3$ | 0.12 | 0.02 | 0.0 | 25 | 38.7 |
| Homogeneous Pyrex | 0.84 | 0.18 | 0.0 | 3.5 | 51.6 |
| $Bi_2O_3/MoO_3$[a] | 1.66 | 0.40 | 0.0 | 37.0 | 34.6 |

$T = 723$ K. Flowrates: $Ar = 0.064\ cm^3\ s^{-1}$; $C_3H_6 = 0.019\ cm^3\ s^{-1}$; $O_2 = 3.76 \times 10^{-4}\ cm^3\ s^{-1}$; Total Pressure = 1 Torr.
[a]The $Bi_2O_3$ catalyst was placed upstream from the $MoO_3$ catalyst.
Reproduced by permission of Academic Press.

radicals were detected in varying amounts depending on the catalyst that was used. Those catalysts which showed a propensity to form 1,5-hexadiene released more radicals into the gas phase than those over which acrolein was formed. When $Bi_2O_3$ was placed upstream of $MoO_3$ in the reactor there was a marked reduction in the amount of radicals detected at the exit of the reactor and an increase in acrolein production. There data are consistent with the work of Gryzbowska *et al.* (1977) whereby radical species do not react with lattice oxygen in $Bi_2O_3$, hence they are more readily released into the gas phase with the formation there of 1,5-hexadiene. The proposed network is outlined in Scheme 4.3. In the presence of the $MoO_3$ component of the catalyst the radical species formed can react with lattice oxygen leading to acrolein.

Compelling evidence that the radical allyl species which forms is important in the catalysis and not just a spectator species comes from measurement of the KIE for allyl radical formation from $C_3H_6$ and $C_3D_6$ over $Bi_2O_3$ catalysts. These data were already presented in Table 3.3 (Driscoll *et al.*, 1987) and show the same KIE associated with acrolein formation from propene in more conventional studies.

The most coherent overall picture of the mechanism of propene oxidation has been presented by Grasselli (1985) and is shown in Scheme 4.4. The mechanism involved abstraction of a $CH_3$ hydrogen from propene by the bismuth oxide part of the active site, with the formation of a $\pi$-allyl species interacting with the Mo centre. Oxygen insertion occurs next, and the $\sigma$-allyl species which forms is more activated for removal of the second hydrogen species than the carbene which would form without oxygen insertion. The final steps involve product desorption and reoxidation of the active site via oxygen transfer from the underlying solid solution in the case of the multi-component catalysts.

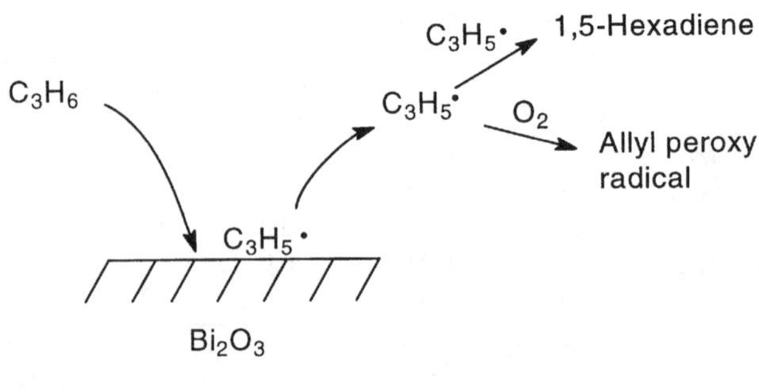


**Scheme 4.3** Schematic diagram for the reactions of allyl radicals species over $Bi_2O_3$ (Driscoll *et al.*, 1987) (reproduced by permission of Academic Press).

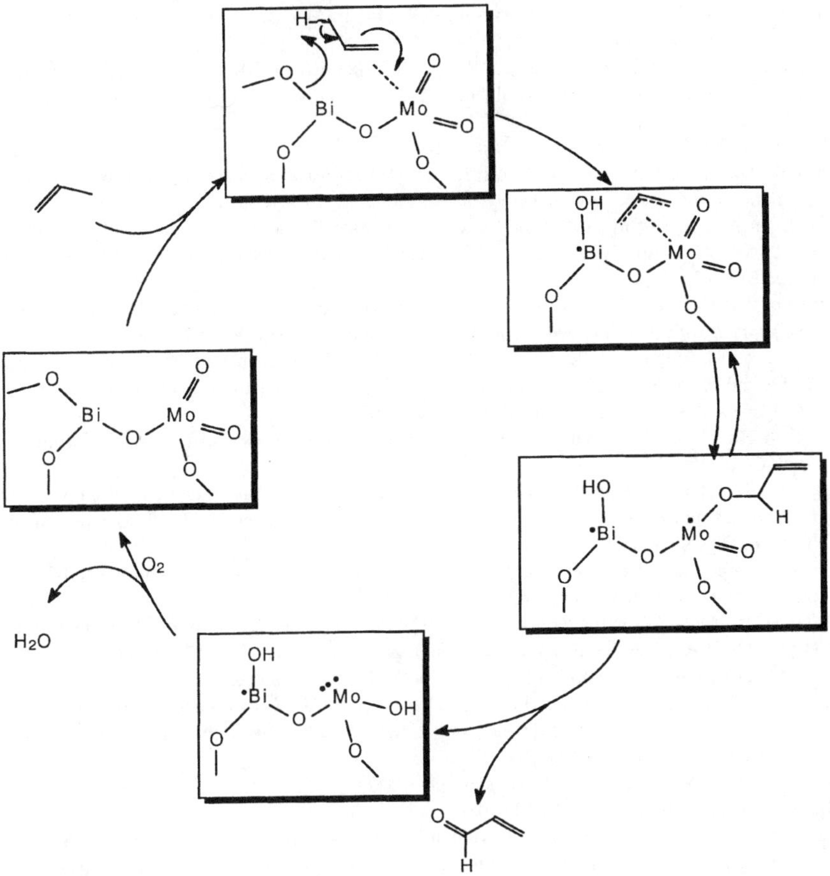


**Scheme 4.4** Schematic diagram of the mechanism of acrolein formation from propene over bismuth molybdate catalysts (Burrington *et al.*, 1983) (reproduced by permission of Academic Press).

## REFERENCES TO CHAPTER 4

Adams CR and Jennings TJ (1963) Investigation of the mechanism of catalytic oxidation of propylene to acrolein and acrylonitrile. *J. Catal.* **2**: 63–68.

Adams CR and Jennings TJ (1964) Mechanistic studies of the catalytic oxidation of propylene. *J. Catal.* **3**: 549–558.

Bielanski A and Haber J (1979) Oxygen in catalysis on transition metal oxides. *Catal. Rev.-Sci. Eng.* **19**(1): 1–41.

Bradzil JF, Suresh DD and Grasselli RK (1980) Redox kinetics of bismuth molybdate ammoxidation catalysts. *J. Catal.* **66**: 347–367.

Bruchner A and Baerns M (1997) Selective gas-phase oxidation of polycyclic aromatic hydrocarbons on vanadium oxide-based catalysts. *Appl. Catal. A General* **157**: 311–334.

Burrington JD, Kartisek CT and Grasselli RK (1980) Aspects of selective oxidation and ammoxidation mechanisms over bismuth molybdate catalysts. *J. Catal.* **63**: 235–254.

Burrington JD, Kartisek CT and Grasselli RK (1983) Mechanism of nitrogen insertion in ammoxidation catalysis. *J. Catal.* **81**: 489–498.

Callaghan JL, Grasselli RK, Milberger EC and Streicker HA (1970) Oxidation and ammoxidation of plopylene over bismuth molybdate catalysts. *Ind. Eng. Chem. Prod. Res. Dev.* **9**(2): 134–142.

Cassidy F (1999) Thesis, University of Limerick.

Driscoll DJ, Campbell KD and Lunsford JH (1987) Surface-generated gas-phase radicals: Formation, detection and role in catalysis. *Ad. Catal.* **35**: 139–186.

Efstathiou AM, Papageorgiou D and Verykios XE (1994) The selective oxidation of methane to C—2-hydrocarbons over Li+-doped $TiO_2$—Catalytic and mechanistic studies. *Stud. Surf. Sci. Catal.* **81**: 217–222.

Etzkorn WJ, Kurland JJ and Neilsen WD (1991) Acrolein and derivatives. In *Kirk–Othmer Encyclopedia of Chemical Technology*. John Wiley and Sons, New York, pp 232–251.

Gaigneaux EM, Ruiz P and Delmon B (1996) Further on the mechanism of the synergy between $MoO_3$ and $\alpha$-$Sb_2O_4$ in the selective oxidation of isobutene to mathacrolein: Reconstruction of $MoO_3$ via spillover oxygen. *Catal. Today* **32**: 37–46.

Grasselli RK (1985) Selectivity and activity factors in bismuth molybdate oxidation catalysts. *Appl. Catal.* **15**: 127–139.

Grasselli RK, Heights G and Hardman HF (1972) Process for the manufacture of isoprene from isoamylenes and methyl butanols and catalyst therefor. US Patent 3 642 930, assigned to the Standard Oil Co Ohio.

Grzybowska B, Haber J and Janas J (1977) Interaction of allyl iodide with molybdate catalysts for the selective oxidation of hydrocarbons. *J. Catal.* **49**: 150–163.

He D-H H, Ueda W and Moro-oka Y (1992) Promotion effect of molybdate support on $Bi_2Mo_3O_{12}$ catalyst in the selective oxidation of propylene. *Catal. Lett.* **12**: 35–44.

Hearne G and Adams M (1948) US Patent 2 451 485, Shell Development Co.

Horowitz HS, Blackstone CM, Sleight AW and Teufer G (1988) V–P–O catalysts for oxidation of butane to maleic anhydride. Influence of $(VO)_2H_4P_2O_9$ precursor morphology on catalytic properties. *Appl. Catal.* **38**: 193–210.

Idol JD (1959) US Patent 2 904 580, Standard Oil Co Ohio.

Knapsak AG (1962) Dutch patent 1 125 901.

Keulks GW and Krenzke LD (1976) The role of lattice oxygen in selective oxidation of propylene over binary oxide catalysts. In *Proceedings of the 6th International Congress on Catalysis*. The Chemical Society, London, pp 806–818.

Krenzke LD and Keulks GW (1980a) The catalytic oxidation of propylene:VI. Mechanistic studies utilizing isotopic tracers. *J. Catal.* **61**: 316–325.

Krenzke LD and Keulks GW (1980b) The catalytic oxidation of propylene: VIII An investigation of the kinetics over $Bi_2Mo_3O_{12}$, $Bi_2MoO_6$ and $Bi_3FeMo_2O_{12}$. *J. Catal.* **64**: 295–302.

Lunsford JH (1989) The role of surface-generated gas-phase radicals in catalysis. *Langmuir* **5**: 12–16.

Matsuura I, Schut R and Hirakawa K (1980) The surface structure of the active bismuth molybdate catalyst. *J. Catal.* **63**: 152–166.

Moro-oka Y, He D-H and Ueda W (1991) Catalyst oxide support interaction to prepare multi-functional oxidation catalysts. In *Structure-Activity Relationships in Heterogeneous Catalysis*, Grasselli RK and Sleight AW (Eds). Elsevier Science, Amsterdam, pp 57–66.

Moro-oka Y and Ueda W (1994) Multicomponent bismuth molybdate catalyst: A highly functionalized catalyst system for the selective oxidation of olefin. *Adv. Catal.* **40**: 233–273.

Miura H, Otsubo T, Shirasaki T and Morikawa Y (1979) Tracer studies of catalytic oxidation by bismuth molybdate: II Propylene reduction of labelled catalysts and catalytic oxidation of propylene. *J. Catal.* **56**: 84–87.

Monnier JR and Keulks GW (1981) The catalytic oxidation of proplene. IX The kinetics and mechanism over $\beta$-$Bi_2Mo_2O_9$. *J. Catal.* **68**: 51–66.

Otsubo T, Miura H, Morikawa Y and Shirasaki T (1975) Tracer studies of catalytic oxidation by bismuth molybdate. I. Hydrogen reduction of labeled catalysts. *J. Catal.* **36**: 240–243.

Peacock JH, Parker AJ, Ashmore PG and Hockey JA (1969) The oxidation of propene over bismuth oxide, molybdenum oxide and bismuth molybdate catalysts. *J. Catal.* **15**: 398–406.

Ponceblanc P, Millet JMM, Coudurier G, Legendre O and Vedrine JC (1992) Solid–solid phase equilibria in the binary system $CoMoO_4$–$FeMoO_4$ and the effect of $Fe^{III}$ on the phase equilibria. *J. Phys. Chem.* **96**: 9462–9465.

Sachtler WMH (1963) Mechanism of catalytic oxidation of propylene to acrolein and of isomeric butenes to butadiene. *Rec. Trav. Chim.* **82**: 243–245.

Sleight AW, Aykan K and Rogers DB (1975) New non-stoichiometric molybdate, tengstate and vanadate catalysts with the scheelite-type structure. *J. Solid State Chem.* **12**: 231–236.

Sleight AW (1977) Crystal chemistry and catalytic properties of oxides with the scheelite structure. In *Advanced Materials in Catalysis. Materials Science and Technology Series*, Burton JJ and Garten T (Eds). Academic Press, New York, Chapter 6, pp 181–208.

Snyder TP and Hill Jr CG (1989) The mechanism of partial oxidation of proplene over bismuth molybdate catalysts. *Catal. Rev. Eng. Sci.* **31**: 43–95.

Topsoe N-Y, Topsoe H, Sorensen O, Clausen BJ and Candia R (1984) IR and AEM evidence for the edge location of the promoter atoms in Co–Mo–S type structures. *Bull. Soc. Chim. Belg.* **93**: 727–733.

Ueda W, Asakawa K, Chen C-L, Moro-oka Y and Ikawa T (1986) Catalytic properties of tricomponent metal oxides having the scheelite structure. I. Role of bulk diffusion of lattice oxide ions in the oxidation of propylene. *J. Catal.* **101**: 360–368.

Voge HH, Wagner CD and Stevenson DP (1963) Mechanism of propylene oxidation over cuprous oxide. *J. Catal.* **2**: 58–62.

Wolfs MWJ and Batist PhA (1974) The selective oxidation of 1-butene over multicomponent molybdate catalysts: Influence of various elements on structure and activity. *J. Catal.* **32**: 25–36.

Wragg RD, Ashmore PG and Hockey JA (1973) On oxygen mobility during catalytic oxidations. *J. Catal.* **28**: 337–337.

Zabetakis, MG (1965) *Flammability Characteristics of Combustible Gases and Vapours*. US Dept of the Interior Bureau of Mines US Government Printing Office, Washington, DC 20402.

# 5

# Oxidative Activation of Alkanes: Oxidation of *n*-butane to Maleic Anhydride


5.1 Introduction 133
5.2 Technologies for the Production of Maleic Anhydride 135
5.3 The Preparation, Structure and Performance of Vanadium Phosphorus Oxide (VPO) Catalysts 139
5.3.1 Catalyst Preparation and Phase Composition 139
5.3.2 Oxidation State of Vanadium in the Working Catalyst 144
5.3.3 The Surface P : V Ratio and Texture 146
5.3.4 Influence of P : V Ratio on Catalyst Performance 147
5.4 Activation of *n*-Butane 149
5.5 The Nature of the Active Site and the Reaction Mechanism 153
5.5.1 The Reaction Network 153
5.5.2 The Location of Active Sites and the Role of Catalyst Microstructure 153
5.6 Conclusions 156
References to Chapter 5 157



Summary

Maleic anhydride features among the first 100 organic chemicals by bulk, with the major part of its production directed towards the polymer industry. It is currently the only commercial catalytic oxidation process which uses an alkane as feedstock. The catalyst used is always an unsupported vanadium phophorus oxide with a platelet type particle shape, which can be used in a variety of fixed and mobile bed reactors. The reaction network involves the formation of butene, 1,3-butadiene and furan as intermediates. Primary activation of *n*-butane occurs at a methylene C—H bond, via an active site which is located on the basal plane of a vanadyl pyrophosphate catalyst particle.

This chapter describes

- Technology for the production of maleic anhydride,
- The preparation, structure and performance of vanadium phosphorus oxide catalysts,
- Activation of *n*-butane,
- The nature of the active site and the reaction mechanism.

## 5.1 INTRODUCTION

The simultaneous presence of an electron acceptor double bond and an anhydride functional group confers unique properties on maleic anhydride and allows it to be used in a wide range of polymer types, including homopolymers, copolymers, graft copolymers and step-growth polymers. Its main use in the polymer industry is in the production of unsaturated polyesters (Culbertson, 1987). It also finds use as a lube oil additive, corrosion inhibitor, in emulsifiers and detergents and as a precursor in the agrichemical industry for the manufacture of herbicides. Production details and principal uses are presented in Table 5.1.

Manufacture of maleic anhydride is predominantly by the oxidation of *n*-butane over vanadium phosphorus oxide (VPO) catalysts. Table 5.2 summarizes details of this reaction with an indication of catalyst composition and reaction conditions. This process, which is practised with a variety of reactor types, started to replace the benzene oxidation route in the 1970s and was the dominant technology by the mid-1980s (Emig and Martin, 1987). A feature of the selective oxidation of *n*-butane is that the reaction (see Table 5.2) involve

**Table 5.1** Annual production and principal uses of maleic anhydride (*Chemical Week*, 1997)

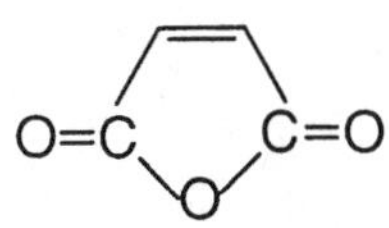


| *Location* | *Production/ktonne yr$^{-1}$* | *Uses* | |
|---|---|---|---|
| America | 315 | Unsaturated polyesters | 63 % |
| Europe | 100 | Oil additives | 12 % |
| Pacific | 96 | Other | 21 % |
| Total | 500 | | |

**Table 5.2** Summary information for *n*-butane oxidation (Cox and Pitcher, 1970; Zabetakis, 1965)

Selective oxidation (14 electron)

$n\text{-}C_4H_{10} + 3.5O_2 \longrightarrow C_4H_2O_3 + 4H_2O \qquad \Delta G_{298\,K} = -1356\ \text{kJ mol}^{-1}$

Total oxidation (26 electron)

$n\text{-}C_4H_{10} + 6.5O_2 \longrightarrow 4CO_2 + 5H_2O \qquad \Delta G_{298\,K} = -2704\ \text{kJ mol}^{-1}$

| Catalyst | Support | Additive | Temp/K | Pressure/bar |
|---|---|---|---|---|
| $(VO)_2P_2O_7$ | None[a] | None | 650 | 1 |

| | | $LFL_{298}$[b] | $UFL_{298}$[b] | AIT/K |
|---|---|---|---|---|
| | $n\text{-}C_4H_{10}$ | 1.8 | 8.4 | 678 |
| | CO | 12.5 | 74 | |

[a]Mixed with silica for fluidized bed operation.
[b]Lower and upper flammability limits at 298 K and the auto ignition temperatures.

the removal of eight hydrogen atoms and the insertion of three oxygen atoms to form the product. This reaction represents a deep (14 electron) oxidation, and in this way it contrasts with many other selective oxidations which are usually restricted to not more than four electrons. The reaction to maleic anhydride is thermodynamically favoured, but the route to total oxidation is even more strongly favoured.

Figure 5.1 summarizes many literature reports of the performance of VPO catalysts for *n*-butane oxidation in terms of a selectivity–conversion plot and clearly shows the upper limit that can be achieved in terms of optimum selectivity. In many commercial operations *n*-butane oxidation is carried out above 80 % conversion and at a limiting selectivity of about 65 %.

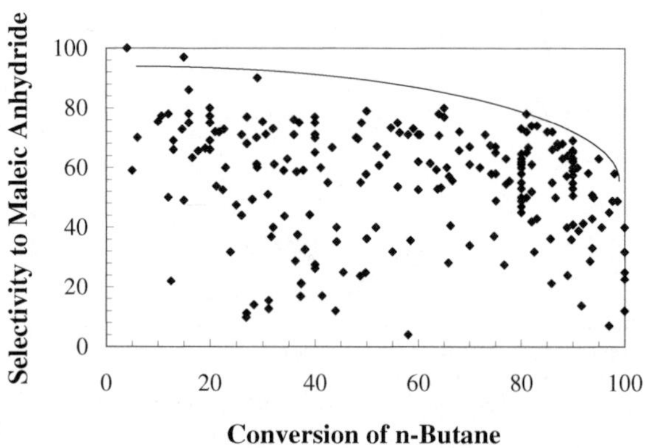


**Figure 5.1** Multiple selectivity–conversion plot for *n*-butane selective oxidation to maleic anhydride over a range of catalysts and in a variety of reaction conditions (Batiot and Hodnett, 1996) (reprinted by permission of Elsevier Science).

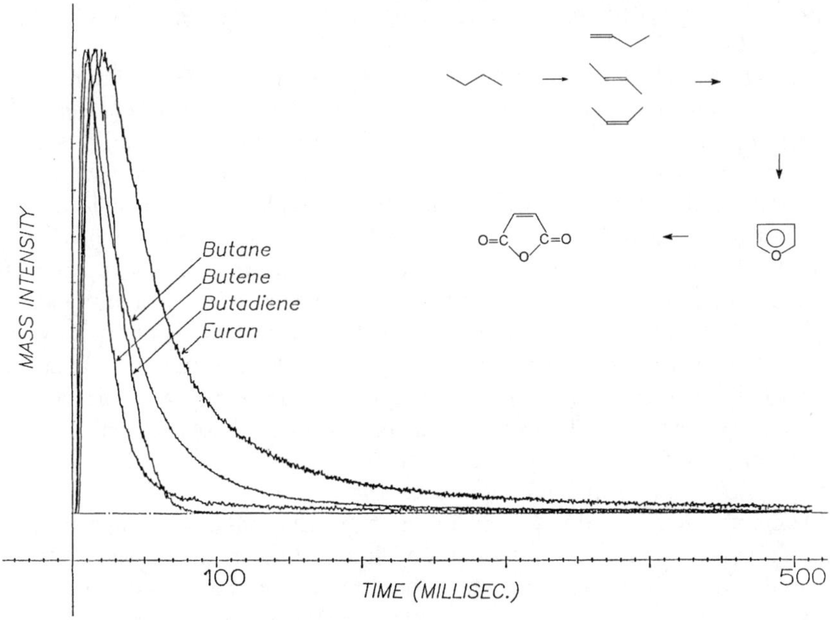


**Figure 5.2** Evolution of reaction intermediates during the oxidation of *n*-butane to maleic anhydride in a TAP reactor study (Gleaves *et al.*, 1988) (reproduced by permission of Marcel Dekker).

Studies in the 1970s (Ai *et al.*, 1970) had already established for butene oxidation that 1,3-butadiene and furan were intermediates in this reaction, and although these intermediates are not detected in the usual experimental condition when *n*-butane is used as the feed stock, the most modern work with TAP reactors has confirmed this as the reaction network. Figure 5.2 presents the results of a TAP reactor study showing the evolution of *n*-butane, butenes, 1,3-butadiene and furan from a VPO catalyst and a schematic diagram of the reaction network. A point of continuing controversy is whether all these steps occur at a single active site on the VPO catalyst or over a number of locations spread across the surface (Gleaves *et al.*, 1988).

## 5.2 TECHNOLOGIES FOR THE PRODUCTION OF MALEIC ANHYDRIDE

The original plant design for the production of maleic anhydride from *n*-butane involved the use of multitubular fixed bed reactors (Burnett *et al.*, 1987). These reactor types still dominate maleic anhydride production and the VPO catalysts must be capable of offering stable activity at greater than 75 % conversion of *n*-butane lean feeds with selectivities to maleic anhydride well above 50 %. Typical inlet and outlet compositions for this type of reactor are presented in Table 5.3.

An inherent feature of fixed bed reactor operation is the requirement to operate below the lower flammability limit (see Section 2.3.2). Operation above the upper flammability limit is impractical because of the need to recycle unreacted *n*-butane and the amount of heat which would have to be dissipated in the reactor. Drawbacks of operating with lean *n*-butane feeds are the cost of pumping large quantities of diluent nitrogen through the reactor and the maleic anhydride recovery system and the requirement to extract maleic anhydride from a stream of which it constitutes less than 1 mol%. In practice, this means that the maleic anhydride collection system must be very efficient and usually represents a much larger part of the overall plant than the catalytic reactor. A typical flowsheet is presented in Figure 5.3. The maleic anhydride collection system involves a solvent, often dibutyl phthalate, into which the cooled maleic anhydride dissolves in the absorber. It is then isolated and purified by distillation (Burnett *et al.*, 1987)

Attempts to reduce the amount of inert material passing through maleic anhydride plants have inspired the use of fluidized bed reactors. These plants operate with nominal *n*-butane inlet concentrations of about 4 vol%, i.e. well within the flammability limits. Here the *n*-butane and air are separately injected directly into the fluidized bed reactor. The extreme turbulence within the reactor causes inlet and outlet gas compositions to become equalized. Provided that the *n*-butane and air are separately fed into the reactor, richer *n*-butane compositions are possible with this type of reactor. For this type of reactor a special attrition resistant form of the VPO catalyst has been developed. Details of the operation of these plants are sparse, but a plan layout for the ALMA process is presented in Figure 5.4.

**Table 5.3** Typical reactor inlet and outlet compositions for *n*-butane oxidation in a fixed bed reactor
(Burnett *et al.*, 1987)

| *Component* | *Reactor inlet/mol%* | *Reactor outlet mol%* |
|---|---|---|
| $O_2$ | 20.3 | 14.8 |
| $N_2$ | 76.4 | 76.4 |
| CO | 0.00 | 0.93 |
| $CO_2$ | 0.00 | 0.85 |
| $H_2O$ | 1.50 | 7.27 |
| $n$-$C_4H_{10}$ | 1.77 | 0.44 |
| $C_4H_2O_3$ | 0.00 | 0.89 |
| | | |
| Conversion | | 75 % |
| Selectivity | | 67 % |
| Yield | | 50 % |

Reproduced by permission of Elsevier Science.

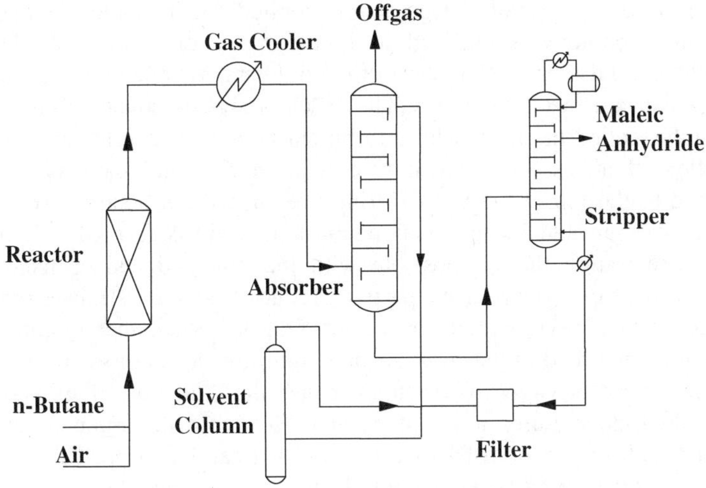


**Figure 5.3** Maleic anhydride process flowsheet: fixed bed reactor with Monsanto solvent collection system (Burnett *et al.*, 1987) (reproduced by permission of Elsevier Science).

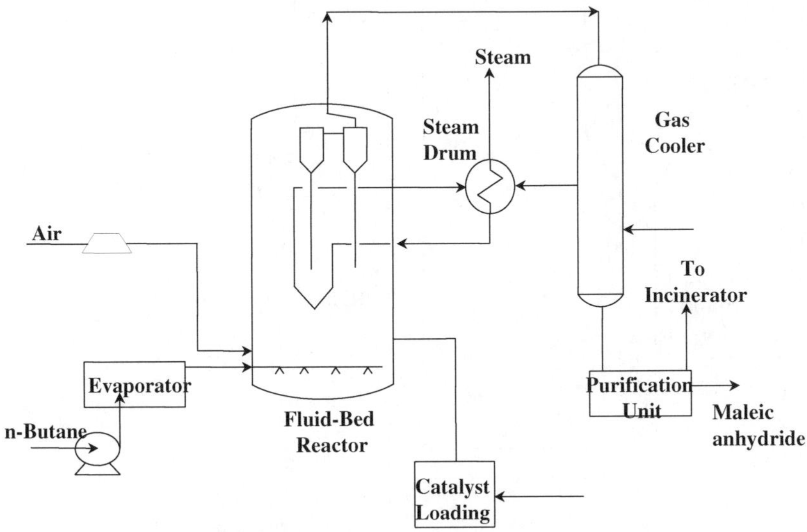


**Figure 5.4** ALMA fluidized bed process for the production of maleic anhydride from *n*-butane (Contractor and Sleight, 1987) (reproduced by permission of Elsevier Science).

The most recent reactor design to be applied to the maleic anhydride production technology is outlined in Figure 5.5 (Contractor and Sleight, 1987). The central idea owes much to the fluid catalytic cracking technology in widespread use in oil refineries. The VPO catalyst is contacted in a long vertical tube, called the riser, with a turbulent flow of pure *n*-butane. The *n*-butane flow fluidizes the catalyst (see Section 8.2) and causes it to be transported to the top of the riser. During this time the *n*-butane reacts with the lattice oxygen and the product maleic anhydride is formed at a much higher concentration than is possible with the fluidized bed or fixed bed reactors. At the top of the riser the partially reduced catalyst is transferred to a regenerator where it is reoxidized by a flow of air or oxygen, then re-contacted with the *n*-butane feed in the riser. At first sight this does not seem to be the ideal reaction for this type of reactor because the oxidation of *n*-butane to maleic anhydride requires a lot of oxygen (3.5 mole per mole of maleic anhydride) and the amount of lattice oxygen that can be stripped from VPO catalysts is very limited (see below). However, this technology has been commercialized by the DuPont company with the aid of a specially formulated VPO catalyst that is resistant to attrition losses (Contractor and Sleight, 1987).

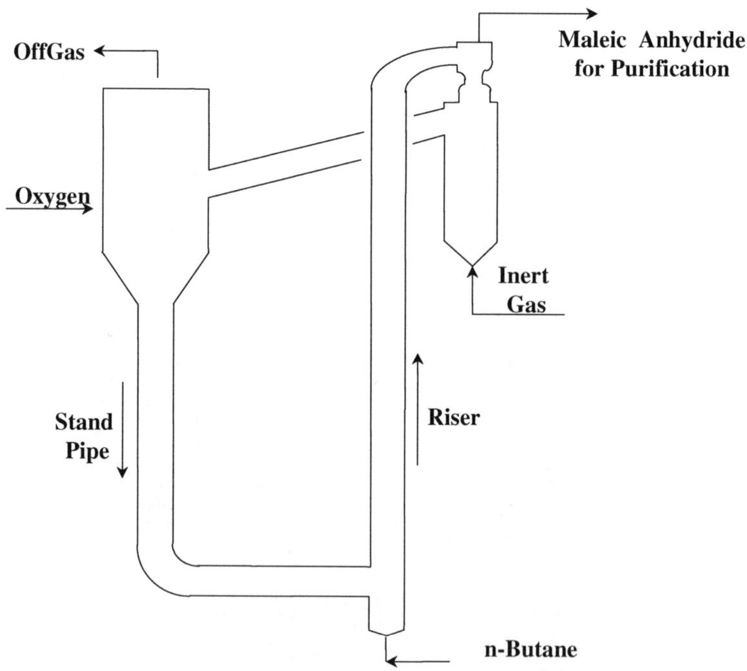


**Figure 5.5** DuPont's circulating solids riser reactor (Contractor and Sleight, 1987) (reproduced by permission of Elsevier Science).

## 5.3 THE PREPARATION, STRUCTURE AND PERFORMANCE OF VANADIUM PHOSPHORUS OXIDE CATALYSTS

Maleic anhydride can be manufactured in varying quantities and at various selectivities from C-4, C-5 or C-6 hydrocarbons, and there are many reports in the literature concerning catalysts that produce a small amount from a variety of feedstocks. Here only the structure of VPO catalysts capable of operation close to that required for commercial practice (Burnett *et al.*, 1987) will be described. Historically, commercial VPO catalysts were first prepared in an aqueous medium, but later preparation methods were developed based on the use of alcohols, which yielded better catalysts (Hodnett, 1985). Both types of preparation methods will be discussed below.

### 5.3.1 Catalyst preparation and phase composition

The original preparation methods for VPO catalysts involved reducing $V_2O_5$ in an aqueous solution of HCl, followed by addition of the phosphorus component, usually as 85 % *ortho*-$H_3PO_4$. The solid precursor was isolated by evaporation of the solvent. These catalysts were largely composed of $(VO)_2P_2O_7$, in which vanadium is in the 4+ oxidation state. However, catalysts prepared in this way often featured other crystalline phases, such as those listed in Table 5.4, many of which have vanadium in the 5+ oxidation state (Bordes, 1987). In general, catalysts prepared in aqueous media are not as active or selective as the modern catalysts described below which are prepared in an organic medium.

The most active and selective VPO catalysts are predominantly (>90 %) vanadyl pyrophosphate, $(VO)_2P_2O_7$, which is prepared by reduction of $V_2O_5$ by an alcohol to a partially soluble $V^{4+}$ complex or suspension, followed by addition of phosphoric acid. The precursor, namely $(VO)_2H_4P_2O_9$, (sometimes written as $VOHPO_4{\cdot}0.5H_2O$ and referred to as a hemihydrate) is isolated by precipitation. A typical preparation procedure is given below. Horowitz *et al.* (1988) report that when this method is used the bulk P : V ratio of the precursor never exceeds 1.1 : 1 irrespective of the solution P : V ratio from which the precursor is isolated. However, the crystallinity of the sample improves with higher solution P : V ratios, as shown by the sharper lines shapes observed by X-ray diffraction in Figure 5.6. The final catalysts is formed by careful thermal treatment of the hemihydrate precursor, in an *n*-butane–air flow. During this treatment the final catalyst form evolves and the selectivity to maleic anhydride increases as the amount of $(VO)_2P_2O_7$ increases. The thermal treatment is prolonged (up to one week) but appears to be crucial to the production of a very selective catalyst (Hutchings *et al.*, 1994). The final catalyst can exhibit a variety of particle shapes and sizes, but a clustered and distorted platelet type crystalline habit shown in Figure 5.7 is common (Horowitz et al. 1988). Small amounts of other phases (crystalline

**Table 5.4** Phases detected in VPO catalysts (Bordes, 1987)

| *Phase* | *Unit cell* | *Cell parameters/nm* |
|---|---|---|
| $VO(PO_3)_2$ | Tetragonal | $a = 1.096$ |
| | | $c = 0.425$ |
| $(VO)_2H_2P_2O_9$ or $VOHPO_4.0.5H_2O$ | Orthorhombic | $a = 0.742$ |
| | | $b = 0.960$ |
| | | $c = 0.569$ |
| $(VO)_2P_2O_7$ | Orthorhombic | $a = 0.773$ |
| | | $b = 1.658$ |
| | | $c = 0.957$ |
| $VOPO_4.2H_2O$ | Tetragonal | $a = 0.620$ |
| | | $c = 0.741$ |
| $\beta$-$VOPO_4$ | Orthorhombic | $a = 0.777$ |
| | | $b = 0.614$ |
| | | $c = 0.697$ |
| $\alpha_I$-$VOPO_4$ | Tetragonal | $a = 0.620$ |
| | | $c = 0.411$ |
| $\alpha_{II}$-$VOPO_4$ | Tetragonal | $a = 0.601$ |
| | | $c = 0.443$ |
| $\delta$-$VOPO_4$ | Orthorhombic | $a = 0.642$ |
| | | $b = 0.626$ |
| | | $c = 0.909$ |
| $\gamma$-$VOPO_4$ | Monoclinic | $a = 0.964$ |
| | | $b = 1.533$ |
| | | $c = 1.662$ |
| | | $\beta = 93.04°$ |
| $VO(H_2PO_4)_2$ | Tetragonal | $a = 0.895$ |
| | | $c = 0.797$ |



and amorphous) have also been detected by a variety of physicochemical techniques (Volta, 1996; Abon and Volta, 1997) . However, the final catalyst closely resembles a simple binary oxide of vanadium and phosphorus, namely ($VO_2P_2O_7$), but the role of minor phases and the small excess phosphorus over vanadium usually present continue to be investigated.

*Preparation of vanadium phosphorus oxide catalysts*

(Horowitz *et al.* 1988)

Synthesis of $(VO)_2H_4P_2O_9$, the most commonly cited precursor of vanadium phosphorus oxide catalysts, is carried out by reacting $V_2O_5$ and organic solvents, often alcohols typically at about 373 K in a flask equipped with a mechanical stirrer and reflux condenser. The reactant P : V atomic ratio can be varied in the range 0.8 and 3.5 by use of the appropriate amount of 85 % $H_3PO_4$ but the P : V

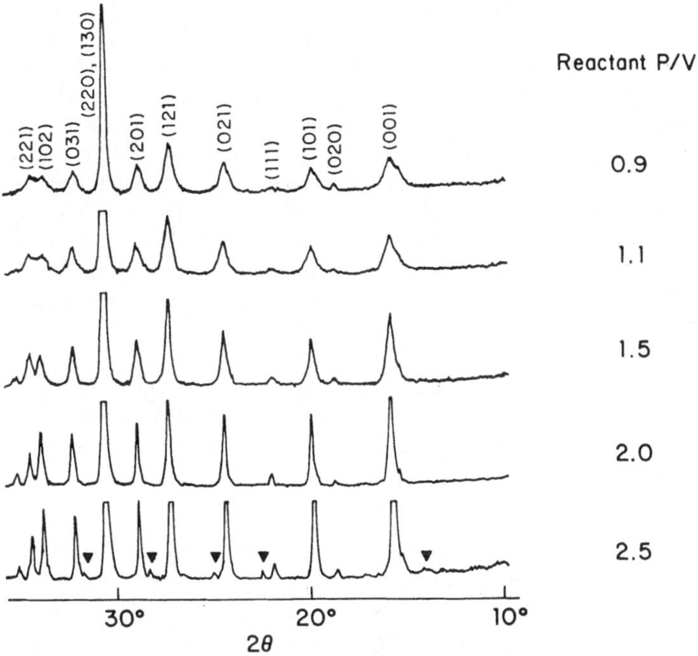


**Figure 5.6** Influence of solution P : V ratio on the crystallinity of $(VO)_2H_4P_2O_9$ prepared in organic media (Horowitz *et al.* 1988) (reproduced by permission of Elsevier Science).

**Figure 5.7** Scanning electron micrographs of VPO catalysts showing the distorted platelet type morphology often encountered with this type of catalyst (Horowitz *et al.*, 1988) (reproduced by permission of Elsevier Science).

ratio which appears in the precursor is never more than 1.1 : 1 when the method described below is used. A typical synthesis involved the reaction of 11.81 g $V_2O_5$ (0.150 mole) and 17.88 g 85 % $H_3PO_4$ (0.156 mole) in 236 ml isobutanol and 23.6 ml benzyl alcohol. After completion of the reaction, $(VO)_2H_4P_2O_9$ precipitates were recovered by vacuum filtration, washed in excess solvent, air dried for 1 h, and then dried at 383 K in a vacuum oven for 12 h. Formation of the final catalyst is achieved by heating the precursor in the catalytic reactor at a rate of 5 K $min^{-1}$ up to 683 K in a flow of 1.5 % *n*-butane–10 % oxygen. Conversion of *n*-butane and selectivity to maleic anhydride continue to change long after the reactor reaches 683 K. In practice these conditions are retained for 24–48 h by which time the conversion and selectivity have usually reach a steady state.

The minimum level of characterization involves analysis by X-ray diffration, elemental analysis for phosphorus and vanadium contents and analysis of the oxidation state of vanadium by redox titrimetry. In addition, BET surface area analysis by nitrogen adsorption is usually carried out.

X-ray diffraction patterns of the precursor hemihydrate and of the final catalysts both exhibit anisotropic line broadening consistent with the presence of a platelet type particle morphology. Inspection of Figure 5.8 reveals considerable broadening of the (010) reflection of the precursor and of the topotactically related (020) reflection of $(VO)_2P_2O_7$ for certain preparation methods. Indeed the more active and selective VPO catalysts tend to exhibit this type of line broadening (Centi *et al.* 1988). It is consistent with a platelet type particle shape, shown in Scheme 5.1, where the basal plane corresponds to the (020) surface of $(VO)_2P_2O_7$ (Okuhara and Misono, 1993). This basic shape is usually distorted or twisted as shown in Figure 5.7. An aspect of the platelet type crystal shape will be important below in locating the active site for this reaction, namely the distinction between the basal and edge planes.

Disorder in the crystalline structure of VPO catalysts has often been cited in the literature (Cavani *et al.*, 1985; Horowitz *et al.*, 1988; Bordes, 1987). The nature of the disorder has never been fully explained. In the past Busca *et al.* (1986b) have postulated that distortion in the platelet habit of $(VO)_2P_2O_7$ results in strain, which changes the Lewis acid strength of surface $V^{4+}$ species. This is outlined in Scheme 5.2. Alternatively, Ebner and Thompson (1993) have postulated that vanadyl columnar disorder can account for the variability in line shape and intensity of the (020) XRD reflection of $(VO)_2P_2O_7$.

Nguyen and Sleight have recently prepared highly perfect crystals of $(VO)_2P_2O_7$, and compared their X-ray diffraction patterns to those of real catalysts. Peak broadening in the latter was associated with extended defects, as well as crystallite size and strain. The types of defects were quantitatively correlated with the $V^{5+}$ concentration in the real catalysts. The authors concluded that most of the $V^{5+}$ in VPO catalysts is present in extended defects (Nguyen and Sleight, 1996).

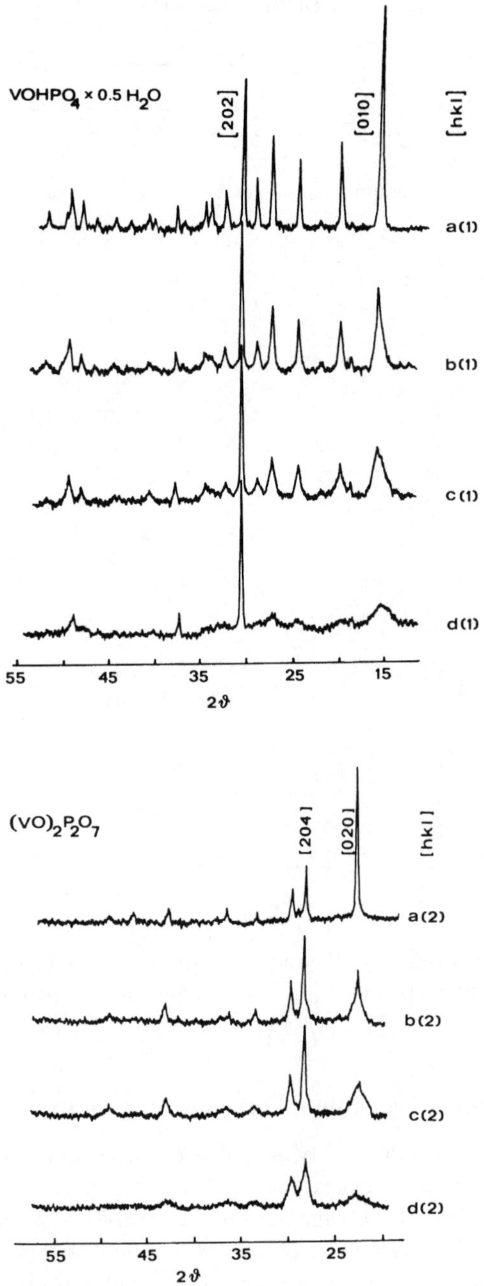


**Figure 5.8** X-ray diffraction patterns of VPO (a) precursors and (b) catalysts showing variability in line width (Busca *et al.*, 1986b) (reproduced by permission of Academic Press).

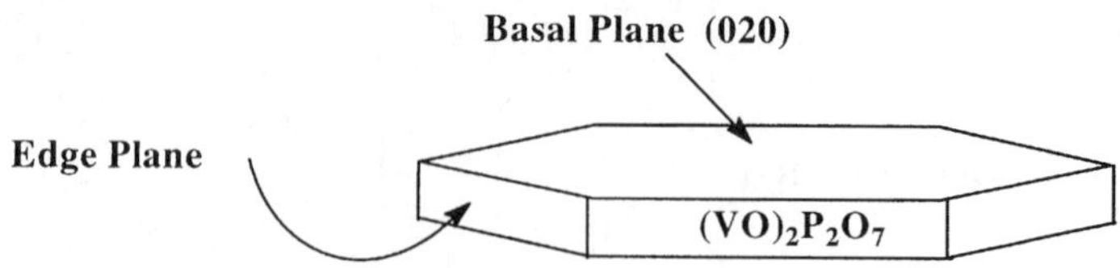


**Scheme 5.1** Schematic diagram of VPO catalyst particle shape.

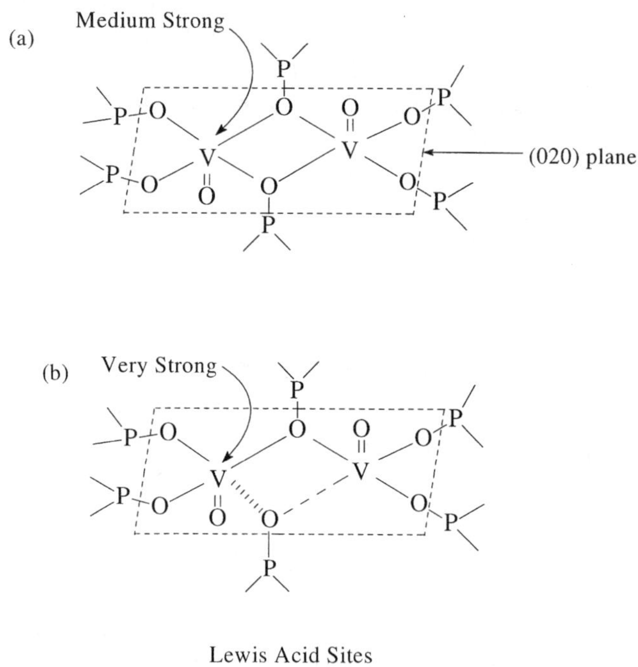


**Scheme 5.2** Schematic diagram of the role of distortion of particle shape on chemical properties of VPO catalysts (Busca *et al.*, 1986c) (reproduced by permission of Elsevier Science).

### 5.3.2 Oxidation state of vanadium in the working catalyst

A consistent feature of VPO catalysts is that during testing in *n*-butane lean conditions, namely with an excess of oxygen in the reactor feed (see Table 5.3), the oxidation state of the vanadium in the catalyst remains very close to 4+ (Cavani *et al.*, 1985, Horowitz *et al.* 1988). About 95 % of the vanadium present in the catalyst is in the 4+ oxidation state. In similar conditions vanadium oxides such as $V_4O_9$ oxidize fully to $V_2O_5$. $V^{4+}$ as the predominant oxidation state appears to be an essential feature to achieve a high selectivity in maleic anhydride, but there is convincing evidence that a small amount of

$V^{5+}$ is necessary for a successful catalyst (Coulson *et al.*, 1997). For catalysts prepared in aqueous solution it is clear that a P : V ratio in excess of 1 : 1 is necessary to render the $V^{4+}$ oxidation state resistant to oxidation (Hodnett *et al.*, 1983). However, for catalysts prepared in organic media there is no evidence to suggest that a high P : V ratio is necessary to achieve a high selectivity. However, there are some indications that the particle size and shape arrived at in the organic preparations may play a role in stabilizing the $V^{4+}$ oxidation state (Duvauchelle *et al.*, 1997). Duvauchelle *et al.* prepared three separate vanadyl pyrophosphate catalysts with differing morphological characteristics. These were described as platelet or mosaic. The shape of the platelet particles is indicated in Scheme 5.1, whereas the mosaic particles were more cubic in appearance. Each exhibited the X-ray diffraction pattern of pure $(VO)_2P_2O_7$, but differences were observed in terms of line shapes, as shown in Figure 5.9. The reactivity of the three catalysts towards oxygen in temperature programmed mode followed by thermogravimetric analysis, where oxidation from the $V^{4+}$ to the $V^{5+}$ state corresponds to a weight gain of ca 5 %, is shown in Figure 5.10. The platelet shaped particles, although the smallest in terms of particle size and the highest in terms of defect concentration, were the least reactive towards oxidation. These results indicate that the thinner the platelets, the stronger the hindrance to oxidation.

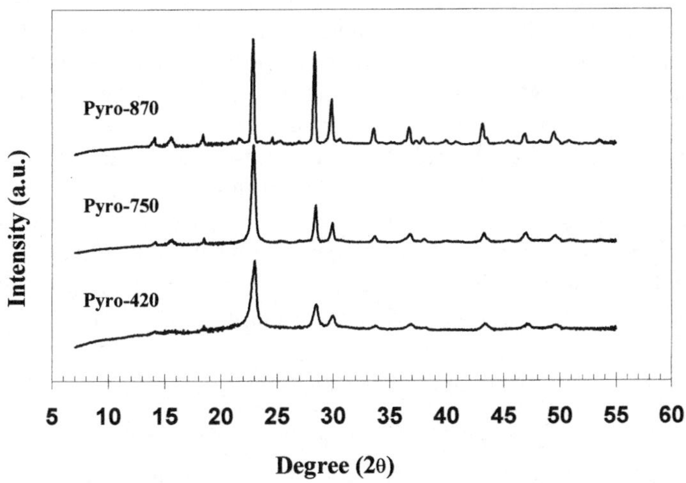


**Figure 5.9** X-ray diffraction pattern of three samples of $(VO)_2P_2O_7$. Pyro 870 has an average particle size greatly in excess of 100 nm and displays a three-dimensional mosaic type morphology. Pyro 750 has an average particle size of ca 100 nm and displays a platelet type morphology. Pyro 420 has an average particle size of 44 nm with platelet morphology. (Duvachelle *et al.*, 1997). Reproduced by permission of DGMK.

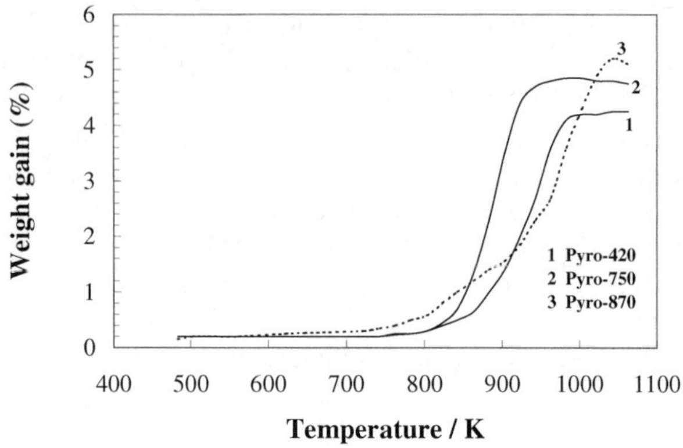


**Figure 5.10** Thermogravimetric analysis of the oxidation of three samples of $(VO)_2$ $P_2O_7$ in pure oxygen (Duvachelle *et al.*, 1997) (reproduced by permission of DGMK).

### 5.3.3 The surface P : V ratio and texture

A large number of studies are available in which the surface P : V ratio has been measured by X-ray photoelectron spectroscopy. Some of these are summarized in Table 5.5

Work in this area prior to about 1993 had relied on the Scofield photo-ionization cross sections as a basis for calibration and all these studies indicated a significant enrichment of the surface in phosphorus, typically showing about twice the amount of phosphorus expected on the basis of the overall bulk composition. However, when calibration standards, such as P–V glasses or organometallic compounds, were used a different picture emerged. The later studies point to a much smaller surface enrichment in phosphorus, of the order of 10 % when the bulk P : V ratio is about 1 : 1. Indeed, the higher P : V ratio measured by Kubias *et al.* (1997), even with calibration standards, may be related to the preparation method used, in particular isolation of the precursor from an aqueous phase by evaporation of the solvent. This method may have deposited phosphorus on the catalyst surface, by impregnation from a solution that was becoming increasingly enriched in phosphorus as the

**Table 5.5** Comparison of bulk and surface P : V ratios

| $P : V_{bulk}$ | $P : V_{surf}$ | *Preparation* | *Calibration* | *Reference* |
|---|---|---|---|---|
| 1.12 : 1 | 2.5 : 1 | aqueous | Scofield | Garbassi *et al.*, 1986 |
| 1.25 : 1 | 1.4 : 1 | aqueous | Scofield | Cornaglia *et al.* 1991 |
| 1.25 : 1 | 2.7 : 1 | organic | Scofield | Cornaglia *et al.* 1991 |
| 1 : 1 | 1.1 : 1 | organic | P–V glass | Okuhara and Misono, 1993 |
| 1 : 1 | 1.08 : 1 | organic | Organometallics | Coulson *et al.*, 1996 |
| 1 : 1 | 1.3-1.5 : 1 | aqueous | P–V glass | Kubias *et al.*, 1997 |

evaporation step proceeded. By contrast in alcohol based preparation methods the precursor is normally isolated by filtration.

Other methods of surface characterization also point to high surface P : V ratios. Delichere *et al.* (1998) used low energy ion scattering (LEIS) to examine the surface composition of a catalysts prepared using an alcohol based method with a bulk P : V ratio of 1.1 : 1. The LEIS technique involves depth profiling by sputtering the catalyst surface with $^{4}He^{+}$ or $^{3}He^{+}$ and monitoring the species expelled into the gas phase in UHV conditions. A surface P : V ratio of ca 2 : 1 was observed by this technique and this value declined to close to 1 : 1 as the sputtering time was increased.

The results of a recent study of the evolution of the surface texture of VPO catalysts, prepared in an organic medium, are presented in Figure 5.11. In this study high resolution electron microscopy was used to image the near surface layers of these catalysts as they were treated in 1.2 % *n*-butane in air at 653 K for up to 23 days (Guliants *et al.*, 1995). Throughout this period these catalysts improved in terms of selectivity to maleic anhydride and in terms of crystallinity as measured by X-ray line width measurements. The fresh catalyst exhibited a surface layer of amorphous material about 1.5 nm thick covering the (020) plane of vanadyl pyrophosphate. The amorphous layer was still present after 8 days but had disappeared after 23 days of treatment. This study is consistent with the surface of the equilibrated catalyst being very close in composition to the underlying bulk phase.

### 5.3.4 Influence of P : V ratio on catalyst performance

Many studies have been undertaken in which the P : V ratios of these catalysts have been varied. The commercial catalysts operate with a slight excess of phosphorus, typically with P : V close to 1.1 : 1 (Horowitz *et al.*, 1988). There is some evidence that part of the phophorus sublimes during normal operation and there are patents which describe methods of replenishing the catalyst with phosphorus without significantly interfering with plant operation (Hodnett, 1985). It is clear from a large number of studies that phosphorus in excess of the 1 : 1 stoichiometric ratio is necessary for the production of selective VPO catalysts prepared in aqueous media. Selectivity to maleic anhydride improves as the P : V ratio increases for catalysts prepared in aqueous media (Hodnett *et al.*, 1983), and this must also be related to the role of excess phosphorus in these catalysts in stabilizing the 4+ oxidation state of vanadium.

There is no clear evidence in the literature that catalysts prepared in organic media require a P : V ratio greatly in excess of the stoichiometric for optimum performance. Figure 5.12 presents yield–conversion plots for a number of VPO catalysts with varying P : V ratios prepared in organic media. Clearly catalysts with P : V ratios less than 1 : 1 do not perform well, especially at high conversions, but neither is optimum performance achieved with catalysts with P : V ratios in excess of 1 : 1.

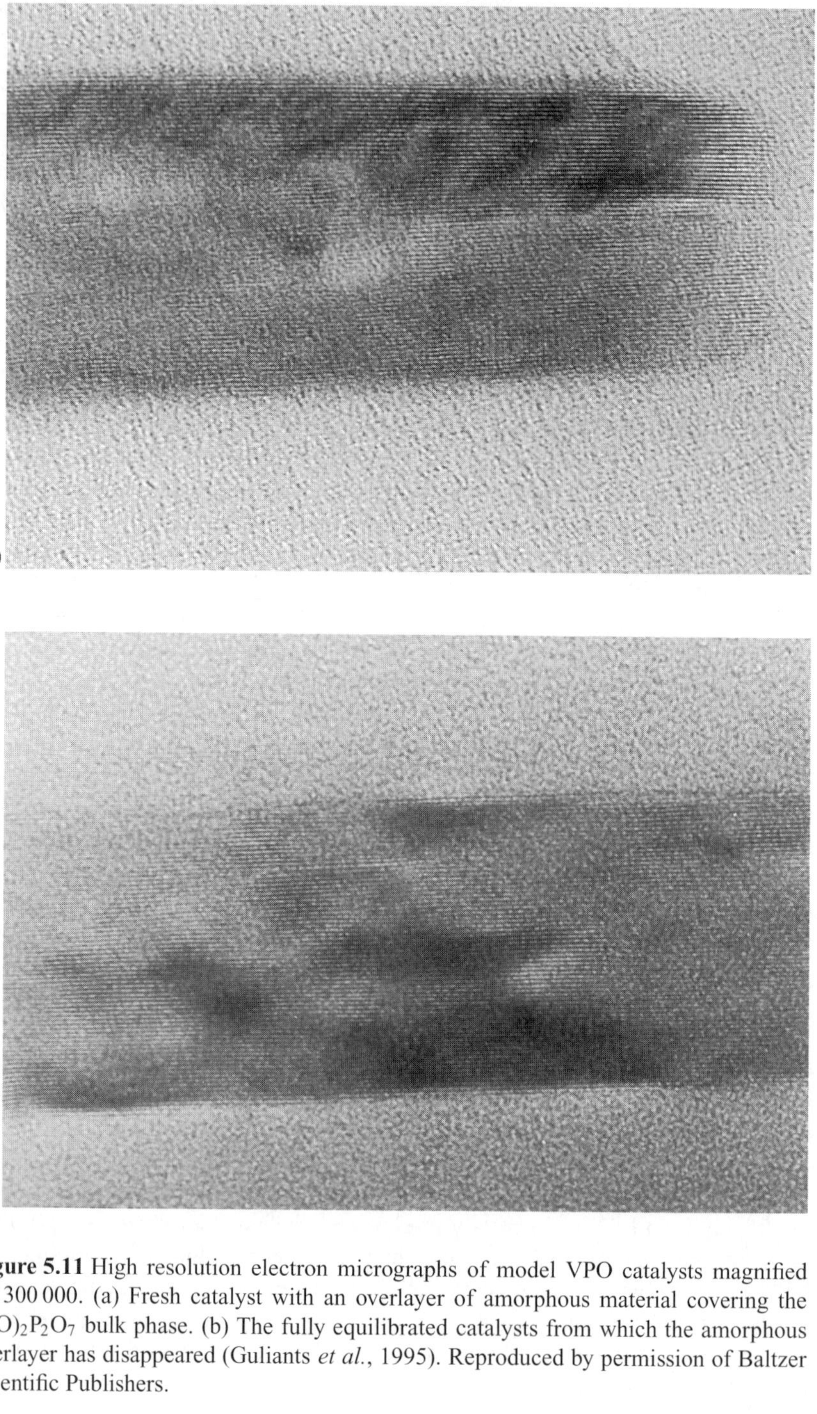

(a)

(b)

**Figure 5.11** High resolution electron micrographs of model VPO catalysts magnified by 300 000. (a) Fresh catalyst with an overlayer of amorphous material covering the $(VO)_2P_2O_7$ bulk phase. (b) The fully equilibrated catalysts from which the amorphous overlayer has disappeared (Guliants *et al.*, 1995). Reproduced by permission of Baltzer Scientific Publishers.

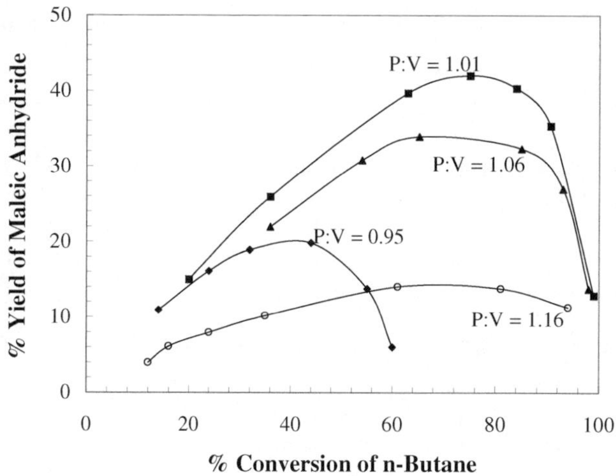


**Figure 5.12** Yield–conversion plots for maleic anhydride production from *n*-butane over VPO catalysts prepared in an organic medium with the P : V ratios indicated (Centi *et al.*, 1988) (reproduced by permission of the American Chemical Society).

## 5.4 ACTIVATION OF *n*-BUTANE

Selective oxidation of *n*-butane to maleic anhydride involves cleavage of eight C—H bonds at various points along the reaction pathway. The question of activation comes down to identifying which is the first C—H bond to be cleaved. Cleavage of a C—C bond will lead to undesirable products. There are two types of C—H bond in *n*-butane, namely methylene and methyl C—H bonds. To identify which of these is involved in the primary activation of the alkane the kinetic isotope effect has been very useful. Pepera *et al.* (1985) carried out a series of oxidation experiments with deuterated *n*-butanes, the results of which are presented in Table 5.6. These data indicate very clearly that the primary activation of the *n*-butane molecule involves rupture of a methylene C—H bond. This is the weaker of the two types of C—H bonds in *n*-butane and represents the point at which the molecule is activated in the direction of selective oxidation.

The preponderance of $V^{4+}$ in the VPO catalyst has made the direct observation of chemisorbed oxygen species such as $O_2^-$ or $O^-$ very difficult because the $V^{4+}$ species is paramagnetic. However, several studies have pointed to the need for a weakly adsorbed surface oxygen species which would be capable of extracting hydrogen from a methylene C—H bond (Centi *et al.*, 1988; Gleaves and Centi, 1993). There is no agreement as to the nature of this oxygen species. On the other hand there have been a number of studies which suggest that the oxygen which appears in the maleic anhydride originates from the lattice, even though the amount of oxygen extractable from a

**Table 5.6** Kinetic isotope effect for *n*-butane oxidation to maleic anhydride over VPO catalysts at 673 K
(Pepera *et al.*, 1985).

| *Components* | $C_x^a$ | $C_y^a$ | *Conv./%* | $k_H/k_D$ |
|---|---|---|---|---|
| $CH_3CH_2CH_2CH_3$ (*x*)<br>$CH_3CD_2CD_2CH_3$ (*y*) | 0.500 | 0.500 | 27.5 | 2.18 |
| $CH_3CH_2CH_2CH_3$ (*x*)<br>$CD_3CH_2CH_2CD_3$ (*y*) | 0.503 | 0.497 | 30.4 | 1.05 |
| $CD_3CH_2CH_2CD_3$ (*x*)<br>$CH_3CD_2CD_2CH_3$ (*y*) | 0.421 | 0.579 | 27.3 | 2.11 |

[a] $C_x$ and $C_y$ are the fractions of component *x* and *y* in the feed.
Reproduced by permission of the American Chemical Society.

VPO catalyst is much less than that available in the bismuth molybdate catalyst discussed in Chapter 4 (Lashier and Schrader, 1991). The first evidence to indicate that lattice oxygen was involved came from experiments in which *n*-butane was passed over a series of VPO catalysts in the absence of gas phase oxygen. Maleic anhydride was formed in these circumstances until extractable oxygen was exhausted (Hodnett and Delmon, 1984).

The TAP reactor technique has been used by Rodemerck *et al.* (1997) to distinguish the roles of lattice and gas phase oxygen during *n*-butane oxidation. Table 5.7 indicates the fraction of $^{18}O$ in furan, maleic anhydride and carbon dioxide formed over a VPO catalyst exposed to *n*-butane and $^{18}O_2$ in a TAP reactor. In these studies the TAP pulse size is much smaller than the lattice oxygen available from the surface layer of the catalyst, so that the reactor outlet can be monitored at a stage in the reaction where most of the surface sites have been involved in less than one turnover. The data in Table 5.7 indicate that nearly all the oxygen incorporated into the oxygenated products, including carbon dioxide, comes from the lattice.

On a longer timescale the $^{16}O$ oxide surface layer is replenished with $^{18}O$ and the results are less striking. This feature was demonstrated in an earlier study wherein Pepera *et al.* (1985), using a conventional reactor, found that with $^{18}O_2$ as oxidant over VPO catalysts the reaction product comprised initially $C_4H_2{}^{16}O_3$, but changed rapidly to $C_4H_2{}^{18}O_3$ as the first few layers of $^{16}O_L$ were depleted and replaced by $^{18}O_L$. The results are consistent with the first few monolayers only of lattice oxygen being involved in the reaction and demonstrate that oxygen species deeper within the solid do not participate actively in the reactions. It appears then that a key feature of these catalysts is that the surface layer of lattice oxygen is quite reactive but good selectivity is achieved only if significant migration of lattice oxygen from the bulk to the surface is restricted. Presumably this arises because an overabundance of reactive oxygen in the surface layers would favour total oxidation reactions.

**Table 5.7** Initial distribution of oxygen isotopes in the reaction products of *n*-butane with $^{18}O_2$ over a VPO catalyst
(Rodemerck *et al.*, 1997)

| *Product* | *Molecular ion* | *Number of $^{18}O$ atoms in product* | *Isotope distribution %* |
|---|---|---|---|
| Furan | $m/e$ 68 | 0 | 93 |
| | $m/e$ 70 | 1 | 7 |
| | % $^{18}O$ | | 7 |
| Maleic anhydride | $m/e$ 98 | 0 | 85 |
| | $m/e$ 100 | 1 | 15 |
| | $m/e$ 102 | 2 | 0 |
| | $m/e$ 104 | 3 | 0 |
| | % $^{18}O$ | | 5 |
| Carbon dioxide | $m/e$ 44 | 0 | 66 |
| | $m/e$ 46 | 1 | 25 |
| | $m/e$ 48 | 2 | 9 |
| | % $^{18}O$ | | 22 |



This model of oxygen movement within VPO catalysts is presented in Scheme 5.3 (Hodnett and Delmon, 1984). The near surface layers are shown as a mixture of $V^{4+}$ and $V^{5+}$ in the working catalyst, and there are interchanges in oxidation states as oxygen is added to and taken from these layers. The underlying bulk remains in the $V^{4+}$ oxidation state and, although essential to having a successful catalyst, does not appear to participate directly in the reaction.

The importance of having an unreactive bulk in VPO catalysts is clearly shown in Figure 5.13 for a series of VPO catalysts with identical P : V ratios, but all prepared by different methods (Busca *et al.*, 1986a and b). When these catalysts were exposed to oxygen (Figure 5.13(a)) they showed nearly identical initial rates of oxidation, corresponding to the oxidation of the near surface layers. For two of the VPO catalysts (nos 2 and 3 in Figure 5.13) the

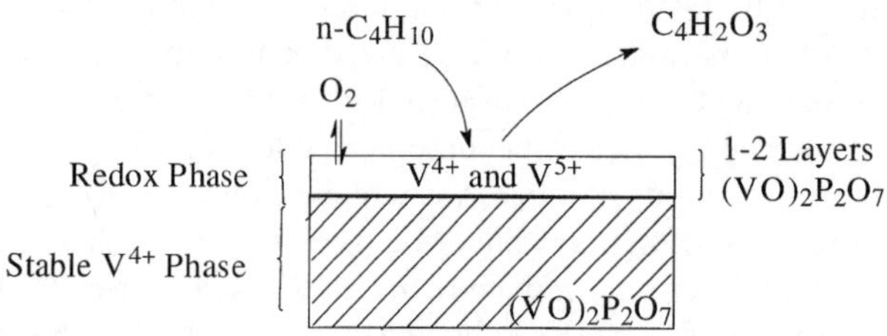


**Scheme 5.3** Schematic diagram of oxygen transport and incorporation during *n*-butane oxidation.

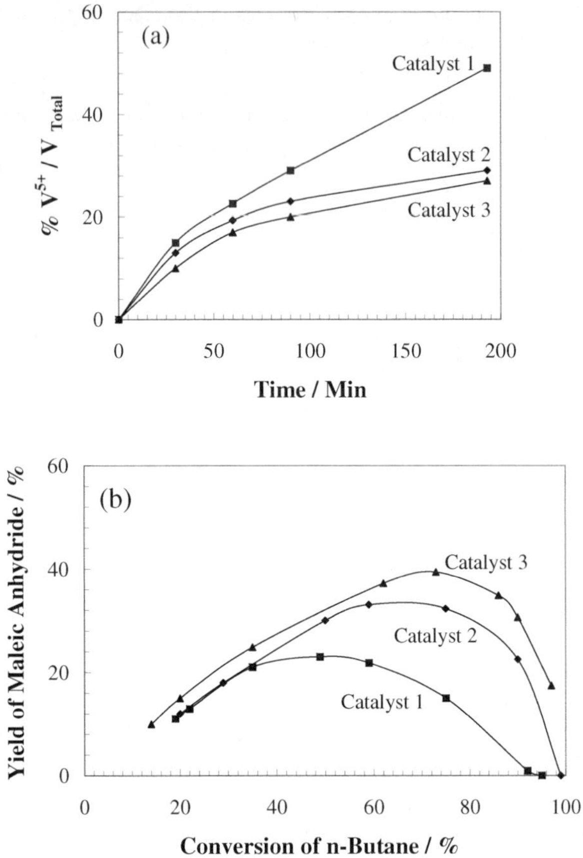


**Figure 5.13** Relationship between VPO (a) oxidizability and (b) selectivity to maleic anhydride for a series of catalysts with identical P : V ratios, but prepared by different methods (Busca *et al.*, 1986a) (reproduced by permission of Academic Press).

rate of oxidation slowed abruptly at a $V^{4+}/V_{tot}$ ratio of about 0.20, indicating that oxygen could not be transported into the interior of these catalysts, so oxidation could not proceed. However, the interior of the third catalyst (no 1 in Figure 5.13) could be more readily oxidized. The implication here is that during reaction oxygen could be transported from the interior of one of these catalysts to the surface layers and become available there for reaction. When tested for catalytic reactivity the catalyst which readily oxidized in the bulk showed the lowest yield of maleic anhydride (see Figure 5.13(b)).

A reactive surface layer on a reduced core shown schematically in Scheme 5.3 (Hodnett and Delmon, 1984) can be achieved for VPO catalysts prepared in aqueous media in the presence of an excess of phosphorus. The mechanism whereby this is achieved for catalysts prepared in organic media may be related to the shape of VPO catalyst particles prepared using alcohols (Duvauchelle *et al.*, 1997). The exact structure of the surface layer has not been fully elucidated. Some workers argue, particularly on the basis of NMR and Raman spectroscopies, that $V^{5+}$ phases such as $\delta$- or $\alpha_{II}$-$VOPO_4$ (see Table 5.4) may be present during catalyst testing (Abon and Volta, 1997; Zhang *et al.*, 1993).

## 5.5 THE NATURE OF THE ACTIVE SITE AND THE REACTION MECHANISM

### 5.5.1 The reaction network

The number of individual steps required for the selective oxidation of *n*-butane into maleic anhydride makes elucidation of the reaction mechanism very difficult, but Trifiro and co-workers (Centi *et al.*, 1988) have made a very useful contribution by proposing a mechanism which identifies a role for each surface species, i.e. $V^{3+}$, $V^{4+}$ and $V^{5+}$, in oxidative dehydrogenation, allylic oxidation, 1–4 oxygen insertion and electrophilic oxygen insertion, all of which are necessary steps in the oxidation of *n*-butane to maleic anhydride.

For butene oxidation these workers observed a correlation between the overall oxidation state of the catalyst and the final product formed. When the catalyst was reduced such that it contained 5 % $V^{3+}$, 1,3-butadiene was the favoured product; when the $V^{4+}$ component was maximized at 97 %, maleic anhydride was the favoured product; and when the $V^{5+}$ component exceeded 5 %, $CO_2$ became the main product. These considerations led to the reaction mechanism outlined in Scheme 5.4, in which *n*-butane and butene are both activated on $V^{4+}$ centres. A butadiene–$V^{3+}$ centre is then established from which butadiene can desorb if insufficient oxygen is available. Further oxidation of the butadiene requires the presence of $V^{5+}$ sites, but their overabundance favours total oxidation products.

An important point addressed by this mechanism is the role of VPO catalysts in miminizing the loss of maleic anhydride product through over-oxidation. Specifically, although $V^{5+}$ is seen as necessary for the oxidation of 1,3-butadiene, too much $V^{5+}$ on the surface results in combustion of the maleic anhydride. The role of phosphorus in excess of the 1 : 1 ratio was seen as limiting the surface concentration of $V^{5+}$ species.

### 5.5.2 The location of active sites and the role of catalyst microstructure

When *n*-butane is fed over a VPO catalyst, reaction intermediates, such as butene, 1,3-butadiene and furan are capable of desorbing from the reactive surface site(s), as shown by TAP reactor studies (see Figure 5.2), but do not normally appear in the gas phase, indicating that all further reaction to maleic

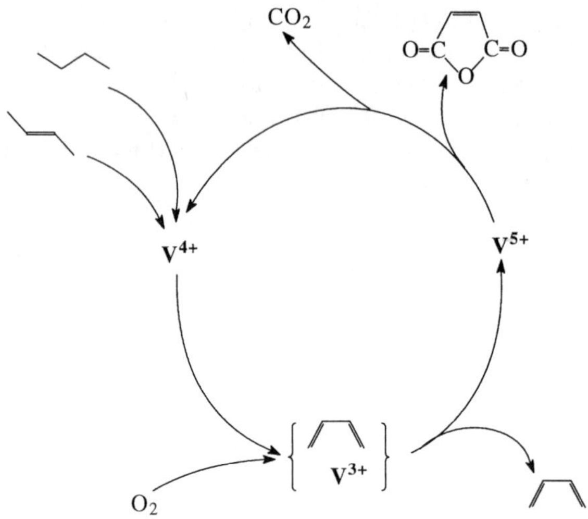


**Scheme 5.4** Generalized scheme for *n*-butane oxidation to maleic anhydride over VPO catalysts (Centi *et al.*, 1988).

anhydride and $CO_x$ occurs more readily than desorption (Centi *et al.*, 1984; Gleaves *et al.*, 1988). This feature points to the need to describe a single active site, with oxygen transport from this site and hydrogen transport from the site as postulated by Pepera *et al.* (1985).

There is now general agreement that VPO catalysts prepared in organic media form platelet or distorted platelet type crystals (see Figure 5.7) and the active sites involved in *n*-butane oxidation to maleic anhydride over VPO catalysts are located on the basal (020) plane of vanadyl pyrophosphate crystals. Two literature sources are most often cited as background in support of this hypothesis: These are (i) the fact that catalysts comprising $(VO)_2P_2O_7$ with some disorder or broadening in the (020) reflection are generally very active and selective for *n*-butane oxidation as indicated by appropriate selectivity–conversion curves (Cavani *et al.*, 1985; Centi *et al.*, 1988; Busca *et al.*, 1986a, b and c), and (ii) the work of Horowitz *et al.* (1988) which showed enhanced performance (again in terms of selectivity–conversion curves) by $(VO)_2P_2O_7$ catalysts which had been wet milled. This process caused stacked $(VO)_2P_2O_7$ platelets (see Figure 5.7) to become detached from one another through the shear forces brought to bear during the milling process, so increasing the (020) specific surface area. When the catalyst which was subjected to this treatment was compared with the non-milled catalyst a considerable boost in selectivity was observed (Figure 5.14) . There appears to be general agreement that $(VO)_2P_2O_7$ with a large specific (020) surface area is desirable because the active sites for *n*-butane oxidation are located on

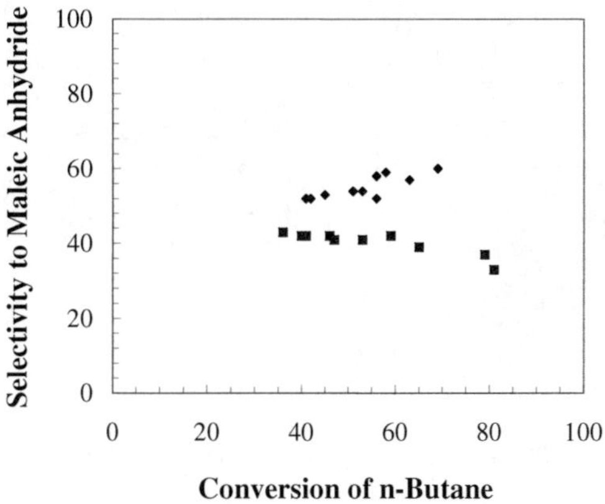


**Figure 5.14** Selectivity–conversion plots for a VPO catalyst following normal preparation (■) and following wet milling to increase the (020) surface area (◆) (Horowitz *et al.* 1988) (reproduced by permission of Elsevier Science).

this surface plane. For this reason some workers have described this reaction as structure sensitive (Zhang *et al.*, 1993)

A basic difficulty now arises in describing the active site: there are no physico-chemical techniques which can be used to characterize, on a molecular scale, the structure of an active site. Hence the tendency with VPO catalysts is to take the (020) plane of $(VO)_2P_2O_7$ as it would occur in the bulk of this oxide and use it as a starting model for the active site (Centi *et al.*, 1988; Ziolkowski *et al.*, 1990). This is the approach of Bordes and co-workers in applying the crystallogaphic model of active sites (CMAS) to VPO catalysts (Ziolkowski, 1983). The model essentially seeks the lowest energetic pathway for the reaction through a series of calculations based on the topography and energy field at each crystal face. These calculations predict that the (020) cleavage plane only is active and selective for *n*-butane oxidation and that $C_{2v}$ symmetry is imposed on the reacting molecule by active site(s) on this plane, as depicted in Scheme 5.5 when this model is applied to *n*-butane oxidation over $(VO)_2P_2O_7$. The first step is hydrogen abstraction from a terminal methyl group by a surface lattice oxygen species, apparently associated with a $V^{4+}$ species. The model predicts all the subsequent steps of the reaction and envisages the presence of $V^{5+}$ species on demand through their incorporation into the vanadia octahedra without distortion of the $(VO)_2P_2O_7$ structure (Bordes, 1993).

Ebner and Thompson argue that pyrophosphate termination of the predominantly (020) plane of $(VO)_2P_2O_7$ results in surface clefts with up to four accessible vanadium centres per cleft and in which the pendant pyrophosphate groups act as acceptor centres for hydrogen from the substrate (see

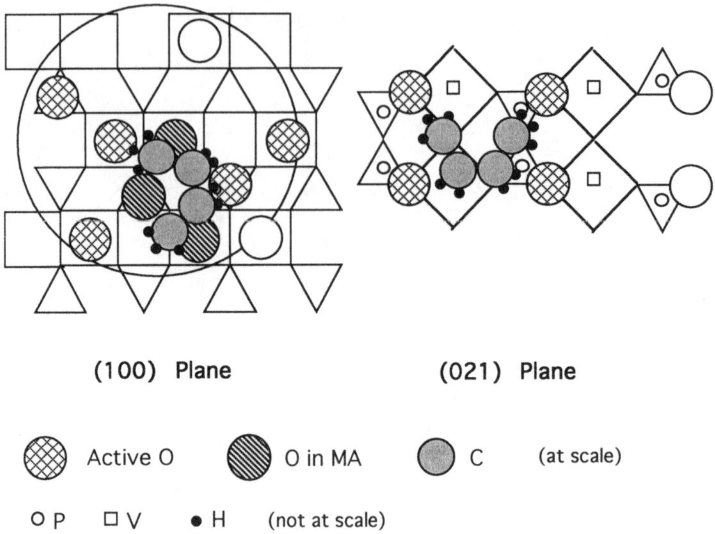


**Scheme 5.5** Adsorption and reaction of *n*-butane on $(VO)_2P_2O_7$ with conservation of $C_{2v}$ symmetry (Bordes, 1993) (reproduced by permission of Elsevier Science).

Scheme 5.6). This hypothesis also fulfils the active site isolation or dilution feature of selective oxidation catalysis as introduced by Grasselli (1983). The idea of active site isolation had been mooted in the past for VPO catalysts (Hodnett *et al.*, 1983), but usually on the basis of there being an excess of inactive phosphorus on the catalyst surface. The model of the active site postulated by Ebner and Thompson (1993) is based on the cleavage structure of the (020) plane, and is consistent with the more recently measured surface P : V ratios for VPO catalysts prepared in organic media (see Table 5.5).

## 5.6 CONCLUSIONS

Selective oxidation of *n*-butane to maleic anhydride is a commercial process which depends for its operation on the structure of active sites located on the basal plane of $(VO)_2P_2O_7$. This catalyst, prepared with an appropriate composition and via an alcohol based preparation method can be used in fixed and fluid bed reactors and is the only catalysts capable of giving high conversions of *n*-butane and high selectivities to maleic anhydride. The VPO catalyst is probably best described as a binary oxide of vanadium and phosphorus, and over 90 % of the vanadium is in the 4+ oxidation state in the working catalyst. During operation the surface layers only are capable of redox behaviour and activation of *n*-butane occurs at one of the methylene C—H bonds in *n*-butane.

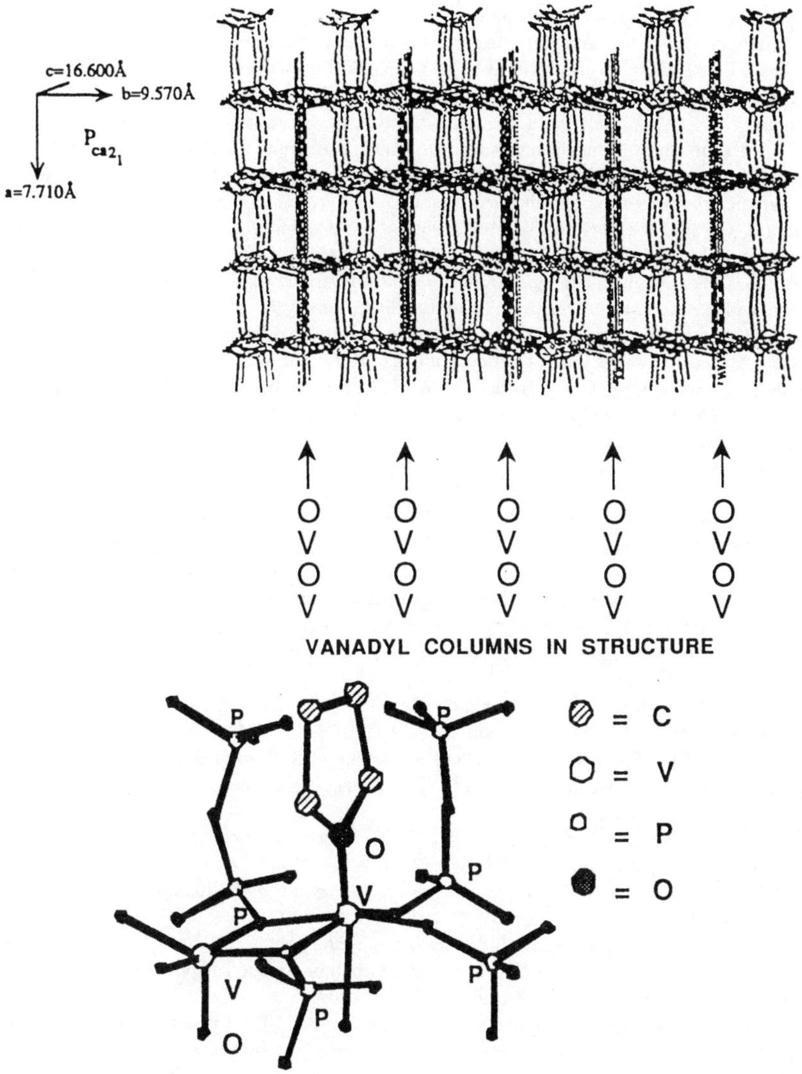


**Scheme 5.6** (a) Pyrophosphate termination of the (020) plane of $(VO)_2P_2O_7$ and (b) a view of the proposed active site (Ebner and Thompson, 1993) (reproduced by permission of Elsevier Science).

## REFERENCES TO CHAPTER 5

Abon M and Volta JC (1997) Vanadium phosphorus oxides for *n*-butane oxidation to maleic anhydride. *Appl. Catal. A: General* **157**: 173–193.

Ai M, Boutry P, Montarnal R and Thomas G (1970) Cinetique d'oxydation des butenes en anhydride maleic sur $V_2O_5$–$P_2O_5$. II Interpretation et exploitation des resultats experimentaux. *Bull. Soc. Chim. Fr.* 2783–2789.

Batiot C and Hodnett BK (1996) The role of reactant and product bond energies in determining limitations to selective catalytic oxidations. *Appl. Catal. A General* **137**: 179–191.

Bordes E (1987) Crystallochemistry of V–P–O phases and applications in catalysis. *Catal. Today* **1**: 499–524.

Bordes E (1993) Nature of the active and selective sites in vanadyl pyrophosphate, catalyst of oxidation of *n*-butane, butene and pentane to maleic anhydride. *Catal. Today*, **16**: 27–38.

Burnett JC, Keppal RA and Robinson WD (1987) Commercial production of maleic anhydride by catalytic processes using fixed bed reactors. *Catal. Today* **1**: 537–584.

Busca G, Cavani F, Centi G and Trifiro F (1986a) Nature and mechanism of formation of vanadyl pyrpohosphate active phase in *n*-butane selective oxidation. *J. Catal.* **99**: 400–414

Busca G, Centi G, Lorenzelli V and Trifiro F (1986b) Surface acidity of vanadyl pyrophosphate, active phase in *n*-butane selective oxidation. *J. Phys. Chem.* **90**: 1337–1344.

Busca G, Centi G and Trifiro F (1986b) *n*-Butane selective oxidation on vanadium-based oxides: Dependence on catalyst microstructure. *Appl. Catal.* **25**: 265–272.

Busca G, Cavani F, Centi G and Trifiro F (1986c) Nature and mechanism of formation of vanadyl pyrophosphate: Active phase in *n*-butane selective oxidation. *J. Catal.* **99**: 400–414.

Cavani F, Centi G, Manenti I and Trifiro F (1985) Catalytic conversion of $C_4$ hydrocarbons on vanadium–phosphorus oxides: Factors influencing the selective 1-butene oxidation. *Ind. Eng. Chem. Prod. Res. Dev.* **24**: 221–226.

Centi G, Fornassari G and Trifiro F (1984) On the mechanism of *n*-butane oxidation to maleic anhydride: Oxidation in oxygen stoichiometry controlled conditions. *J. Catal.* **89**: 44–51.

Centi G, Trifiro F, Ebner JR and Franchetti VM (1988) Mechanistic aspects of maleic anhydride synthesis from C-4 hydrocarbons over phosphorus vanadium oxide. *Chem. Rev.* **88**: 55–80.

*Chemical Week* (1997) (http:\\www.chemweek.com\marketplace\prod_focus.html).

Contractor RM and Sleight AW (1987) Maleic anhydride from C-4 feedstocks using fluidized bed reactors. *Catal. Today* **1**: 587–607.

Cornaglia LM, Caspani C and Lombardo EA (1991) Physocochemical characterization and catalytic behavior of VPO formulations. *Appl. Catal.* **74**: 15–27.

Coulson GW, Bare SR, Kung H, Birkeland K, Bethke G K, Harlow R, Herron N and Lee PL (1997) The kinetic significance of $V^{5+}$ in *n*-butane oxidation over vanadium phosphates. *Science* **275** (Jan 10): 191–193.

Coulson GW, Thompson EA and Herron N (1996) Characterization of VPO catalysts by x-ray photoelectron spectroscopy. *J. Catal.* **163**: 122–129.

Cox JD and Pitcher G (1970) *Thermochemistry of Organic and Organometallic Compounds*. Academic Press, New York.

Culbertson BM (1987) Maleic anhydride uses in resins and polymers. *Catal. Today* **1**: 609–629.

Delichere P, Bere KE and Abon M (1998) Vanadyl pyrophosphate catalysts: Surface analysis by XPS and LEIS, *Appl. Catal. A: General.*, **172**, 295–309.

Duvauchelle N, Zanthoff H, Hannour F and Bordes E (1997) Crystallite shape of vanadyl pyrophosphate related to its oxygen capacity characterized by vacuum transient experiments (TAP). In *Proceedings of the DGMK Conference $C_4$ Chemistry—Manufacture and Uses of $C_4$ Hydrocarbons*. Tagungsbericht 9705, Aachen.

Ebner JR and Thompson MR (1993) An active site hypothesis for well crystallized vanadium–phosphorus oxide catalyst systems. *Catal. Today* **16**: 51–60.

Emig G and Martin F (1987) Economics of maleic anhydride production from C-4 feedstocks. *Catal. Today* **11**: 477–498.

Garbassi F, Bart JCJ, Montino F and Petrini G (1985) Preparation and characterization of vanadium–phosphorus oxidation catalysts. *Appl. Catal.* **16**: 271–287.

Gleaves JT and Centi G (1993) The reaction mechanism of alkane selective oxidation on vanadyl pyrophosphate catalysts. Features gleaned from TAP reactor transient response studies. *Catal. Today* **16**: 69–78.

Gleaves JT, Ebner JR and Kuechler TC (1988) Temporal analysis of products (TAP): A unique catalyst evaluation system with sub millisecond time resolution. *Catal. Rev. Sci. Eng.* **30**: 49–116.

Grasselli RK (1983) Selective oxidation and ammoxidation catalysis: History of catalyst design. In *Surface Properties and Catalysis by Non-Metals*, Bonnelle JP, Delmon B and Derouane E (Eds). Reidel, Dordrecht, pp 273–288.

Guliants VV, Benziger JB, Sundaresan S, Yao N and Wachs IE (1995) Evolution of the active surface of the vanadyl pyrophosphate catalysts. *Catal. Lett* **32**: 379–386.

Hodnett BK (1985) Vanadium phosphorus oxide catalysts for the selective oxidation of C-4 hydrocarbons to maleic anhydride. *Catal. Rev. Sci. Eng.* **27**: 373–424.

Hodnett BK and Delmon B (1984) Influence of reduction pretreatments on the activity and selectivity of vanadium–phosphorus oxide catalysts for *n*-butane partial oxidation. *Ind. Eng. Chem. Fundam.* **23**: 465–470.

Hodnett BK, Permanne Ph and Delmon B (1983) Influence of P : V ratio on the phase composition and catalytic activity of vanadium phosphate based catalysts. *Appl. Catal.* **6**: 231–244.

Horowitz HS, Blackstone CM, Sleight AW and Teufer G (1988) V–P–O catalysts for oxidation of butane to maleic anhydride. Influence of $(VO)_2H_4P_2O_9$ precursor morphology on catalytic properties. *Appl. Catal.* **38**: 193–210.

Hutchings GJ, Desmartinchomel A, Olier R and Volta JC (1994) Role of the product in the transformation of a catalyst to its active state. *Nature* **368**: 41–45.

Kubias B, Richter F, Papp H, Krepel A and Kretschmer A (1997) The nature of the active site of the $(VO)_2P_2O_7$ catalyst: An investigation of the chemical composition and dynamics of the catalyst surface. In *Proceedings 3rd World Congress on Oxidation Catalysis*, Grasselli RK, Oyama ST, Gaffney AM and Lyons JE (Eds). Elsevier Science, Amsterdam.

Kung HH (1992) Kinetic analysis of a generalized catalytic selective oxidation reaction. *J. Catal.* **134**: 691–701.

Lashier ME and Schrader GL (1991) Reactive lattice oxygen sites for $C_4$-hydrocarbon oxidation over $\beta$-$VOPO_4$. *J. Catal.* **128**: 113–125.

Nguyen PT and Sleight AW (1996) Vanadium phosphate catalysts: Ideal structure, real structure and stability region. In *Heterogeneous Hydrocarbon Oxidation, ACS Symposium Series* **638**, Warren BK and Oyama ST (Eds). American Chemical Society, Washington, DC, pp 236–248.

Okuhara T, and Misono M (1993) Key reaction steps and active surface phase of vanadyl pyrophosphate for selective oxidation of butane. *Catal. Today* **16**: 61–67.

Pepera MA, Callahan JL, Desmond MJ, Milberger EC, Blum PR and Bremer NJ (1985) Fundamental study of the oxidation of butane over vanadyl pyrophosphate. *J. Am. Chem. Soc.* **107**(17): 4883–4892.

Rodemerck U, Kubias B, Zanthoff H-W and Baerns M (1997) The reaction mechanism of the selective oxidation of butane on $(VO)_2P_2O_7$ catalysts: The role of oxygen in the reaction chain to maleic anhydride. *Appl. Catal. A: General* **153**: 203–216.

Volta JC (1996) Dynamic processes on vanadium phosphorus oxides for selective alkane oxidation. *Catal. Today* **32**: 29–36.

Zabetakis MG (1965) *Flammability Characteristics of Combustible Gases and Vapors*. US Dept of the Interior, Bureau of Mines, US Government Printing Office, Washington, DC 20402.

Zhang Y, Sneeden RPA and Volta JC (1993) On the nature of active sites of the VPO catalysts for *n*-butane oxidation to maleic anhydride. *Catal. Today* **16**: 39–49.

Ziolkowski J (1983) Advanced bond strength model of active sites on oxide catalysts. *J. Catal.* **84**: 317–332.

Ziolkowski J, Bordes E and Courtine P (1990) Oxidation of butane and butene on the (100) face of $(VO)_2P_2O_7$: A dynamic view in terms of the crystallochemical model of active sites. *J. Catal.* **122**: 126–150.

# 6

# Epoxidation of Alkenes: Reactivity of Electrophilic Oxygen Species


6.1 Introduction 161
6.2 Technology for Ethylene Oxide Production 162
6.3 Catalyst Composition, Structure and Performance 164
6.4 Reaction Kinetics and Network 170
6.5 The Nature of the Oxygen Species on and under the Silver Surface 174
6.5.1 Oxygen on the Pure Silver Surfaces 174
6.5.2 Interaction and Interconversion of Oxygen Species on the Alkali Metal Promoted Silver Surfaces 177
6.6 Reactivity of Oxygen Species on Silver 179
6.6.1 The Role of Chlorine in Determining the Relative Amounts of Adsorbed and Subsurface Oxygen in Silver 185
References to Chapter 6 187



Summary

Ethylene oxidation to ethylene oxide is by far the biggest epoxidation reaction in commercial operation. The reaction is carried out at elevated pressure over a silver catalyst supported on a non-porous alumina, promoted by alkali metals and moderated by the addition of a few ppm of a chlorocarbon to the reactor feed gas. A feature is the ability of the oxygen species produced at the silver surface to attack the electron rich C=C bond, hence the term electrophilic oxygen. The promoters and moderators act by inhibiting sintering of the supported silver phase and by boosting the selectivity to ethylene oxide.

This chapter describes

- Technology for ethylene oxide production,
- Catalyst composition, structure and performance,
- Reaction kinetics and network,

- The nature of the oxygen on and under the silver surface,
- Reactivity of oxygen species on silver.

## 6.1 INTRODUCTION

Ethylene oxide, manufactured from ethylene feedstock, always features among the top 50 bulk organic chemicals and in turn it serves as feedstock for the manufacture of ethylene gylcol and other commodity chemicals. It is produced commercially over a silver catalyst, promoted by alkali metals and supported on a non-porous form of $\alpha$-alumina. The catalyst is further modified by the addition of small amounts (ppm) of chlorinated hydrocarbons to the reaction feed. This process was introduced by Union Carbide in 1937 and by Shell in 1958 to replace the practice of ethylene oxide production via the chlorhydrin process (Berty, 1983).

Epoxidation represents one of the mildest forms of oxidation possible. It is unusual in that it requires activation of a C=C double bond without rupture of both bonds in the linkage. Ethylene epoxidation to ethylene oxide is a two electron oxidation for the reaction as written in Table 6.1 and is one of the least exothermic of commercial oxidation reactions. In principal, the reaction is atom efficient but there is a very strong thermodynamic driving force towards the total oxidation reaction. A further technological difficulty is that the product, ethylene oxide, has no upper flammability limit, as indicated in Table 6.1.

The performance of a wide range of catalysts is presented in Figure 6.1 in terms of a multiple selectivity–conversion plot (Cassidy, 1999). This plot gathers together literature reports of selectivity–conversion performance for a

**Table 6.1** Summary information for ethylene epoxidation

Selective oxidation (2 electron)

$C_2H_4 + 0.5O_2 \longrightarrow C_2H_4O$ $\quad \Delta G_{298\,K} = -404\ kJ\,mol^{-1}$

Total oxidation (12 electron)

$C_2H_4 + 3O_2 \longrightarrow 2CO_2 + 2H_2O$ $\quad \Delta G_{298\,K} = -1315\ kJ\,mol^{-1}$

| Catalyst | Support | Additive | Temp/K | Pressure/bar |
|---|---|---|---|---|
| Ag | $\alpha$-$Al_2O_3$ | Alkali metal | 470–570 | 15–25 |
| 5–15 wt% | (0.03–1 $m^2\,g^{-1}$) | 0.005–0.05 wt% | | |

| | $LFL_{298}$[a] | $UFL_{298}$[a] | AIT/K |
|---|---|---|---|
| $C_2H_4$ | 2.7 % | 36 % | 763 |
| $C_2H_4O$ | 3.6 % | 100 % | — |

[a]% in air at 1 bar.

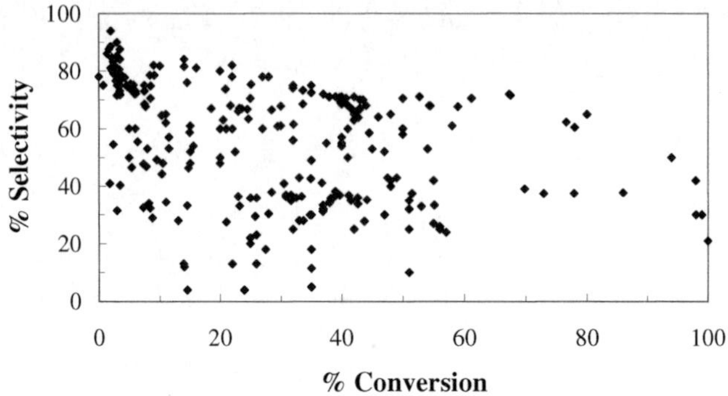


**Figure 6.1** Multiple selectivity–conversion plot for ethylene epoxidation to ethylene oxide over a range of catalysts and in a variety of experimental conditions (Cassidy, 1999).

wide range of catalysts operating in different experimental conditions. A feature of this plot is that no catalyst appears to exist which gives 100 % selectivity to the epoxide even at very low conversions, but selectivity to the epoxide above 80 % can be achieved for conversions up to 60 %.

## 6.2 TECHNOLOGY FOR ETHYLENE OXIDE PRODUCTION

Ethylene oxide is manufactured by the direct oxidation of ethylene over a promoted silver catalyst in multitubular fixed bed reactors operating in recycle conditions. The majority of modern plants operate with pure oxygen although a number of older plants still use air. A typical modern plant produces about 250 ktonne $yr^{-1}$.

The process can be divided into three sections, namely the reaction system, ethylene oxide recovery and ethylene oxide purification. A simplified flow sheet for an oxygen based direct oxidation process for ethylene oxide production is shown in Figure 6.2 and typical feed gas compositions and reactor operating conditions are shown in Table 6.2. Clearly the reactor operates above the LFL and oxygen is the limiting reagent. Unreacted ethylene is recycled. (Dever *et al.*, 1994).

The reactor is usually of the shell and tube type comprising several thousand mild steel or stainless steel tubes, 20–50 mm inside diameter. Tube lengths are in the range 6–12 m and the $Ag/\alpha\text{-}Al_2O_3$ catalyst is in the form of 3–10 mm diameter particles.

The oxygen based process uses almost pure oxygen. This reduces the amount of inerts introduced into the stream and allows a more complete recycle of unreacted ethylene. Oxygen and ethylene are separately introduced into the recycling stream and the reactor effluent is passed through a heat

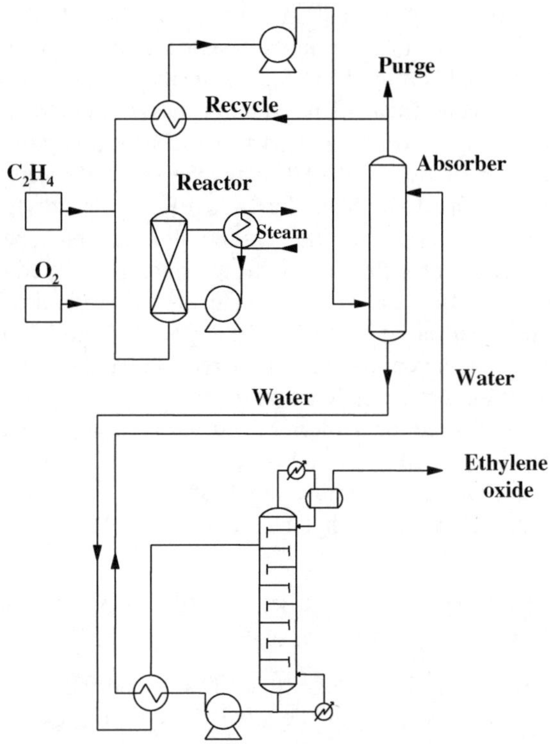


**Figure 6.2** Simplified flow sheet for an oxygen based, direct oxidation process for ethylene oxide production from ethylene (Dever *et al.*, 1994) (reproduced by permission of John Wiley & Sons).

**Table 6.2** Operating conditions for the oxidation of ethylene to ethylene oxide over promoted silver catalysts (Dever *et al.*, 1994; Porcelli, 1981; Zomerdijk and Hall, 1981)

| *Parameter* | *Range* |
|---|---|
| $C_2H_4$/mol% | 20–35 |
| $O_2$/mol% | 4–8 |
| $CO_2$/mol% | 5–10 |
| $C_2H_6$/mol% | 0–1 |
| $C_2H_4Cl_2$/ppm | 1–10 |
| Temperature/K | 490–510 |
| Pressure/MPa | 2–3 |
| GHSV/$h^{-1}$ | 2000–4500 |
| Conversion of $C_2H_4$/% | 8–12 |
| Selectivity/% | 75–82 |

Reproduced by permission of John Wiley and Sons.

exchanger and into a water absorption column. A typical high capacity plant may have several reactor–absorber units in series. Ethylene oxide is recovered in the water stream and is directed to the purification unit. Unreacted ethylene is directed to the recycle stream. This stream is passed through a $CO_2$ removal unit (not shown in Figure 6.2) and a part of the stream is purged. When the selectivity to ethylene oxide is 70–80 % there is still a substantial amount of carbon dioxide generated. In fact, the ratio of recycle ethylene to carbon dioxide is about 2 : 1. Some of the carbon dioxide must be removed to maintain its concentration below 15 vol%, at which point it adversely effects catalyst performance. Because of the recycle, argon, a significant impurity in the oxygen source, can build up to 30–40 mol%. This problem is overcome by introducing a process vent for part of the recycled gases.

In the purification unit, ethylene oxide is removed by steam stripping following feeding into the top section of a desorber unit and the almost pure water generated is recycled to the adsorber units. Further purification of the ethylene oxide is required to remove small amounts of carbon dioxide, and traces of aldehydes, ethane and ethylene (Dever *et al.*, 1994).

## 6.3 CATALYST COMPOSITION, STRUCTURE AND PERFORMANCE

A basic description of the preparation of a supported silver catalyst is presented below. Although long and detailed, the system is extremely sensitive to small amounts of impurity so that extreme caution and care must be exercised in its preparation and handling.

### Preparation of a supported silver catalyst

The following description was taken from UK patent GB 2 043 481 B assigned to Union Carbide Corporation (Bhasin *et al.*, 1984)

A catalyst containing 11.14 wt% Ag, 0.00477 wt% Cs and 0.00158 wt% K was prepared as hereinafter described on an alpha-alumina carrier 'A' having the following chemical composition and physical properties.

| *Chemical composition of carrier 'A'* | | *Physical properties of carrier 'A'* | | *Pore size, μm* | *% TPV* |
|---|---|---|---|---|---|
| $\alpha$-$Al_2O_3$ | 98.5 wt% | Surface area | 0.03 $m^2\ g^{-1}$ | 0.1–1.0 | 1.5 |
| $SiO_2$ | 0.74 wt% | | | 1.0–10.0 | 38.5 |
| $CaO$ | 0.22 wt% | Pore volume | 0.50 $cm^3\ g^{-1}$ | 10.0–30.0 | 20.0 |
| $NaOH$ | 0.16 wt% | | | 30–100 | 32.0 |
| $Fe_2O_3$ | 0.14 wt% | Packing density | 0.70 g $m^{-1}$ | >100 | 8.0 |
| $K_2O$ | 0.04 wt% | | | | |
| $MgO$ | 0.03 wt% | Median pore | diameter | 21 μm | |

The carrier was impregnated under vacuum with a solution of silver salts and alkali metals. Assuming that all of the silver in the impregnating solution contained in the pores of the carrier is deposited upon the carrier, approximately 20 wt% silver in solution is necessary to prepare a catalyst containing about 11 wt% silver on the catalyst.

*Impregnating solution preparation*

80.16 g of ethylenediamine was mixed with 176 g of distilled water. 60.06 g of anhydrous oxalic acid was then added slowly to the solution at 296 K while continuously stirring. During this addition, the solution temperature rose to about 313 K. 147.25 g of silver oxide powder was then added to the diamine–oxalic acid–water solution. Finally, 29.30 g of monethanolamine, 13.26 g of aqueous Cs solution as hydroxide (0.00444 g Cs/ml soln, or 0.0589 g Cs), 34.28 g of aqueous K solution as carbonate (0.0005669 K/ml soln, or 0.0194 g K) and 120.6 g of distilled water were added to complete the solution. The solution was filtered and 0.9 g of undissolved silver was recovered.

*Catalyst preparation*

125 g of Carrier 'A' was impregnated in a 12 in long × 2 in I.D. glass cylindrical vessel equipped with a side arm and suitable stopcocks for impregnating the carrier under vacuum. The impregnating vessel containing the carrier was evacuated to approximately 5 cm of mercury pressure for about 20 min after which the impregnating solution was slowly added to the carrier by opening the stopcock between the separatory funnel and the impregnating vessel until the carrier was completely immersed in solution, the pressure within the vessel being maintained at approximately 5 cm of mercury. Following addition of the solution, the vessel was opened to the atmosphere. The carrier then remaining immersed in the impregnating solution at ambient conditions for about 1 h, and thereafter drained of excess solution for about 30 min. The impregnated carrier was then heat treated to effect reduction of silver salt and deposition of cesium and potassium ions on the catalyst surface. The impregnated carrier was spread out in a single layer on a 7 cm wide endless stainless steel belt (spiral weave) and transported through a 5 cm × 2 cm square heating zone for 2.5 min. The heating zone was maintained at 773 K by passing hot air upward through the belt and about the catalyst particles. By chemical analysis, this catalyst was found to contain 10.88 % Ag, in close agreement with the calculated value.

The support used is always a form of $\alpha$-$Al_2O_3$ with a very low surface area. The silver loading tends to be above 10 wt%, well above the amount required to give a monolayer covering of the support. The alkali metal promoter tends to be less than 0.05 wt%. It is not added seperately, as might be expected if it was to be deposited preferentially on the surface of the silver.

The silver metal loading at about 10 wt% is much more than is required for monolayer coverage of the $\alpha$-$Al_2O_3$ surface. This point is illustrated by the SEM photographs shown in Figure 6.3. Prior to calcination the $\alpha$-$Al_2O_3$

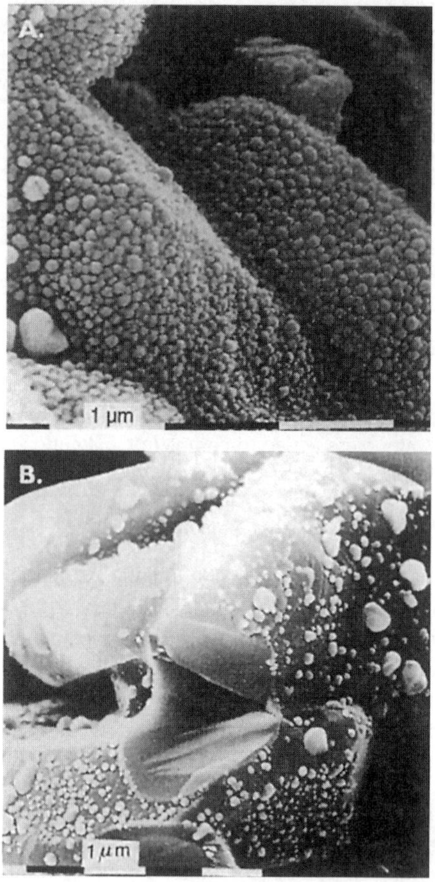


**Figure 6.3** Scanning electron microscopy photographs of silver catalysts supported on $\alpha$-$Al_2O_3$: (a) fresh; (b) calcined. (De Jong, 1998). Reproduced by permission of Baltzer Scientific Publishers.

surface is covered by silver particles about 0.1 μm in diameter. Following calcination the silver particles' size increases further and appears to spread fairly evenly over the support. Table 6.3 shows a comparison of BET and silver metal surface areas for a range of alumina supports with a variety of silver loadings. The former were measured by argon adsorption and the latter by oxygen chemisorption. Provided the sintering temperature is not excessive and the silver loading is sufficient then the BET and silver metal surface areas are almost identical, consistent with the spreading of silver multilayers over the alumina support.

This image of the supported silver catalysts has been confirmed by AES and ISS analyses, each of which showed strong attenuation of the aluminium signal and emergence of a silver signal as the predominant feature when the

**Table 6.3** Comparison of BET and silver metal surface areas for a range of supported silver catalysts
(Verykios *et al.*, 1980)

| *Silver loading/wt%* | *Sintering temperature/K* | $S_{BET}/m^2$ $(g\ cat)^{-1}$ | $S_{Ag}/m^2$ $(g\ cat)^{-1}$ |
|---|---|---|---|
| 13.9 | — | 0.63 | 0.62 |
| 13.9 | 673 | 0.63 | 0.56 |
| 5.6 | — | 0.97 | 0.59 |
| 5.6 | 673 | 0.97 | 0.46 |
| 5.6 | 973 | 0.97 | 0.26 |
| 1.0 | — | 0.20 | 0.16 |
| 1.0 | 673 | 0.20 | 0.14 |
| 1.0 | 1023 | 0.20 | 0.09 |
| 6.4 | — | 3.03 | 1.20 |
| 6.4 | 683 | 3.03 | 0.85 |
| 6.4 | 1123 | 3.03 | 0.30 |

Reproduced by permission of Academic Press.

bare support is compared with the catalyst. Figure 6.4 illustrates this effect for ISS and Auger data (Minihan and Hoflund, 1996a) all of which are consistent with the presence of a thick layer of silver or silver particles on the $\alpha$-$Al_2O_3$ support. This study also showed that even during operation, and especially

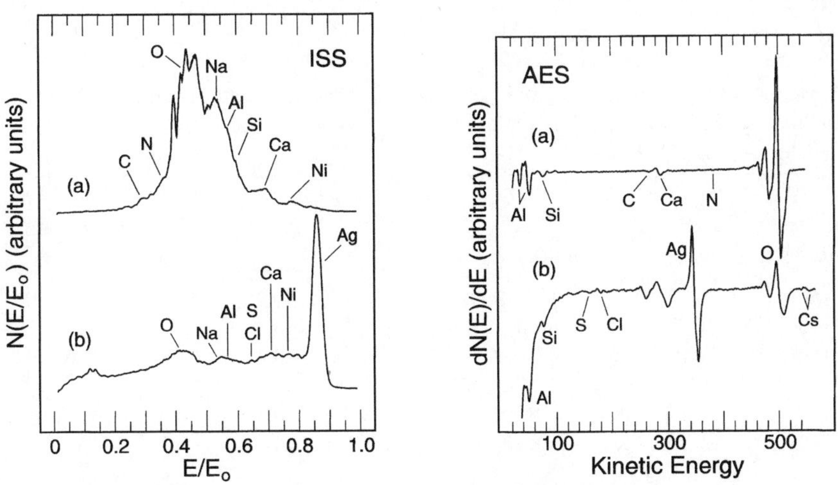


**Figure 6.4** (A) ISS and (B) Auger spectra obtained from (a) an $\alpha$-alumina support and (b) a caesium-promoted Ag/$\alpha$-alumina catalyst (Minihan and Hoflund, 1996a). Reproduced by permission of Academic Press.

during deactivation, the silver tends to agglomerate into larger crystals, leaving areas of the support uncovered by silver (Minihan and Hoflund, 1996b).

The influence of silver particle size on the catalytic properties and the role of added caesium is illustrated by the data in Table 6.4 (Bukhtiyarov *et al.*, 1997). These data show greater selectivity to ethylene oxide associated with larger silver particles. In addition, added caesium inhibits the total oxidation reaction without significantly affecting the rate of the selective oxidation reaction for all silver particle sizes tested.

A chlorine source is added to the feed gas in very small amounts because, like caesium in Table 6.4, it inhibits the total oxidation reaction without significantly affecting the epoxidation reaction. In general, laboratory scale testing of epoxidation catalysts tends to be carried out at atmospheric pressure or slightly above. Typical operating conditions have been described as (Minihan *et al.*, 1997): temperature 543 K; total pressure 18 bar; GHSV = 8000 $h^{-1}$; 30 wt% ethylene; 8 wt% oxygen; 6 wt% $CO_2$; 0.5 wt% $C_2H_6$; 10 ppm ethyl chloride. The chlorine source reacts with silver according to equation (6.1), generating a surface partially covered by AgCl:

$$Ag + C_2H_4Cl_2 + 0.25O_2 \longrightarrow C_2H_3Cl + AgCl + 0.5H_2O \qquad (6.1)$$

The exact surface chlorine content is very critical to the operation. This is shown in Figure 6.5 for epoxidation over a silver (110) surface treated so as to produce different surface chlorine coverages. Too much chlorination stops all reactions, so that the TOF for ethylene oxide and carbon dioxide formation

**Table 6.4** Influence of mean particle size and added caesium on the rates of epoxidation and total oxidation over a series of $Ag/Al_2O_3$ catalysts (Bukhtiyarov *et al.*, 1997)

| *Silver particle size/nm* | *Wt% Cs* | *Rate of $CO_2$ formation $\times 10^{-17}$/* molec $cm^{-2}$ $s^{-1}$ | *Rate of $C_2H_4O$ formation $\times 10^{-17}$/molec $cm^{-2}$ $s^{-1}$* | *Selectivity to $C_2H_4O$/%* |
|---|---|---|---|---|
| 160 | — | 2.5 | 0.2 | 8 |
| 160 | 0.1 | 1.7 | 0.3 | 15 |
| 160 | 0.2 | 0.4 | 0.4 | 50 |
| 560 | — | 20 | 6.3 | 24 |
| 560 | 0.2 | 12 | 6.5 | 35 |
| 1000 | — | 7.0 | 6.2 | 45 |
| 1000 | 0.2 | 4.0 | 6.5 | 61 |

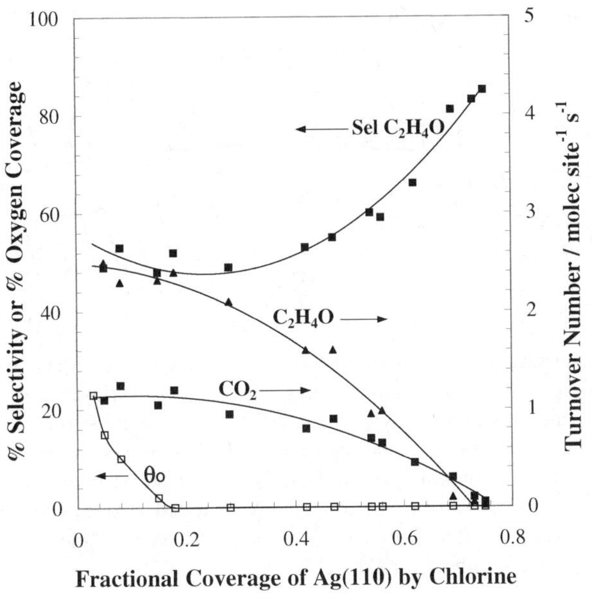


**Figure 6.5** Oxygen coverage ($\theta_O$), ethylene oxide and carbon dioxide turnover frequencies and selectivity to ethylene oxide as a function of chlorine precoverage of the silver (110) surface (Campbell and Paffett, 1984a and b) (reproduced by permission of Elsevier Science).

goes to zero as the surface coverage by chlorine nears completion. However, the silver surface can be about one third covered by chlorine without any big changes in TOF or selectivity. Thereafter, there is a tradeoff between increasing selectivity and diminishing activity (Campbell and Paffett, 1984b).

Berty (1983) has proposed that alkali metals, such as barium, and the addition of a small amount of ethane can reverse the surface chlorine reactions, or more particularly when these additives are present in appropriate amounts they can prevent overchlorination. Overchlorination can occur because the addition of ppm quantities of a chlorinated hydrocarbon to the ethylene epoxidation feed is difficult to control. Ethane can reverse overchlorination of silver via equation (6.2):

$$AgCl + 0.25O_2 + C_2H_6 \longrightarrow C_2H_5Cl + 0.5H_2O + Ag \qquad (6.2)$$

and barium, added as barium carbonate can act as a sink for areas of the silver surface that have been overchlorinated (equation (6.3)):

$$2AgCl + BaCO_3 \longrightarrow BaCl_2 + 2Ag + 0.5O_2 + CO_2 \qquad (6.3)$$

These concepts are represented Scheme 6.1.

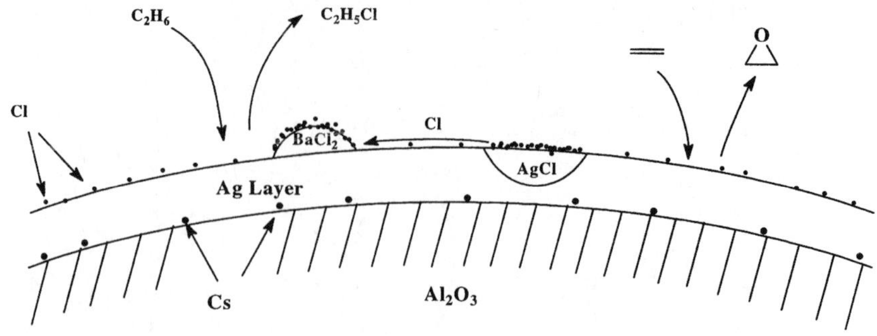


**Scheme 6.1** Schematic representation of the role of promoters and gas phase additives in the operation of silver epoxidation catalysts. The silver overlayer is shown with the Cs promoter located predominantly at the interface between the silver and the alumina support (Minihan *et al.*, 1997). Small particles of $BaCl_2$ are shown which act as sinks for those areas of the silver catalyst that have been over chlorinated (shown as AgCl). These sections of the silver surface are inactive. Added ethane also moderates the surface chlorine concentration. Ethylene epoxidation proceeds efficiently on those parts of the surface that are lightly covered by chlorine.

## 6.4 REACTION KINETICS AND NETWORK

A feature of ethylene epoxidation is that no catalyst exists which gives 100 % selectivity even at close to zero conversion (see Figure 6.1). In fact, only very good catalysts achieve 85 % selectivity even at low conversions. This means that the direct route for ethylene oxidation to carbon oxides and water must account for up to 15 % of the ethylene consumed even at low conversions:

$$C_2H_4 + 0.5O_2 \longrightarrow C_2H_4O \qquad 85\,\% \tag{6.4}$$

$$C_2H_4 + O_2 \longrightarrow 2CO_2 + 4H_2O \qquad 15\,\% \tag{6.5}$$

When ethylene and ethylene oxide are adsorbed onto silver and heated in an oxidizing atmosphere, $CO_2$ evolves at 432 K when ethylene is the adsorbate and at 463 K when ethylene oxide is the adsorbate (Van Santen and Kuipers, 1987). $CO_2$ desorbs from the silver surfaces at 383 K. This implies that the desorption temperatures quoted above for $CO_2$ derived from ethylene or ethylene oxide are a consequence of reaction just prior to desorption. A clear implication is that in the reaction conditions ethylene oxide is more resistant to total oxidation than ethylene. In this regard, Berty (1983) has estimated for catalysts operating close to industrial conditions ratios of $k_1 : k_2 : k_3$ of $6 : 0.2 : 1$ in Scheme 6.2, namely that the predominant reaction route is the selective oxidation of ethylene to ethylene oxide ($k_1$), with six times less ethylene being directly oxidized to carbon oxides and water ($k_3$). In these conditions, the consecutive oxidation of ethylene oxide to carbon oxides

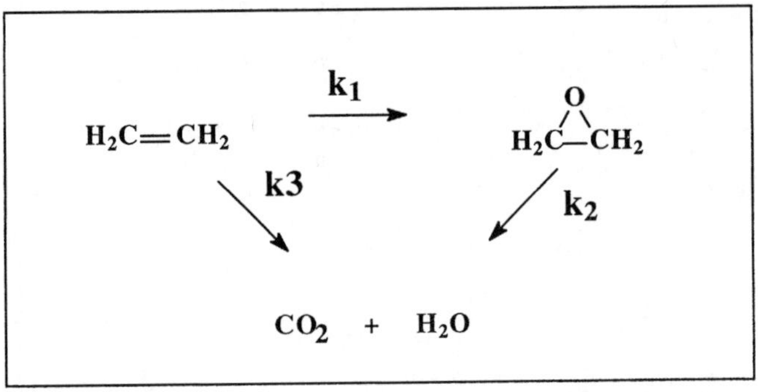


**Scheme 6.2** Simplified reaction network for ethylene epoxidation over silver catalysts.

and water ($k_2$) is very small. Ethylene oxide combustion only becomes important at higher conversions where the gas phase concentration of ethylene oxide increases to an extent that it is the predominant gas phase species. It should be recalled that ethylene oxide production units operate in the conversion range 8–12 % (see Table 6.2).

An interesting set of kinetic isotope effects has been observed for ethylene epoxidation over supported silver catalysts, whereby the ratio of $k_1 : k_3$ at 523 K in Scheme 6.2 increased by a factor of 5 when $C_2H_4$ was replaced with $C_2D_4$. In fact the rate of epoxidation increased slightly with $C_2D_4$, whereas the rate of total oxidation decreased. The conversion levels being investigated were small so that the total combustion product was coming predominantly from total oxidation of ethylene. These data were interpreted in terms of C—H bond activation being the slow step in ethylene combustion (Van Santen *et al.*, 1980; Grant and Lambert, 1985; Cant and Hall, 1978). Little is known about the reaction mechanism for the combustion reaction after the initial activation of the C—H bond, but a number of reports indicate that the species observed on the surface by infrared spectroscopy are identical to those which appear when acetaldehyde is oxidized in similar conditions (Force and Bell, 1975).

A detailed kinetic study of ethylene epoxidation over supported silver catalysts has been undertaken by Klugherz and Harriott (1971) and the results are presented in Figures 6.6–6.8. The reaction products in this case have an inhibiting effect on the rate of ethylene oxide formation as shown in Figure 6.6. Rather than attempt to extrapolate or correct results back to zero conversion, it was decided that all rates would be compared at the same partial pressure of the three main products,ethylene oxide, carbon dioxide and water, ($P_p$), of 0.01 atm (Klugherz and Harriott, 1971).

A prominent feature of Figure 6.7 is that the rate passes through a maximum with increasing ethylene pressure at constant oxygen pressure. The maximum shifts to higher ethylene pressures as the oxygen pressure increases.

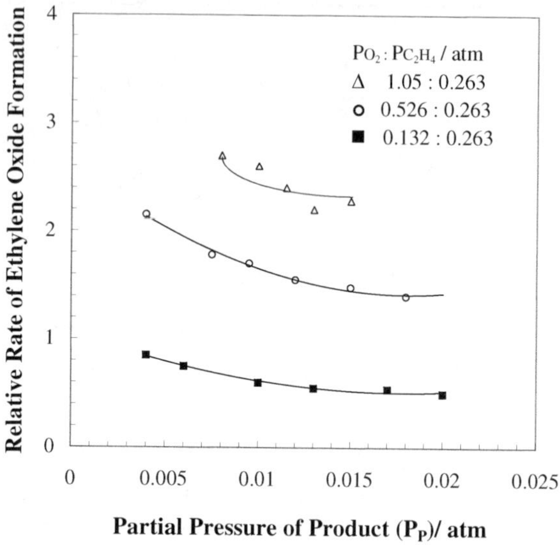


**Figure 6.6** Effect of product ($C_2H_4O + CO_2 + H_2O$) partial pressure on the relative rate of ethylene oxide formation at the ethylene and oxygen partial pressures indicated over a supported silver catalyst (8.1 % $Ag/\alpha$-$Al_2O_3$) at 493 K (Klugherz and Harriott, 1971). (Relative rate is defined as the observed rate in the specified reaction conditions divided by the rate at $P_E = 0.263$ atm; $P_O = 0.263$ atm; $P_P = 0.01$ atm and $T =$ 493 K.) Reproduced by permission of the AIChE.

A relationship of this type between reaction rate and reactant pressure is typically associated with a Langmuir–Hinshelwood mechanism for bimolecular reactions. In this mechanism reaction occurs between two adsorbed species competing for sites on the surface of the catalyst. The reaction rate passes through a maximum for the optimum surface coverage by both reactants. Additional data to support this hypothesis were presented by Kagawa *et al.* (1981) who showed that when 300 Torr of ethylene was exposed to an oxygen saturated silver surface at 298 K, no adsorption occurred, indicating that silver sites not covered by oxygen species are necessary for ethylene adsorption and that the ethylene molecules were incapable of displacing the preadsorbed oxygen.

Figures 6.8 shows that the rate of epoxidation increases with increasing oxygen pressure. At the lowest ethylene pressures, the rates increase only slightly as the oxygen pressure is increased, indicating that the surface is fully covered by oxygen at low ethylene partial pressures. At moderate ethylene pressures the reactions are nearly first order in oxygen. Most significantly, at the two highest ethylene pressures the reactions are greater than first order in oxygen.

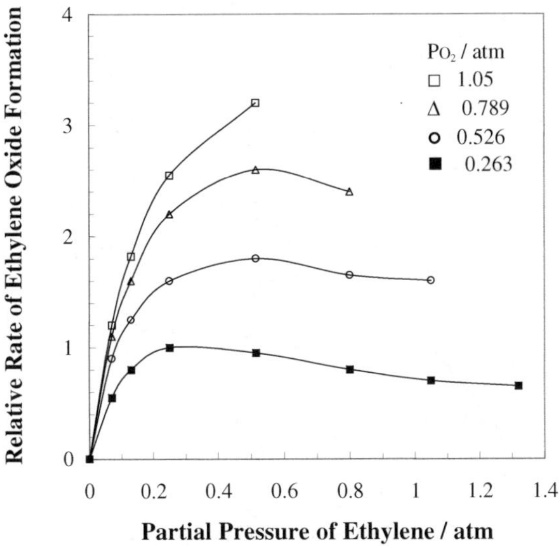


**Figure 6.7** Effect of ethylene pressure on the relative rate of ethylene oxide formation over a supported silver catalyst (8.1 % $Ag/\alpha\text{-}Al_2O_3$) at 493 K for the fixed oxygen partial pressures indicated. $P_p$ equal to 0.01 atm (Klugherz and Harriott, 1971). (Relative rate is defined as the observed rate in the specified reaction conditions divided by the rate at $P_E = 0.263$ atm; $P_O = 0.263$ atm; $P_P = 0.01$ atm and $T = 493$ K). Reproduced by permission of the AIChE.

The following rate law was developed to describe these results (Klugherz and Harriott, 1971):

$$r = \frac{kP_E P_O^{3/2}}{(1 + K_E P_E + \sqrt{K_O P_O} + K_P P_P)^2 (1 + \sqrt{K_S P_O})^2} \tag{6.6}$$

where $K_E$, $K_O$, $K_S$ and $K_P$ are the equilibrium constants for ethylene, molecular oxygen, atomic oxygen and reaction products present at the appropriate pressures of ethylene ($P_E$), oxygen ($P_O$) and products ($P_P$).

The dependence on oxygen pressure is unusual. The observed reaction order is greater than unity at high ethylene pressures and low oxygen pressures. The overall conclusion from this study was that reaction takes place between adsorbed ethylene and an adsorbed oxygen species on active sites formed by the dissociative adsorption of oxygen on metallic silver. Ethylene and oxygen, as well as the reaction products, compete for adsorption sites on the catalyst surface, which was considered to be similar to silver oxide. The concentration of active sites depends on the oxygen partial pressure. Such a mechanism can explain the rate maxima with respect to ethylene pressure and the orders in oxygen greater than unity (Klugherz and Harriott, 1971).

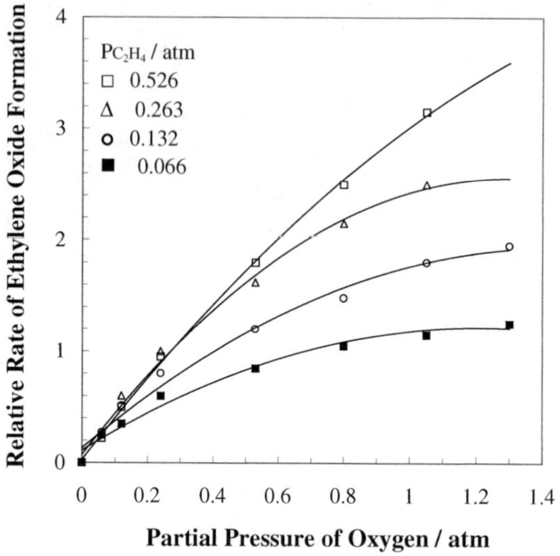


**Figure 6.8** Effect of oxygen pressure on the relative rate of ethylene oxide formation over a supported silver catalyst (8.1 % $Ag/\alpha\text{-}Al_2O_3$) at 493 K for the fixed ethylene partial pressures indicated. $P_p$ equal to 0.01 atm. (Klugherz and Harriott, 1971). (Relative rate is defined as the observed rate in the specified reaction conditions divided by the rate at $P_E = 0.263$ atm; $P_O = 0.263$ atm; $P_P = 0.01$ atm and $T = 493$ K.) Reproduced by permission of the AIChE.

## 6.5 THE NATURE OF THE OXYGEN SPECIES ON AND UNDER THE SILVER SURFACE

### 6.5.1 Oxygen on the pure silver surfaces

Because of its industrial significance a great deal of fundamental research work has been carried out on the modes of interaction of oxygen with single crystals of pure silver and with the alkali earth modified analogues. This work has demonstrated that there are three forms of oxygen present on or just below the surface of the silver. Each type of oxygen species is readily characterized by its desorption temperature in TPD. The exact amounts of each species present is determined by the reaction temperature, the exposed silver surface and the presence or absence of chlorine or alkali metals.

*Adsorbed molecular oxygen on the silver surface*

A good deal of evidence has been accumulated from single crystal work to indicate that when the (110) or (111) plane of pure silver is exposed to oxygen at low temperatures (about 100 K) an adsorbed molecular oxygen species forms (van Santen and Kuipers, 1987). This species is reported to carry an average charge of minus 1.7. The surface $O_2^{1.7-}$ species does not generate a LEED pattern characteristic of an ordered overlayer. When the silver sample is heated, some of this oxygen desorbs at 173 K, as shown in the TPD profile presented in Figure 6.9. The remaining oxygen appears to be converted into a form of atomic oxygen.

There is an important difference in sticking coefficients for oxygen between the (110) and (111) surfaces of silver. These values are shown in Table 6.5. Effectively oxygen adheres much more strongly to the silver (110) surface. At 173 K most of the adsorbed oxygen stays on the silver (110) surface, whereas most of it desorbs from the silver (111) surface. That part which desorbs at 173 K does not appear to dissociate at any time prior to desorption: when a mixture of $^{16}O_2$ and $^{18}O_2$ was contacted with silver at 113 K, no $^{18}O^{16}O$

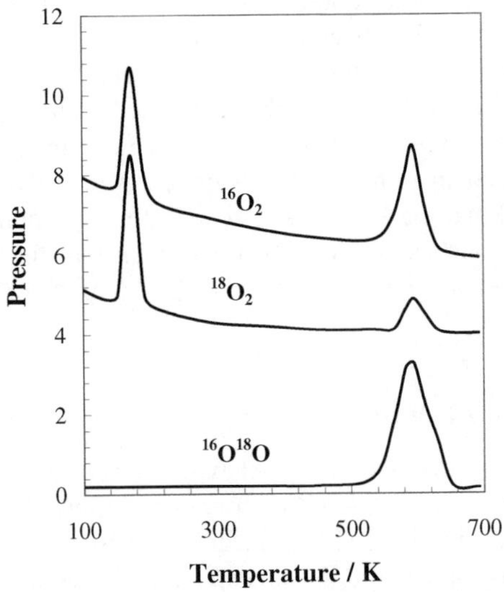


**Figure 6.9** TPD profile of oxygen from the Ag(110) surface following exposure to 220 L $^{18}O_2$ and 720 L $^{16}O_2$ at 113 K (L stands for a Langmuir which is defined as exposure of a surface to sufficient adsorbent to achieve monolayer coverage) (Backx *et al.*, 1981) (reproduced by permission of Elsevier Science).

**Table 6.5** Sticking coefficients for oxygen on silver surfaces (van Santen and Kuipers, 1987)

| *Surface* | *Sticking coefficient (temperature/K)* | |
|---|---|---|
| | *Molecular oxygen* | *Atomic oxygen* |
| Ag (110) | $7 \times 10^{-4}$ (130) | 5–30 $\times 10^{-4}$ (—) |
| Ag (111) | $5 \times 10^{-6}$ (147) | $10^{-6}$ (490) |

species were observed in the products which desorbed at 173 K, as shown in Figure 6.9 (Backx *et al.*, 1981).

### *Atomic oxygen on silver*

Increasing oxygen exposure to the silver (110) surface at room temperature generates a 2 × 1 ordered LEED surface overlayer due to monoatomic surface oxygen (Rovida and Pratesi, 1975). However, it is believed that the surface monoatomic oxygen layer is not intact, but that islands form for coverages in the range $0.5 < \theta < 0.7$. Desorption of this form of oxygen from the silver (110) face is characterized by a desorption peak at ca 600 K, as shown in Figure 6.9.

On the silver (111) face an ordered 4 × 4 superstructure is established above 398 K but this structure could also be described as a surface layer of $Ag_2O$. The lattice mismatch between the (111) faces of silver and silver oxide is just 0.5 %. In fact when angular resolved XPS analysis was undertaken the intensity of the peak at 529.8 eV, characteristic of monoatomic oxygen, decreased strongly for increasing off-normal angles, an indication that the oxygen was embedded in rather than adsorbed on the silver surface. Some workers have argued that the low sticking coefficient for monoatomic oxygen in the silver (111) face (see Table 6.5) must mean that the slow step is the adsorption of molecular oxygen on the surface (van Santen and Kuipers, 1987).

### *Oxygen under the surface of silver*

Backx *et al.* (1981) observed curious behaviour when they ran a number of cycles of oxygen adsorption at room temperature followed by TPD analysis. A first contact of a clean silver (111) surface with oxygen followed by TPD analysis gave a peak at ca 600 K, as expected. When a second cycle of adsorption followed by TPD was undertaken, the desorption peak at ca 600 K was 70 % larger, and continued to increase for a further six cycles. These data are consistent with some of the adsorbed oxygen being retained and the formation of another type of oxygen which in fact leads eventually to a desorption peak at 880 K (Backx *et al.*, 1981) . At no stage does this form of oxygen lead to a structure with long range order, hence LEED structures were never observed. In addition, this structure forms in conditions where $Ag_2O$ is not stable as indicated by the phase diagram presented as Figure 6.10 (Rovida *et al.*, 1974). This form of oxygen is described as subsurface oxygen and will be referred to again below when we consider the influence of alkali metal additives on oxygen adsorption. A final feature is that, when subsurface oxygen is formed from $^{16}O_2$ and the system is subsequently exposed to $^{18}O_2$, exchange occurs, but evidently just below 573 K, the temperature where monoatomic oxygen desorbs (van Santen and Kuipers, 1987).

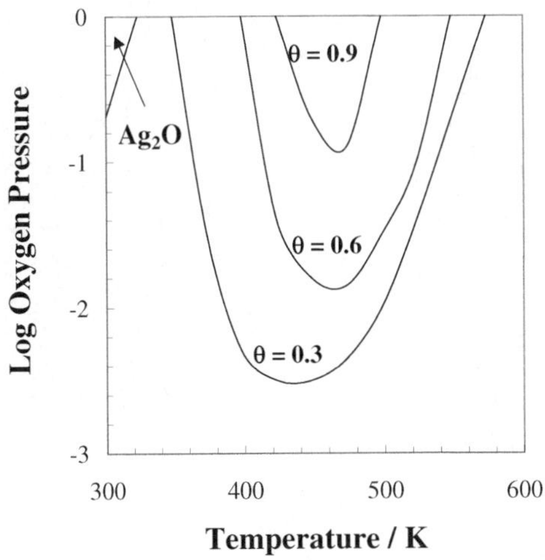


**Figure 6.10** Stability diagram for oxygen adsorbed or absorbed on Ag(111). The stable region for $Ag_2O$ is shown and the contour lines correspond to $\theta$ values of 0.3, 0.6 and 0.9 (Rovida *et al.*, 1974) (reproduced by permission of Elsevier Science).

### 6.5.2 Interaction and interconversion of oxygen species on the alkali metal promoted silver surfaces

Although the analysis of $Ag/\alpha$-$Al_2O_3$ epoxidation catalysts indicates that most of the alkali metals may not be on the silver surface but at the interface between the silver overlayer and the support (Minihan *et al.*, 1997), the role of alkali metals on the surface of silver has been studied extensively by a wide variety of surface science techniques. The phase diagram for K, Rb and Cs adsorbed on Ag(111) has been published (Leatherman and Diehl, 1996). This study rules out the formation of islands of alkali metal on this face of silver. Each alkali metal forms hexagonal overlayers which increase in density uniformly as coverage is increased. The stability of the overlayers increases also with coverage with the melting temperatures for the overlayer changing, typically from 40 K at $\theta = 0.11$ to 250 K at $\theta = 0.33$ for potassium on Ag(111). A number of the observed structures are presented in Figure 6.11 for increasing coverages by alkali metal. A point of caution is that none of these structures was observed above 423 K, which is well below the temperature of operation of silver based epoxidation catalysts.

Some alkali metals appear to play a role as shuttle between atomic and subsurface oxygen. When the silver (100) face was modified with potassium the adsorbed atomic form of oxygen was no longer stable and transformed more readily into subsurface oxygen. The adsorbed form of atomic oxygen then only appeared when the subsurface region was saturated. An additional feature was that potassium also tended to dissolve into the silver bulk, but it could be removed by heating above 873 K (Kitson and Lambert, 1981).

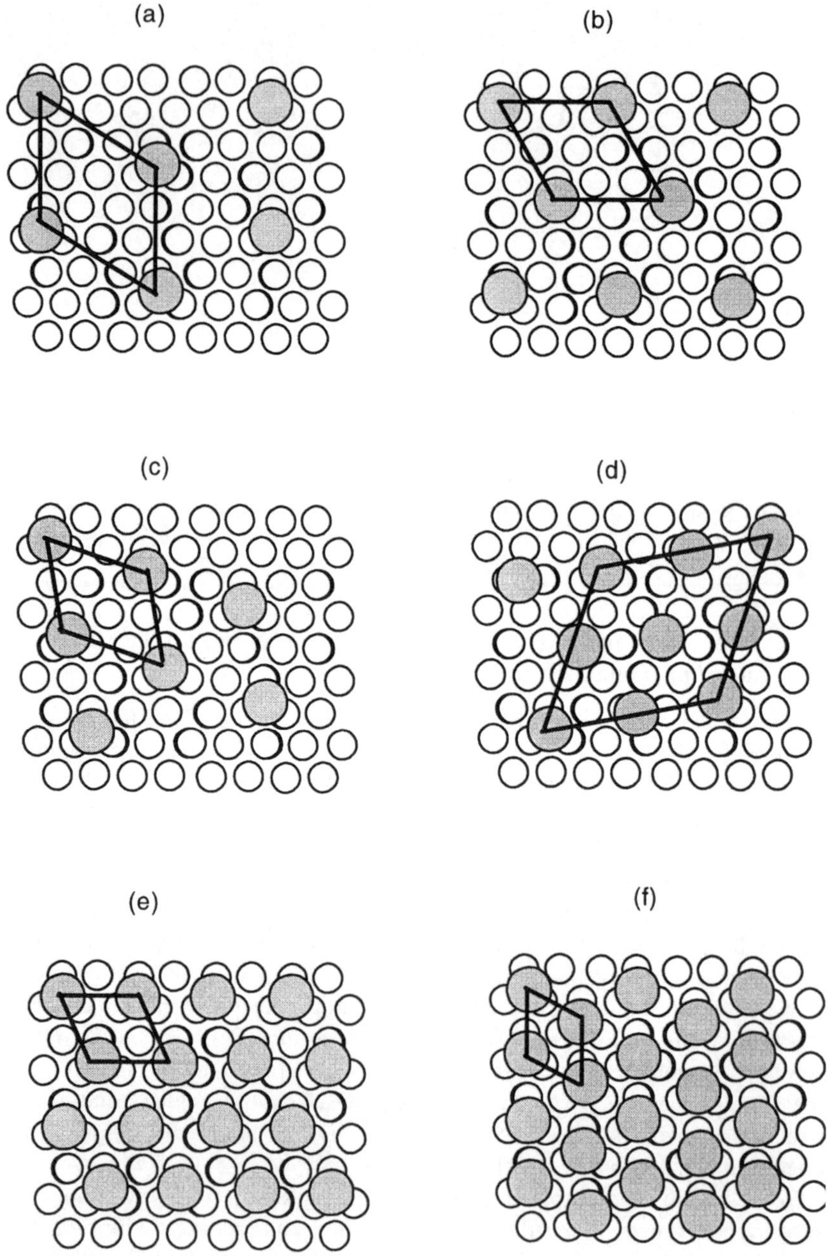


**Figure 6.11** Proposed commensutate structures for alkalis on Ag(111): (a) $(2\sqrt{3} \times 2\sqrt{3})R30°$; (b) $p(3 \times 3)$; (c) $(\sqrt{7} \times \sqrt{7})R19.1°$; (d) $(\sqrt{21} \times \sqrt{21})R10.8°$; (e) $p(2 \times 2)$; (f) $(\sqrt{3} \times \sqrt{3})R30°$ (Leatherman and Diehl, 1996). Reproduced by permission of the American Physical Society.

Briggs *et al.* (1977) also point to this function for sodium on Ag(111). In fact these workers suggest that sodium catalyses the otherwise activated oxygen adsorption and they measured an Na : O ratio of 1 : 1 for all sodium coverages up to saturation. These workers also concluded that oxygen speciation is not drastically affected by the addition of sodium to the silver (110) face, although Garfunkel *et al.* (1985) report that the adsorbed molecular oxygen state could be populated and remained stable up to 300 K when sodium was present on the surface.

## 6.6 REACTIVITY OF OXYGEN SPECIES ON SILVER

Table 6.6 summarizes catalytic data for the epoxidation of a variety of olefins over promoted $Ag/Al_2O_3$ catalysts where oxygen is the oxidant (Monnier, 1997). A striking feature of these data is that the presence of a so called allylic C—H bond in the substrate eliminates the possibility of achieving a high selectivity in epoxidation. This is because the reaction is directed towards cleavage of the allylic C—H bond rather than addition of oxygen. For example epoxidations of ethylene, 1,3-butadiene, styrene and its pyridine analogue are all possible at good selectivitites and reasonable conversions. However, when allylic hydrogens in the form of $CH_3$ groups are present the selectivity to the epoxide becomes zero, because a labile C—H bond has been introduced into the molecule which is readily cleaved, so directing the reaction away from the epoxidation pathway. These data also serve to illustrate that the concept of electrophilic or nucleophilic oxygen species on oxidation catalysts must be judged against the nature of the hydrocarbon species being oxidized. Clearly the so called electrophilic oxygen present on the surface of silver catalysts is not sufficiently electrophilic to be directed away from the very labile methyl C—H bonds in propene.

Figure 6.12 presents a full multiple selectivity–conversion plot for the epoxidation of propene with molecular oxygen, constructed in the same manner as Figure 6.1 for ethylene epoxidation. Clearly, no catalyst has been developed which will give a good selectivity at a reasonable conversion. Although there are not enough data available in the literature to generate full selectivity–conversion plots for the reactions in Table 6.6 it is apparent even with the limited amount of data available that there is a relationship between substrate structure and the ultimate selectivity that can be achieved in epoxidation reactions involving molecular oxygen. To illustrate this point further the bond dissociation energies of the weakest C—H bond in the substrates listed in Table 6.6 are presented in Table 6.7 and these values are plotted against the selectivity in Figure 6.13.

There appears to be a limiting value of bond dissociation enthalpy below which epoxidation is not favoured when molecular oxygen is the oxidant. (The situation with $H_2O_2$ as oxidant is treated in Chapter 10.) The presence of a C—H bond in the alkene with a bond energy less than 400 kJ $mol^{-1}$ renders

**Table 6.6** Catalytic activity for olefin epoxidation with molecular oxygen (Monnier, 1997)

| *Reaction* | *Temperature/K* | *Selectivity to epoxide/%* | *Conversion of olefin/%* |
|---|---|---|---|
| 1,3-Butadiene | | | |
| 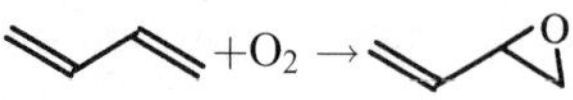 | 523 | 96 | 21 |
| 1-Butene | | | |
| 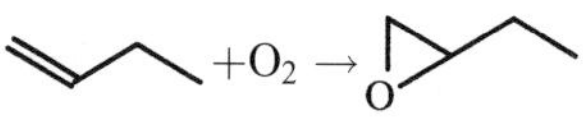 | 523 | 0 | 8 |
| 2-Butene | | | |
| 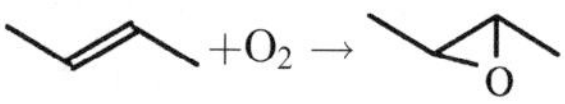 | 523 | 0 | 3.2 |
| 4-Vinyl pyridine | | | |
| 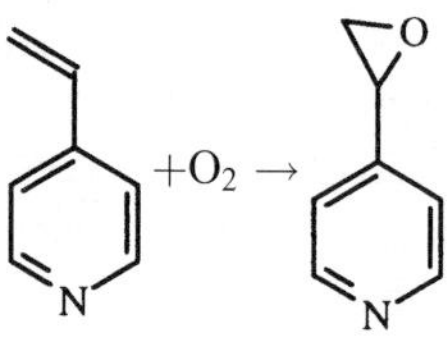 | 518 | 85 | 21 |
| Styrene | | | |
| 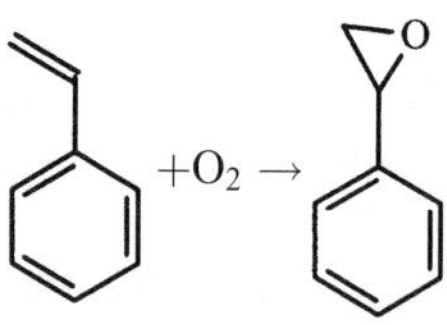 | 523 | 95 | 19 |
| 4-vinyl toluene | | | |
| 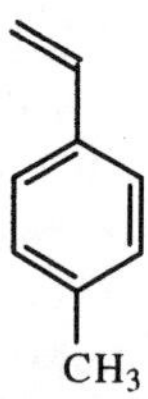 $+O_2 \rightarrow CO_2, H_2O$ | 498 | 0 | 100 |

Reproduced by permission of Elsevier Science.

it prone to other oxidation routes that involve cleavage of the weak C—H bond rather than addition of oxygen across the double bond, lowering the selectivity for epoxidation.

While kinetic studies indicate involvement of adsorbed oxygen species in

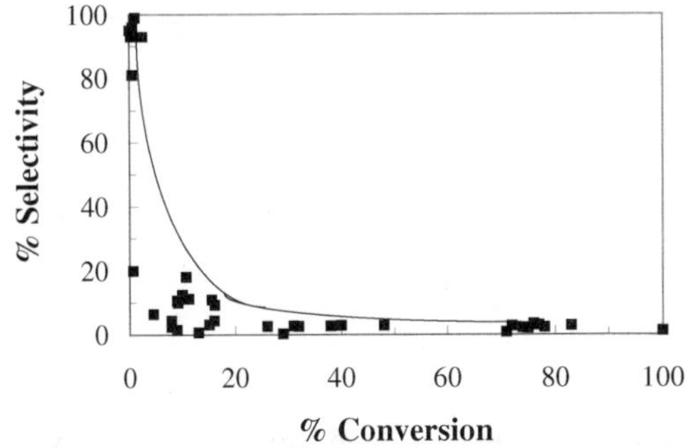


**Figure 6.12** Selectivity–conversion plot for propene epoxidation with oxygen (Cassidy, 1999).

**Table 6.7** Selectivity and bond energy data for olefin epoxidation with molecular oxygen at ca 20 % olefin conversion
(Cassidy, 1999; Monnier, 1997)

| *Olefin* | *Selectivity (%) to epoxide at ca. 20 % olefin conversion* | *Weakest C—H bond in olefin/kJ mol$^{-1}$* |
|---|---|---|
| Ethylene | 85 | 444 |
| Propylene | 10 | 361 |
| 1-Butene | 0 | 345 |
| 2-Butene | 0 | 358 |
| 1,3-Butadiene | 96 | 409 |
| 4-Vinyl-toluene | 0 | 371 |
| 4-Vinyl pyridine | 85 | 411 |
| Styrene | 95 | 410 |

the epoxidation of ethylene they do not shed any light on the nature of these species. However, a great many other techniques have focused on this problem. The essential question is to know which of the three forms of oxygen observed on or just under silver surfaces is the one which is involved in the epoxidation mechanism. Interestingly, some discrepancies still appear between the rate of epoxidation and the rate of oxygen uptake. For oxygen surface coverage ($\theta$) equal to 0.7, Voge measured a rate of oxygen consumption during epoxidation reactions that was 200 times greater than the rate of adsorption (Van Santen and Kuipers, 1987) and the rate of ethylene oxide formation has been reported as significantly higher than the rate of oxygen dissociation at the surface (Imre, 1970). These data probably point to a mechanism whereby oxygen chemisorption is governed by the availability of bare silver sites created by ethylene oxide production and desorption.

The simplest argument put forward to elucidate the nature of the active form of oxygen is that six ethylene molecules could react with six molecules

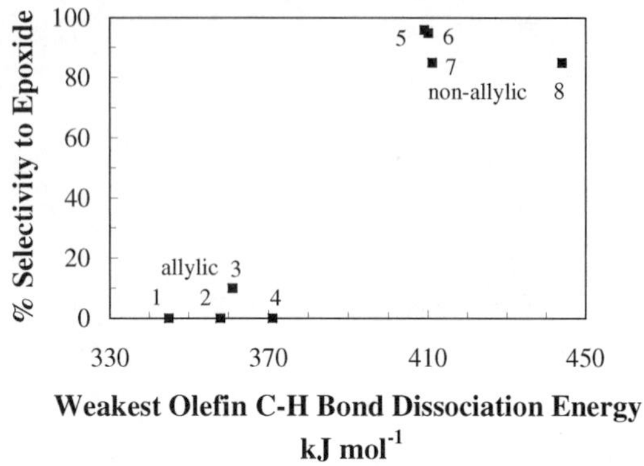


**Figure 6.13** Epoxidation selectivity olefins with molecular oxygen at ca 20 % olefin conversion as a function of the weakest C—H bond in the olefin: 1, but-1-ene; 2, but-2-ene; 3, propylene; 4, 4-vinyl-toluene; 5, 1,3-butadiene; 6, styrene; 7, 4-vinyl pyridine; 8, ethylene (Cassidy, 1999).

of adsorbed oxygen to yield six molecules of epoxide, but this would leave a residue of six atomic oxygen species on the catalyst surface which would have to be removed by combustion of one molecule of ethylene. This argues that the maximum selectivity that could be achieved in ethylene epoxidation is $(6/7) \times 100$ (Sachtler *et al*., 1981):

$$6C_2H_4 + 6O_{2(ads)} \rightarrow 6C_2H_4O + 6O_{(ads)} \tag{6.7}$$

$$C_2H_4 + 6O_{(ads)} \rightarrow 2CO_2 + 2H_2O \tag{6.8}$$

$$7C_2H_4 + 6O_{2(ads)} \rightarrow 6C_2H_4O + 2CO_2 + 2H_2O \tag{6.9}$$

This hypothesis is no longer supported because the 6/7 or 86 % selectivity barrier has now been exceeded with the most modern of catalysts and technologies (Bhasin, 1999). In addition, there is little evidence to suggest that a form of adsorbed molecular oxygen is present on silver surfaces at the reaction temperature. For example, $O_2$(ads) is not stable above 383 K on Ag(111) and 215 K on Ag(110) (Van Santen and Kuipers, 1987).

There is a good deal of evidence to suggest that very high surface coverages of oxygen are necessary to achieve high selectivities in epoxidation, namely that the form of oxygen which selectively reacts with ethylene to give epoxide via a very mild oxidation is formed only at high oxygen coverages. Akella and Lee (1984) have demonstrated that selectivity to ethylene oxide over pure silver increases as the surface coverage by oxygen increases (Figure 6.14).

It is clear from the thermodynamics that the form of oxygen involved is not $Ag_2O$ (Rovida and Pratesi, 1975); it is not stable in typical epoxidation reaction conditions. However, Backx *et al*. (1981) have shown that the O : Ag

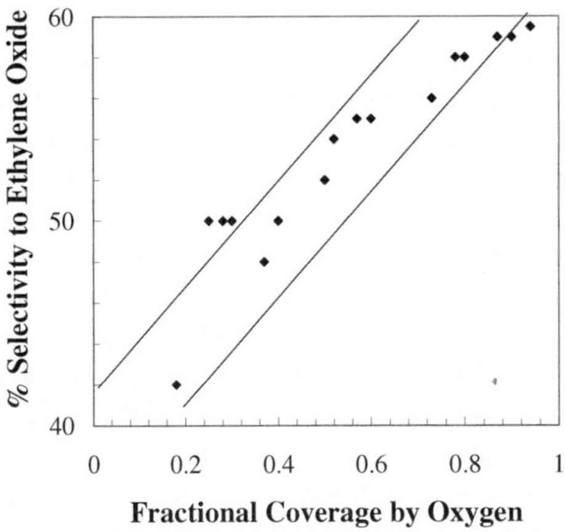


**Figure 6.14** Selectivity to ethylene oxide as a function of oxygen coverage over a range of silver catalysts (Akella and Lee, 1984) (reproduced by permission of Academic Press).

surface ratio must be greater than 0.5 : 1 before epoxide is formed and Grant and Lambert (1983) indicated that some pretreatment in an ethylene–oxygen mixture was necessary before the surface could produce epoxide. Each of these studies was carried out in conditions where subsurface oxygen was generated, pointing to a role for this type of oxygen in the epoxidation mechanism over silver catalysts.

Van Santen and De Groot (1986) studied the role of adsorbed atomic oxygen and subsurface oxygen in epoxidation. Table 6.8 summarizes a number of experiments in which the silver surface was exposed to $^{16}O_2$ to produce

**Table 6.8** Summary of $^{16}O_2$ incorporation and $C_2H_4{}^{16}O$ formation over silver surfaces treated with $^{16}O_2$ prior to reaction with $C_2H_4$ and $^{18}O_2$ (van Santen and de Groot, 1986)

| *Experiment* | *$^{16}O_2$ preadsorbed/ μmol (g Ag)$^{-1}$* | *$^{18}O_2$ initial partial pressure/Torr* | *Maximum $C_2H_4{}^{16}O$ produced$^a$/ μmol (g Ag)$^{-1}$* | *$^{16}O_2$ incorporated/ μmol (g Ag)$^{-1}$* |
|---|---|---|---|---|
| 1 | 1.09 | 0 | 0.12 | 0.51 |
| 2 | 0.94 | 2.6 | 0.19 | 0.43 |
| 3 | 1.08 | 20.4 | 0.42 | 0.9 |
| 4 | 0.46 | 20.4 | 0.12 | 0.28 |
| 5 | 0.46 | 0 | 0 | 0.03 |

Total number of silver surface atoms = 2 μmol (g Ag)$^{-1}$; 1 Torr = 0.58 μmol (g Ag)$^{-1}$

[a]Measured at the temperature where $C_2H_4{}^{16}O$ formation was at a maximum.

Reproduced by permission of Academic Press.

varying amounts of adsorbed and subsurface oxygen. Following evacuation, ethylene or a mixture of ethylene and $^{18}O_2$ was introduced and the evolution from the surface of $C_2H_4{}^{16}O$ and $C_2H_4{}^{18}O$ was followed during temperature programmed reaction in the range 298–598 K. Typical results are presented in Figure 6.15. In experiments 1–3 of Table 6.8 a silver surface was prepared with an $O_s : Ag_s$ ratio of approximately 1 : 1 by treating the silver surface at 373 K with $^{16}O_2$. For this $O_s : Ag_s$ (1 : 1) composition, $C_2H_4{}^{16}O$ evolved at lower temperatures than those required for the formation of $C_2H_4{}^{18}O$. It was established that $^{18}O_2$ did not exchange with $^{16}O$ in the subsurface in the absence of ethylene but it did exchange when ethylene was present. In fact experiments 2 and 3 indicate that the amount of exchange increased in the presence of ethylene as the $^{18}O_2$ partial pressure increased because the amount of $C_2H_4{}^{16}O$ produced increased. The appearance of $C_2H_4{}^{16}O$ in Figure 6.15 at temperatures lower than those required for $C_2H_4{}^{18}O$ is cited as proof that epoxide formation occurs by reaction of ethylene with adsorbed oxygen.

This system was then heated to 573 K so as to remove of all the adsorbed oxygen, leaving the subsurface oxygen. At this point the $O_s : Ag_s$ ratio was 0.5 : 1. When ethylene alone was contacted with this system no epoxide was produced (experiment 5 of Table 6.8) indicating that subsurface oxygen,

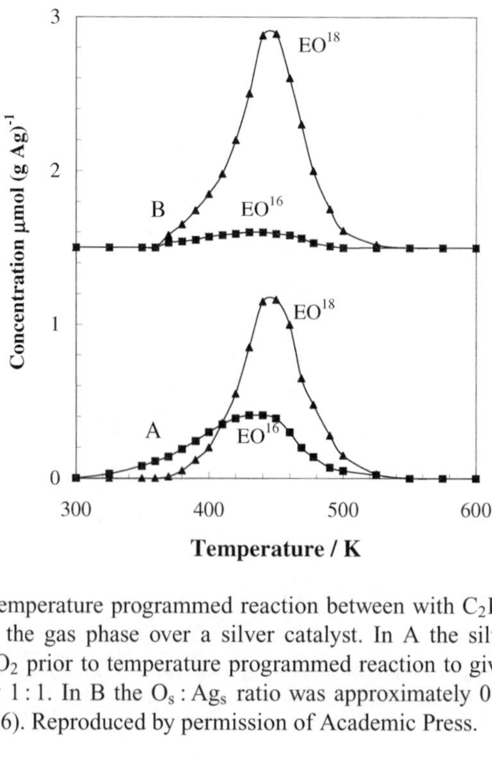


**Figure 6.15** Temperature programmed reaction between with $C_2H_4$ (13 Torr) and $^{18}O_2$ (20.4 Torr) in the gas phase over a silver catalyst. In A the silver catalyst had been exposed to $^{16}O_2$ prior to temperature programmed reaction to give an $O_s : Ag_s$ ratio of approximately 1 : 1. In B the $O_s : Ag_s$ ratio was approximately 0.5 : 1 (van Santen and De Groot, 1986). Reproduced by permission of Academic Press.

although necessary for efficient reaction, was not itself the reactive species. When the mixture of ethylene and oxygen was contacted with the subsurface oxygen saturated silver surface, $C_2H_4{}^{16}O$ and $C_2H_4{}^{18}O$ evolved at identical temperatures during temperature programmed reaction (see Figure 6.14(b) and experiment 4 of Table 6.8). The amount of $C_2H_4{}^{18}O$ greatly exceeded the other isotope. These data were interpreted as evidence that ethylene cannot react directly with subsurface oxygen, but that some of this oxygen became available for reaction with ethylene through exchange with $^{18}O_2$ in the presence of the hydrocarbon. This process is presented schematically in Scheme 6.3 showing no exchange between $^{18}O_2$ and $^{16}O_{subsurface}$ in the absence of ethylene and no direct reaction between subsurface oxygen and ethylene. However, when ethylene and oxygen are both present in the gas phase, exchange between gas phase and subsurface oxygen is possible, probably mediated through the epoxidation reaction itself.

### 6.6.1 The role of chlorine in determining the relative amounts of adsorbed and subsurface oxygen on silver

When chlorine is added to the silver system the oxygen coverage decreases drastically (Campbell and Paffett, 1984b), as does the turnover frequency whereas the selectivity to ethylene oxide increases (Berty, 1983) (see Figure 6.5).

In fact, the presence of chlorine on the silver surface affects the coverage by different oxygen species differently. Kilty *et al.* (1973) identified three adsorbed oxygen species. The first of these had an apparent activation energy of less than 10 kJ mol$^{-1}$; surface chlorine atoms blocked the adsorption of this type of oxygen on a one for one basis which stopped completely when the Cl : Ag ratio was about 0.25 : 1 (Figure 6.16). Adsorption of a second form of oxygen, with an apparent activation energy of 33 kJ mol$^{-1}$, was not disrupted by preadsorbed chlorine until adsorption of the first form of oxygen was completely inhibited; thereafter this form of adsorbed oxygen was more strongly inhibited than the first form. This behaviour is consistent with

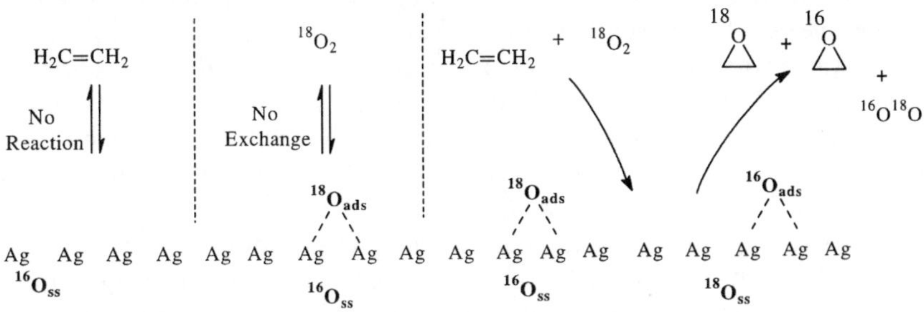


**Scheme 6.3** Schematic diagram of exchange of adsorbed and subsurface oxygen only when ethylene is present in the gas phase.

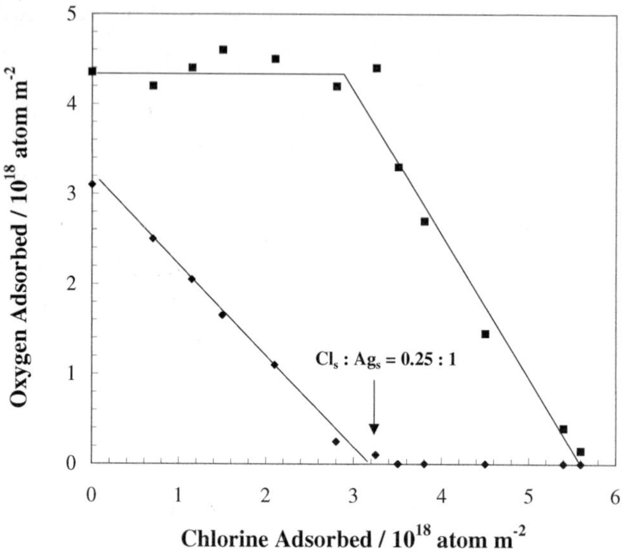


**Figure 6.16** Influence of the amount of preadsorbed chlorine on the adsorption of oxygen on a silver surface (Kilty *et al.*, 1973) (reproduced by permission of Elsevier Science).

preferential poisoning of a specific set of adsorption sites by the initial doses of chlorine, which adsorbs preferentially on that site; then a second set of surface sites are poisoned. Chlorine adsorbs on these sites as a second preference.

There seems to be general agreement that, although adsorbed chlorine decreases the gross amount of adsorbed oxygen, it increases the amount of oxygen that participates in the epoxidation reaction. Van Santen and Kuipers (1987) have proposed a mechanism whereby the presence of subsurface oxygen, or more effectively chlorine, renders the adsorbed atomic oxygen more electrophilic by withdrawing electron density from around the adsorbed oxygen species. The presence of highly charged silver species, such as $Ag^+$ or $Ag^{3+}$, was also seen as beneficial to the creation of electrophilic adsorbed atomic oxygen species. Electrophilic oxygen tends to attack the C═C bond in ethylene whereas the more nucleophilic oxygen tend to activate the C—H bonds leading to total oxidation products. The proposed mechanism is presented in Scheme 6.4. It shows the creation of a surface layer of positively charged silver species, brought about by the presence of oxygen or chloride anions. The positively charged silver species draw electron density away from the adsorbed oxygen, making it more electrophilic. By contrast oxygen adsorbed on metallic silver is seen as being more negatively charged, making it more nucleophilic.

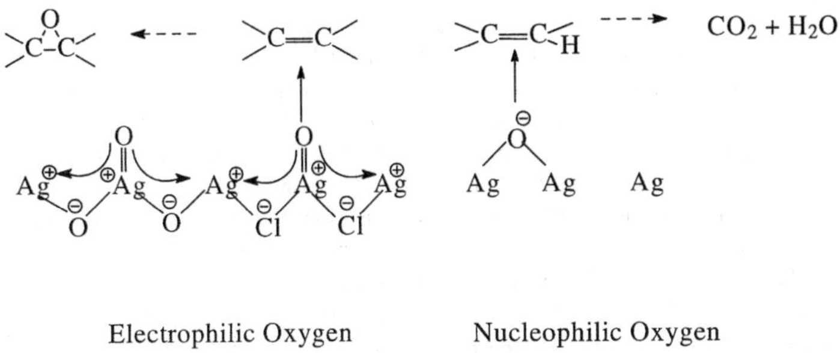


**Scheme 6.4** Schematic representation of the proposed transition state leading to epoxidation. The presence of negatively charged subsurface oxygen or surface chlorine generates a positive charge on the uppermost silver layer. This draws negative charge away from adsorbed atomic oxygen, making it more electrophilic. In the absence of a charged upperlayer of silver the surface oxygen species carries a stronger negative charge and is less inclined to attack the electron rich region of the ethylene molecule.

## REFERENCES TO CHAPTER 6


Akella LM and Lee HH (1984) Selectivity characterization of ethylene oxidation reactions: Oxygen chemisorption. *J. Catal.* **86**: 465–472.

Backx C, de Groot CPM and Biloen P (1981) Adsorption of oxygen on Ag(110) studied by HREELS and TPD. *Surf. Sci.* **104**: 300–317.

Berty JM (1983) *Ethylene Oxide Synthesis in Applied Industrial Catalysis*, Vol. 1, Leach BE (Ed.). Academic Press, New York, pp 207–238.

Bhasin MM (1999) Importance of surface science and fundamental studies in heterogeneous catalysis. *Catal. Lett.* **59**: 1–7.

Bhasin MM, Eligen PC and Hendrix CD (1984) Catalyst composition and process for oxidation of ethylene to ethylene oxide. UK patent GB 2 043 481 B, assigned to Union Carbide Corporation.

Briggs D, Marbrow RA and Lambert RM (1977) An XPS and UPS study of the interaction of oxygen with sodium-dosed Ag(111). *Surf. Sci.* **65**: 314–324.

Bukhtiyarov VI, Prosvirin IP, Kvon RI, Bal'zhinimaev BS and Podgornov EA (1997) XPS and TPD studies of Cs–O complexes on Ag surfaces: single crystal versus supported catalysts. *Appl. Surf. Sci.* **115**: 135–143.

Campbell CT and Paffett MT (1984a) The interactions of $O_2$, CO and $CO_2$ with Ag(110). *Surf. Sci.* **143**: 517–535.

Campbell CT and Paffett MT (1984b) The role of chlorine promoters in catalytic ethylene epoxidation over the Ag(110) surface. *App. Surf. Sci.* **19**: 28–42.

Cant NW and Hall WK (1978) Catalytic oxidation VI. Oxidation of labelled olefins over silver. *J. Catal.* **52**: 81–94.

Cassidy F (1999) Thesis, University of Limerick.

De Jong KP (1998) Assembly of solid catalysts. *CATTECH* **2**: 87–95.

Dever JP, George KF, Hoffman WC and Soo H (1994) Ethylene oxide. In *Kirk–Othmer Encyclopedia of Chemical Technology*, Vol. 9, 4th edition. Wiley, New York, pp 915–959.

Force EL and Bell AT (1975) Infrared spectra of adsorbed species present during the epoxidation of ethylene over silver. *J. Catal.* **38**: 440–460.

Garfunkel EL, Ding X, Dong G, Yang S, Hou X and Wang X (1985) The coadsorption of sodium and oxygen on Ag(100): An XPS, UPS and HREELS study. *Surf. Sci.* **164**: 511–525.

Grant RB and Lambert RM (1983) Mechanism of the silver-catalysed heteogeneous epoxidation of ethylene. *J. Chem. Soc., Chem. Commun.* 662–663.

Grant RB and Lambert RM (1985) A single crystal study of the silver-catalysed selective oxidation and total oxidation of ethylene. *J. Catal.* **92**: 364–375.

Imre VL (1970) Untersuchungen zum Mechanismus der Oxydation van Athylen durch Sauerstoff zu Athylerr\noxid an Silber al katalysator. *Ber. Bunsenges. Phys. Chem.* **74**: 220–226.

Kagawa S, Iwamoto M, Mori H and Seiyama T (1981) Formation of adsorbed oxygen and its reactivity with ethylene over silver catalysts. *J. Phys. Chem.* **85**: 434–439.

Kilty PA, Rol NC and Sachtler WMH (1973) Identification of oxygen complexes adsorbed on silver and their function in the catalytic oxidation of ethylene. In *Proc. 5th Int. Congr. Catalysis*, Hightower JW (Ed.). North-Holland, Amsterdam, pp 929–943.

Kitson M and Lambert RM (1981) Basic studies of the oxygen chemisorption on silver: oxygen, dioxygen and superoxide on potassium-dosed Ag(100). *Surf. Sci.* **109**: 60–74.

Klugherz PD and Harriott P (1971) Kinetics of ethylene oxidation on a supported silver catalyst. *AIChE J.* **17**(4): 856–866.

Leatherman GS and Diehl RD (1996) Phase diagrams and rotated incommensurate phases of K, Rb and Cs adsorbed on Ag(111). *Phys. Rev. B* **53**(8): 4939–4946.

Minihan DM and Hoflund GB (1996a) Study of Cs-promoted, $\alpha$-alumina-supported silver ethylene epoxidation catalyst. I Characterization of the support and as-prepared catalyst. *J. Catal.* **158**: 109–115.

Minihan DM and Hoflund GB (1996b) Study of Cs-promoted, $\alpha$-alumina-supported silver ethylene epoxidation catalyst. II Effect of aging. *J. Catal.* **162**: 48–53.

Minihan DM, Hoflund GB, Epling WS and Schoenfeld DW (1997) Study of Cs-promoted, $\alpha$-alumina-supported silver ethylene epoxidation catalyst. III Characterization of Cs-promoted and non-promoted catalysts. *J. Catal.* **168**: 393–399.

Monnier JR (1997) The selective epoxidation of non-allylic olefins over supported silver catalysts. In *Proceedings 3rd World Congress on Oxidation Catalysis*, Grasselli RK, Oyama ST, Gaffney AM and Lyons JE (Eds). Elsevier Science, Amsterdam, pp 135–149.

Porcelli JV (1981) Ethylene oxidation–exploratory research. *Catal. Rev. Sci. Eng.* **23**: 151–162.

Rovida G, Pratesi F, Maglietta M and Ferroni E (1974) Chemisorption of oxygen on the silver (111) surface. *Surf. Sci.* **43**: 230–256.

Rovida G and Pratesi F (1975) Chemisorption of oxygen on the silver (100) surface. *Surf. Sci.* **52**: 542–555.

Sachtler WMH, Backx C and van Santen RA (1981) On the mechanism of ethylene epoxidation. *Catal. Rev. Sci. Eng.* **23**: 127–149.

Seyedmonir SR, Plischke JK, Vannice MA and Young HW (1990) Ethylene oxidation over small silver crystallites. *J. Catal.* **123**: 534–549.

Sajkowski DJ and Boudart M (1987) Structure sensitivity of the catalytic oxidation of ethene by silver. *Catal. Rev. Sci. Eng.* **29**(4): 325–360.

Van Santen RA and de Groot CPM (1986) The mechanism of ethylene epoxidation. *J. Catal.* **98**: 530–539.

Van Santen RA and Kuipers HPCE (1987) The mechanism of ethylene epoxidation. *Adv. Catal.* **35**: 265–321.

Van Santen RA, Moolhuysen J and Sachtler WMH (1980) The role of C—H bond breaking in the epoxidation of ethylene. *J. Catal.* **65**: 478–480.

Verykios XE, Stein FP and Coughlin RW (1980) Influence of metal crystalline size and morphology on selectivity and activity of ethylene oxidation catalysed by supported silver. *J. Catal.* **66**: 368–382.

Zomerdijk JC and Hall MW (1981) Manufacture of ethylene oxide. *Catal. Rev. Sci. Eng.* **23**: 163–185.

# 7

# Catalytic Combustion


7.1 Introduction 190
7.2 Environmental Applications of Catalytic Combustion 191
7.2.1 The VOC Problem 191
7.2.2 The NOx Problem 192
7.2.3 Operation of Three-Way Catalysts 196
7.3 Principles of Catalytic Combustion 199
7.4 Choice of Materials for the Construction of Combustion Catalysts 202
7.4.1 Chemical Reactivity of the Active Phase 202
7.4.2 Stability of the Active Phase 203
7.4.2 Stability of Support Materials 205
7.4.3 Choice of Support Acid–Base Properties 214
7.5 Poisoning of Combustion Catalysts 216
7.6 Reactivity of the Substrate 218
7.7 Kinetics of Catalytic Combustion Reactions 223
7.8 Nature of the Active Species on Working Precious Metal Catalysts 227
7.8.1 Reactivity of Surface Species in Combustion Catalysis 227
7.8.2 Nature of Oxygen Species in Combustion Catalysis 231
References to Chapter 7 238



Summary

Catalytic combustion is an important element in the worldwide fight against air pollution. One of the best known applications is in the treatment of exhaust gases from petrol driven spark ignition engines, with the so called three-way catalysts. Other applications include the use of catalytic units for natural gas combustion in conditions that avoid by-product nitrogen oxides. Catalytic combustion is also suitable for the elimination of low concentrations of volatile organic compounds in process outlet streams and from indoor air in the work place and in the home. One of the principal features of catalytic combustion is its capability to oxidize very low concentrations of combustibles in air at concentrations below the minimum required to sustain thermal combustion. Reactor configuration is a major design feature in combustion catalysis, principally because of the very high gas throughputs required. Sintering and poisoning of supports and supported phases present major technological challenges, which are particularly severe in these types of catalysts because of the intermittent nature of their operation and the variability of the gas sources that need to be treated.

This chapter describes

- Environmental applications of catalytic combustion,
- Principles of catalytic combustion,
- Choice of materials for the construction of combustion catalysts,
- Poisoning of combustion catalysts,
- Reactivity of the substrate,
- Kinetics of catalytic combustion reactions,
- Nature of the active species on working precious metal catalysts.

## 7.1 INTRODUCTION

Catalytic combustion is the process whereby a combustible is completely oxidized over a heterogeneous catalyst. It has two major applications. These are the combustion of fuels and the oxidation of volatile organic compounds (VOCs). In the former, a fuel such as methane, at a concentration in excess of 2 vol%, is combusted over a heterogeneous catalyst and the hot gases generated are fed to a gas-turbine generator. The process is somewhat similar to conventional combustion, but can operate with leaner feeds and at lower temperatures, offering advantages in terms of the environmental impact. In particular, catalytic combustion represents one possible strategy for the reduction of emissions of NOx from combustion sources.

For the second application, organic compounds present at concentrations less than 1000 ppm are oxidized by passing them over a combustion catalyst. A feature of catalytic combustion is the ability to oxidize at any oxidant : fuel ratio. In practice, this means that catalytic combustion can operate at very low fuel concentrations, where conventional combustion cannot be sustained. Typically, concentrations of hazardous VOCs are present in the air in concentrations that do not support conventional flame combustion but are amenable to destruction through catalytic oxidation.

The origins of the VOC and NOx problems are presented below followed by a detailed description of the state of the art of combustion catalysts. Consideration of chlorinated VOCs will be reserved for Chapter 9.

## 7.2 ENVIRONMENTAL APPLICATIONS OF CATALYTIC COMBUSTION

### 7.2.1 The VOC problem

VOCs are an important class of air pollutant found in the atmosphere at ground level in all urban and industrial centres. Many hundreds of compounds fall into this category. A standard definition for VOCs is any organic compound whose vapour pressure at 298 K is between 1 and 760 Torr. However, this definition is not strictly adhered to in practice and the term has come to mean any carbon-containing compound found in the atmosphere, excluding elemental carbon, carbon monoxide and carbon dioxide. This definition includes methane, as well as oxygenated, chlorinated, sulfur- and nitrogen-containing compounds.

Human activities cause VOCs to be emitted into the atmosphere mainly from motor vehicle exhausts, evaporation of fuels, solvent usage, industrial processes, landfill wastes and agricultural activities. These sources are responsible for 75–95 % of VOC emissions (Crump, 1995). The balance comes from natural biogenic sources such as plants, trees, wild animals, forest and other natural fires and anaerobic processes in bogs and marshes. Isoprene and monoterpenes account for many of the naturally occuring VOCs whereas man-made VOCs include the full range of oxygenated, chlorinated, fluorinated, sulfur- and nitrogen-containing aliphatic and aromatic compounds, many of which carry well documented risks to human health.

Worldwide estimates of VOC emissions are comparable with those for $SO_2$ and NOx. Some comparative global emission figures are presented in Table 7.1. Table 7.2 presents the results of a study of the VOC concentrations in indoor and outdoor air in Avon, England, and indicates typical levels that are encountered in practical situations. Catalytic combustion is one of the few technologies suitable for the treatment of the very low levels of VOCs encountered in indoor air.

**Table 7.1** World pollutant emissions 1986–1992 in kilotonnes (Eurostat Yearbook, 1996)

| *Year* | *$SO_2$* | *NOx* | *CO* | *VOC* |
|---|---|---|---|---|
| 1986 | 46 625 | 45 116 | 145 029 | 33 652 |
| 1988 | 45 100 | 46 737 | 143 043 | 35 124 |
| 1990 | 56 190 | 52 044 | 162 905 | 37 646 |
| 1992 | 43 420 | 41 436 | 116 945 | 30 081 |

**Table 7.2** Concentrations of a selection of VOCs in indoor and outdoor air in Avon, England (Crump, 1995)

| *VOC* | *No of Readings* | *Indoor concentrations/ppb* | | | *No of Readings* | *Outdoor concentrations/ppb* | | |
|---|---|---|---|---|---|---|---|---|
| | | *Mean* | *Min* | *Max* | | *Mean* | *Min* | *Max* |
| Formaldehyde | 1501 | 21 | 1 | 174 | 126 | 2 | nd | 9 |
| Benzene | 1504 | 7 | nd | 16 | 125 | 4 | nd | 14 |
| Toluene | 1504 | 36 | 2 | 1523 | 125 | 10 | 2 | 215 |
| Undecane | 1504 | 12 | nd | 677 | 125 | 1 | nd | 3 |
| Total VOC | 1505 | 340 | 21 | 7130 | 125 | 30 | 4 | 269 |

nd: less than 1 ppb.
Reproduced by permission of the Royal Society of Chemistry.

### 7.2.2 The NOx problem

Oxides of nitrogen (NO and $NO_2$, collectively referred to as NOx) are recognized as major contributors to photochemical smog, acid rain and ozone formation in urban areas, and to ozone depletion in the stratosphere. NOx in air is one of six criteria pollutants monitored by the US Environmental Protection Agency. The others are CO, $O_3$, $SO_2$, Pb and PM10 (particulate matter 10 μm). The results of monitoring in terms of average human exposure for four of the six criteria pollutants, namely CO, $O_3$, NOx and $SO_2$ are presented in Figure 7.1. The data show that, although the levels of exposure to the criteria pollutants have declined in general, NOx levels have remained

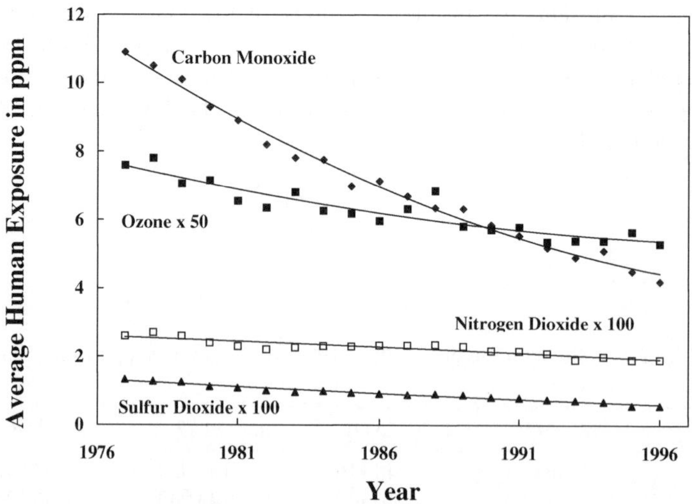


**Figure 7.1** Average human exposure in ppm to the indicated pollutants in the USA between 1975 and 1993 (National air quality and emissions trends report, 1996).

static at about 20 ppb since monitoring began in the mid-1970s (National air quality and emissions trends report, 1996).

Minimizing NOx production during combustion processes is a major current research topic and one in which catalytic combustion can play a role. In the absence of strategies to control this problem, NO formation from a clean fuel (one which does not contain fuel-bound nitrogen, such as pyridine) increases with combustion temperature as shown in Figure 7.2 (Garten *et al.*, 1997). Clearly any process which can maintain combustion temperatures below 1800 K would help to control NOx emissions, and catalytic combustion can operate in this temperature range.

There are three important routes to NOx formation during combustion. These are normally called (i) thermal, (ii) prompt and (iii) fuel. The first two arise from the oxidation of atmospheric nitrogen and the third involves the combustion of nitrogen-bearing hydrocarbons, often called fuel-bound nitrogen, in the fuels (Warnatz *et al.*, 1996).

*Thermal NO*

If the flame temperature is sufficiently high (usually above 1800 K) $N_2$ decomposes and reacts directly with gas phase oxygen according to equations (7.1)–(7.3):

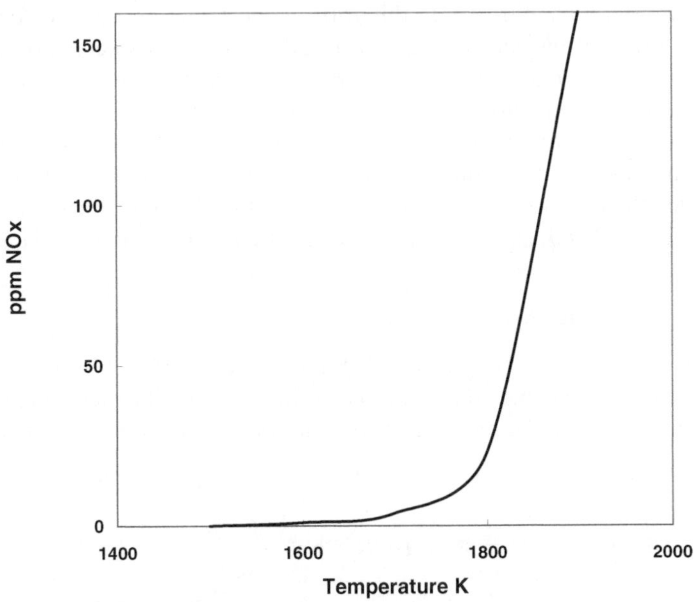


**Figure 7.2** NOx formation as a function of flame temperature for the combustion of methane (Garten *et al.*, 1997) (reproduced by permission of Wiley–VCH).

$$O + N_2 \longrightarrow NO + N \tag{7.1}$$

$$N + O_2 \longrightarrow NO + O \tag{7.2}$$

$$N + OH \longrightarrow NO + H \tag{7.3}$$

High temperatures are required for this reaction because of the stability of the N≡N bond, making step one above the slow step in the overall reaction.

*Prompt NOx*

Prompt NOx forms in regions of flames where there is a high concentration of the $CH^{\bullet}$ radicals. This radical is formed from hydrocarbon fuels only at the flame front and in fuel rich combustion conditions where significant amounts of $C_2H_2$, the precursor to the $CH^{\bullet}$ radical, can form. The principal reactions which occur are shown in Scheme 7.1.

Figure 7.3 shows the influence of oxidant to fuel ratio ($\lambda$) on the amount of thermal and prompt NOx generated; prompt NOx is generated principally in fuel rich flames ($\lambda$ less than 1) and thermal NOx in fuel lean flames ($\lambda$ greater than 1).

*Fuel NO*

Fuel NO is generated from the combustion of fuel-bound nitrogen. Coal contains up to 1 wt% nitrogen in this form while crude oil derived fuels, in spite of hydrodenitrogenation treatment in oil refineries, also contain significant amounts of nitrogen heterocycles. In the combustion process fuel-bound nitrogen is converted into HCN, and then into a number of intermediates which lead to the formation of $NH_3$, NO and $N_2$. A reaction sequence is outlined in Scheme 7.2. Again the oxidant to fuel ratio, $\lambda$, is a major factor in determining the relative amounts of $NH_3$, NO and HCN which form, as shown in Figure 7.4. NO is the major product when sufficient oxygen is available.

*Strategies to minimize NOx emissions*

There are a number of strategies which can be used to reduce the amounts of NOx emitted from combustion operations. Primary reduction methods include staged combustion; in the first stage the operation is run at $\lambda = 0.7$, namely in

**Scheme 7.1** Principal reaction involved in prompt NO formation.

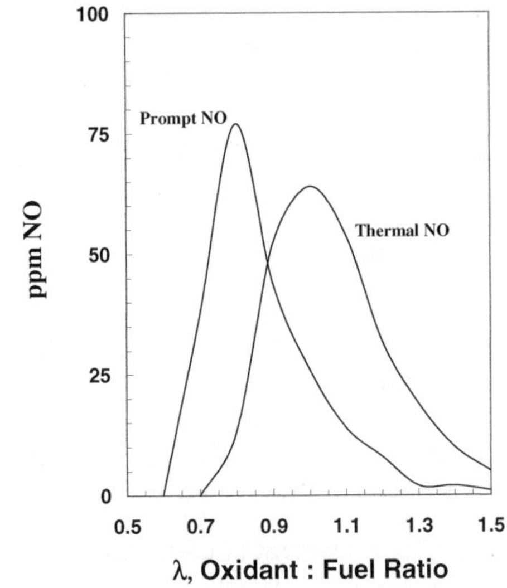


**Figure 7.3** Production of prompt and thermal NOx as a function of the oxidant to fuel ratio. $\lambda = 1.0$ corresponds to the stoichiometric requirement for a given fuel. $\lambda$ is less than unity for a fuel rich mixture and greater than unity for a fuel lean mixture (Warnatz *et al.*, 1996) (reproduced by permission of Springer).

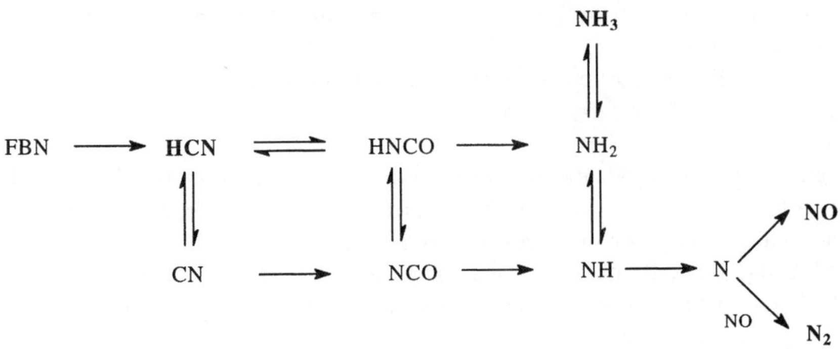


**Scheme 7.2** Reaction sequence for NO production from fuel-bound nitrogen (Warnatz *et al.*, 1996) (reproduced by permission of Springer).

fuel rich conditions to minimize fuel and prompt NOx (see Figures 7.3 and 7.4); the second stage is operated in fuel lean conditions $\lambda > 1.4$, minimizing NOx formation through the thermal mechanism. Injection of inert gases, particularly recycled nitrogen or water vapour serves to reduce the combustion temperature and further to minimize NO formation through the thermal mechanism.

Catalytic combustion can also be used to fight this problem. There are three immediate advantages associated with its use. These are:

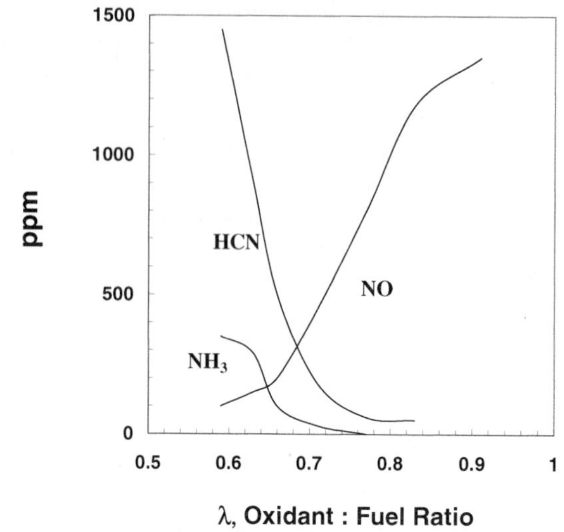


**Figure 7.4** Formation of $NH_3$, NO and HCN at different oxidant: fuel ratios in a propane air flame doped with 2400 ppm $CH_3NH_2$ added to simulate fuel-bound nitrogen (Warnatz *et al.*, 1996) (reproduced by permission of Springer).

- Operation at low temperatures, eliminating NO production through the thermal mechanism;
- Operation in fuel lean conditions, minimizing NO formation through the prompt mechanism;
- A capacity to convert selectively a significant amount of fuel-bound nitrogen into $N_2$ rather than NO.

Figure 7.5 compares catalytic combustion with conventional and advanced low-emission burners. The plot shows how CO and NOx emissions are related and that in many circumstances improving the efficiency of combustion (less CO being emitted) occurs at the expense of higher NOx emissions. The challenge in the future is to reduce both and catalytic technologies will play an important role here.

### 7.2.3 Operation of three way catalysts

One of the first applications of combustion catalysts for the abatement of environmental problems was the so called three way catalysts. These were first introduced in 1975 and are now in widespread use throughout the developed world for the treatment of exhaust gases from petrol or gasoline fueled internal combustion engines. The principal pollutants treated by three way catalysts are unburned hydrocarbons (HC), carbon monoxide (CO) and nitrogen oxides (NO). The amounts of each of the pollutants mentioned above are known to depend on the air to fuel ratio fed to the engine.

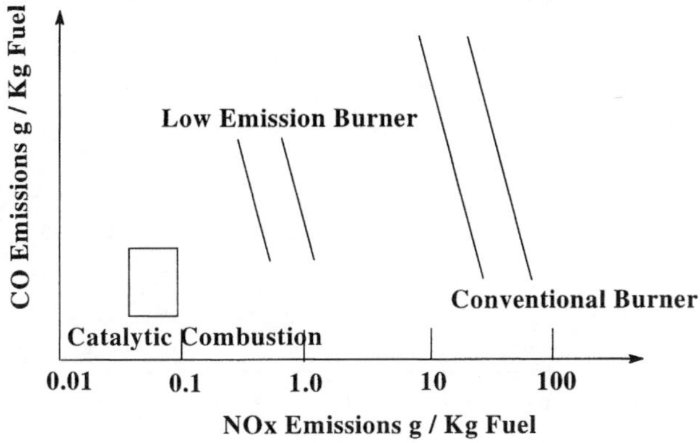


**Figure 7.5** Relationship between CO and NO emissions for conventional and advanced burners and for catalytic combustion.

Figure 7.6(b) shows the exhaust gas composition of a petrol fueled test engine capable of operating at varying air to hydrocarbon ratios. Carbon monoxide is the biggest pollutant in terms of concentration for all except very high values of $\lambda$. For high values of $\lambda$, hydrocarbon emissions persist, in spite of increasing availability of air, because the combustion temperature drops resulting in incomplete combustion.

Modern three way catalysts efficiently remove NO, CO and HC from exhaust gases only in a very narrow window of $\lambda$ values. This effect is shown in Figure 7.6(a). Close to the stoichiometric value for $\lambda$, the catalyst operates efficiently. However, in fuel rich mixtures ($\lambda < 1$), conversion of CO and HC is poor, whereas in fuel lean mixtures ($\lambda > 1$), conversion of NO is poor. This feature requires a very careful control of the fuel to air ratio in automobiles operating with this technology. Some modern three way catalysts have a cerium oxide component which can accommodate a small change in oxygen content of the exhaust gases. This is achieved by alternately acting as an oxygen sponge, when the exhaust gases become oxygen rich, and as a source of oxygen when the exhaust is depleted of oxygen.

A modern development is the so called lean burn petrol engine ($\lambda > 1.3$). This seeks to operate in conditions where very little NO or CO is produced and HC emissions can be reduced by appropriate engine design.

The main reactions occurring on a modern three way catalyst are:

$$CO + 0.5O_2 \longrightarrow CO_2 \tag{7.4}$$

$$CO + NO \longrightarrow 0.5N_2 + CO_2 \tag{7.5}$$

$$C_xH_y + nO_2 \longrightarrow xCO_2 + (y/2)H_2O \tag{7.6}$$

$$C_xH_y + mNO \longrightarrow (m/2)N_2 + (y/2)H_2O + xCO_2 \tag{7.7}$$

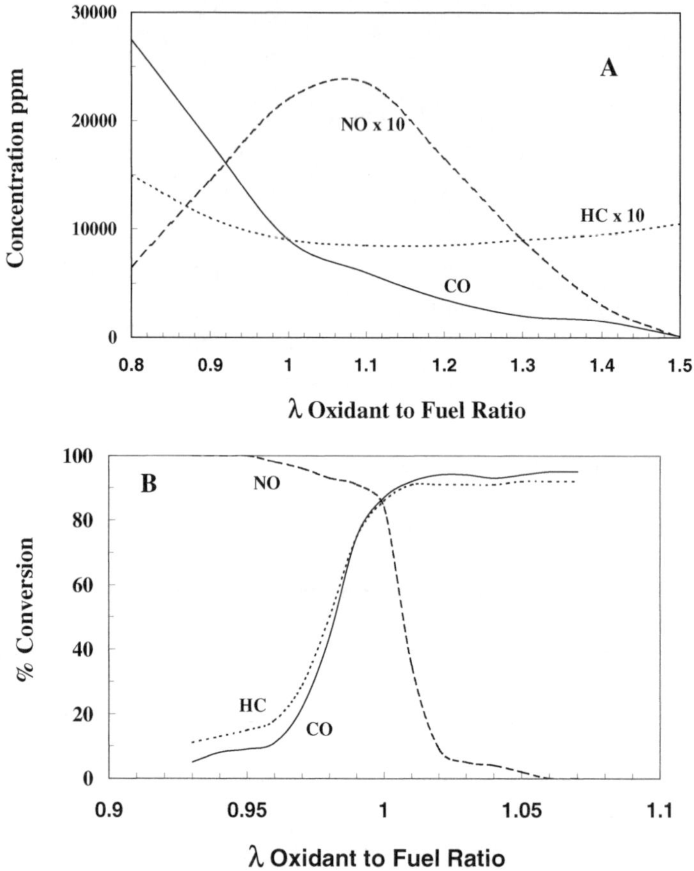


**Figure 7.6** (A) Efficiency of NO, CO and HC removal by a modern three way catalysts as a function of the oxidant to hydrocarbon ratio ($\lambda$). (B) NO, CO and HC emissions as a function of the oxidant to hydrocarbon ratio ($\lambda$) for a spark ignition engine.

All these reactions can be classified as redox with NO replacing oxygen at times as the oxidant. Normally, the catalysts do not operate efficiently during engine warm-up, but during steady state operation the heat input from the hot exhaust gases and the heat of reaction is more than sufficient to maintain efficient operation.

There are two primary requirements for three way catalysts. The first is good oxidizing efficiency in the temperature range 600– 1200 K. Not surprisingly then the principal active phase in three way catalysts is platinum, with smaller amounts of rhodium added ($Pt:Rh = 5–20:1$). In the absence of rhodium, NO is not efficiently converted.

The second requirement is the ability to handle large volumes of exhaust gases without generating a pressure drop across the catalyst bed that would interfere with engine operation. In the earliest three way catalysts the active components were deposited onto alumina beads, but modern devices make use of monolith technologies. These are ceramic devices with a honeycomb structure manufactured from cordierite ($2MgO{\cdot}2Al_2O_3{\cdot}5SiO_2$), coated with a thermally stable alumina support (additives used include La, Ba and Ce) and impregnated with a mixture of platinum and rhodium. The platinum and rhodium contents together amount to 1–1.5 g per litre of reactor volume.

Modern developments include a move to replace platinum with less expensive palladium. This requires a higher metal loading and higher operating temperatures. The latter is achieved by moving the entire catalysts assembly closer to the engine outlet. Attempts at tackling the inactive period during engine startup (cold start) include the use of adsorbents in a bed placed upstream of the catalyst and onto which pollutants would be adsorbed when the system is cold and desorbed after the catalyst comes to its operating temperature (Heck and Farrauto, 1995).

## 7.3 PRINCIPLES OF CATALYTIC COMBUSTION

When a mixture of combustible and oxidant is passed over a catalyst the reaction commences and initially follows the normal Arrhenius type behaviour as the reaction temperature is increased, as shown in Figure 7.7. The temperature at which this process commences depends on the chemical structures of the combustible and the catalyst. For example, combustion of alcohols begins at low temperatures on precious metal catalysts, whereas alkanes require significantly higher temperatures on the same type of catalyst and higher temperatures again on metal oxides.

For a limited temperature range over combustion catalysts, the normal activation energy can be determined. However, the majority of combustion reactions are very exothermic and temperature control within the reactor can be a problem as the percentage conversion increases, even with very lean feeds. Eventually a process called light-off occurs whereby very rapid increases in reaction rates can be achieved, apparently, within a very narrow temperature range. This reaction zone is shown in Figure 7.7.

Beyond this range, diffusion limitations can become apparent because the rate at which the reactants are being consumed is faster than the rate at which they can be delivered to the catalyst surface. At higher temperatures again homogeneous processes become important and the reaction rate can increase once again (Zwinkels *et al.*, 1993) as shown in Figure 7.7.

Temperature measurement and control are difficult problems for all catalytic combustion processes. Table 7.3 lists a number of total oxidation reactions, their associated standard enthalpies of reaction and the calculated adiabatic temperature rises for the complete conversion of 1000 ppm of the organic substrate (Heck and Farrauto, 1995). The data illustrate the substantial adiabatic temperature rise that can be expected during combustion even of

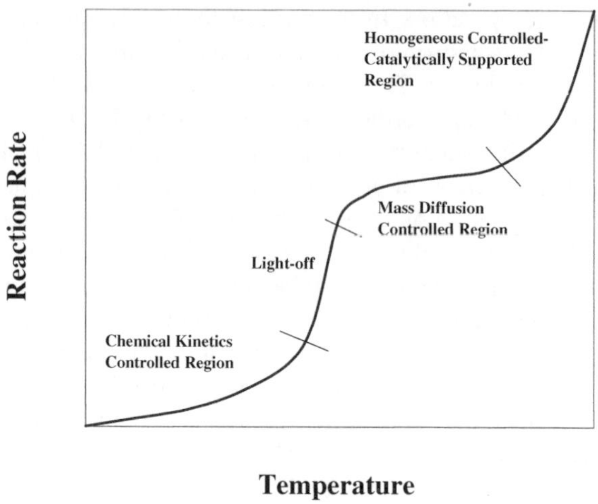


**Figure 7.7** Variation of reaction rate with temperature for a typical combustion process showing the four different reaction zones.

very lean feeds. More generally, adiabatic temperature rises tend to be in the range 15–20 K for each 1000 ppm of $C_1$ fed to the reactor.

A feature of combustion operations is the need to treat very large gas volumes. To minimize plant size this requires very active catalysts that must be capable of stable operation at high temperatures and in the presence of typical catalyst poisons such as sulfur-bearing compounds and trace levels of metals, such as lead, selenium and copper, introduced as VOCs or as dust. A further feature is the need to design a geometric shape that will optimize gas–catalyst contact but minimize pressure drop through the reactor.

A recent innovation has been the monolith or honeycomb catalyst support. These materials have replaced packed bed arrangements in three way auto-exhaust catalysts. The honeycomb shapes constructed of cordierite ($2MgO \cdot 2Al_2O_3 \cdot 5SiO_2$) or metal are manufactured to give a cluster of regularly shaped square, rectangular or corrigated channels that extend through the entire

**Table 7.3** Standard enthalpies of oxidation and adiabatic temperature rise for the combustion reactions indicated (Heck and Farrauto, 1995)

| *Reaction* | *Standard enthalpy of reaction kJ (mol VOC)$^{-1}$* | *Adiabatic temperature rise (K) for 1000 ppm of substrate* |
|---|---|---|
| $HCHO + O_2 \longrightarrow CO_2 + H_2O$ | −519 | 17 |
| $CH_3COCH_3 + 4O_2 \longrightarrow 3CO_2 + 3H_2O$ | −1687 | 57 |
| $CH_3CHOHCH_3 + 4.5O_2 \longrightarrow 3CO_2 + 4H_2O$ | −1873 | 64 |
| $C_6H_5CH_3 + 9O_2 \longrightarrow 7CO_2 + 4H_2O$ | −3768 | 123 |

length of a fixed bed catalytic reactor. One such structure is shown in Scheme 7.3. The channel size is between 1 and 10 mm depending on the application envisaged. The honeycomb defines the basic shape of the catalyst bed and offers near zero resistance to gas flow. An aluminium oxide layer is attached to the insides of the channels in the form of a washcoat and other components, such as precious metals, may then be added by ion exchange or impregnation.

## 7.4 CHOICE OF MATERIALS FOR THE CONSTRUCTION OF COMBUSTION CATALYSTS

### 7.4.1 Chemical reactivity of the active phase

The first factor to be considered in designing a combustion catalyst is the nature of the materials that are to be used. The active phase refers to the material on which the active site is located and on which the principal or kinetically slow step in the overall chemical transformation of the substrate occurs. Typically, Pt and Pd are the active components for the vast majority of combustion applications, although a great deal of work is ongoing in the development of catalysts based on cheaper active components. The high activity of the precious metals is related to their ability to activate $H_2$, $O_2$, C—H and O—H bonds. Figure 7.8 presents a rationale for the wide use of Pt and Pd as active phases, namely that they far outperform other materials as catalysts for combustion reactions (Morooka and Ozaki, 1966). In addition, these metals are generally more resistant to deactivation and can be regenerated when necessary using relatively simple methods (see Section 7.5 below).

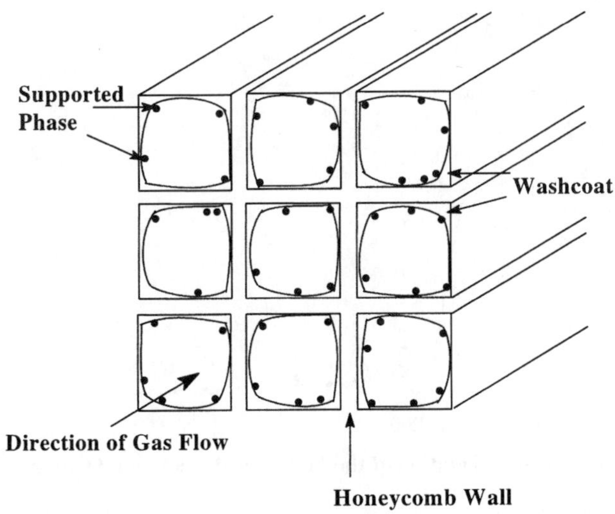


**Scheme 7.3** Cross-section of a ceramic honeycomb structure.

The activity data in Figure 7.8 have been plotted against the enthalpy of formation of the metal oxide. It is questionable whether there is in fact a significant correlation between combustion activity and enthalpy of formation, but the important factor to bear in mind is that, irrespective of whether oxidation rates are expressed per unit of catalyst mass ($mol\,g^{-1}\,s^{-1}$) or, as in Figure 7.8, per unit of surface area ($mol\,m^{-2}\,s^{-1}$), a very similar reactivity pattern emerges (Morooka and Ozaki, 1966).

### 7.4.2 Stability of the active phase

Highly dispersed materials are inherently unstable and will tend towards a state of minimum surface area. The basic thermodynamic driving force towards minimization of surface area is often written as a factor in the New Fundamental equation:

$$dG = V\,dP - S\,dT + \gamma\,dA \tag{7.8}$$

where $V$ is volume, $P$ pressure, $T$ is temperature, $S$ is entropy, $\gamma$ is the surface tension and $A$ is surface area. Creation of a new surface always requires an increase in Gibbs free energy ($G$) so that the direction of spontaneous change is towards lower surface areas. Finely dispersed metals on supports operate in metastable, or kinetically stabilized, regimes which prevent or delay loss of surface area during high temperature operation.

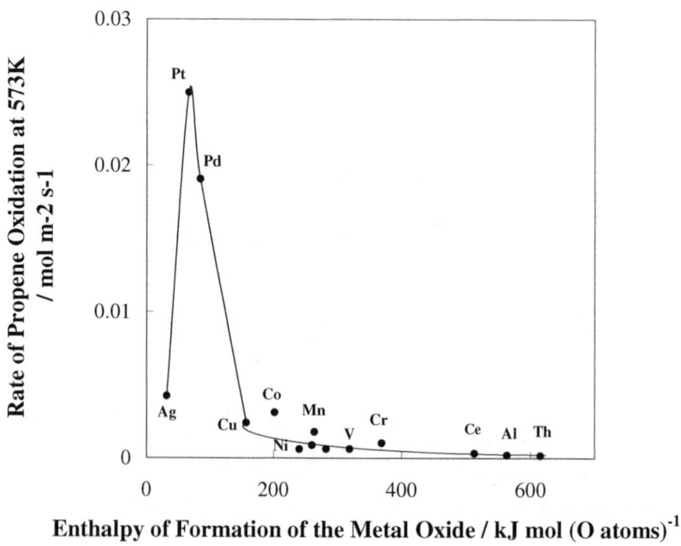


**Figure 7.8** Areal rates of propane oxidation over the metals or metal oxides indicated (Morooka and Ozaki, 1966) (reproduced by permission of Academic Press).

Sintering is a general term to describe a loss in surface area and starts to be important when the operating temperature is about half the melting point (Knozinger and Taglauer, 1997). This is referred to as the Tammann temperature and some values for metals and oxides are given in Table 7.4. At and above the Tammann temperature two phenomena that contribute to sintering start to occur:

- Migration of small crystallites over a support followed by their coalescence, and
- Detachment of atoms from small crystallites and their capture by larger ones.

Sintering can be the result of a combination of these two phenomena. Two dimensional melting is an important part of the latter process, whereby the surface layer of the metal takes on a liquid like behaviour.

Figures 7.9 show an example of one supported phase that does and one that does not sinter at high temperatures. This figure presents a series of TEM observations for individual particles of $PdO/Al_2O_3$ and $Pt–PdO/Al_2O_3$ at a series of sintering temperatures. Heating to 973 K was carried out inside the electron microscope and the appearance of a single catalyst particle was monitored. The $PdO/Al_2O_3$ catalyst sintered in the temperature range 843–863 K. Particle growth was by the migration and coalescence of individual crystallites of PdO. By contrast, this phenomenon was not observed with the $Pt–PdO/Al_2O_3$ catalysts, which appeared to be stable to sintering in the temperature range studied (Narui *et al.*, 1999).

In practice sintering, redispersion and spreading are dynamic processes that tend to drive a system towards its thermodynamic equilibrium. The extent to which sintering will occur in practice depends on a number of factors such as the nature of the metal, the temperature and time of treatment, gas phase

**Table 7.4** Melting points and Tammann temperatures for selected metal and metal oxides (Knozinger and Taglauer, 1997)

| *Compound* | $T_{Melt}/K$ | $T_{Tam}/K$ |
|---|---|---|
| Ag | 1235 | 618 |
| Cu | 1356 | 678 |
| Fe | 1808 | 904 |
| Ni | 1726 | 863 |
| Pd | 1827 | 914 |
| Pt | 2045 | 1023 |
| $MoO_3$ | 1068 | 534 |
| $V_2O_5$ | 963 | 482 |
| $WO_3$ | 1746 | 873 |

Reproduced by permission of Wiley–VCH.

composition and the strength of the metal–support interaction. A typical example is shown in Figure 7.10. Here sintering of metallic nickel on silica and alumina supports in hydrogen at 1023 K is presented (Bartholomew, 1994). The Tammann temperature for metallic nickel is 863 K. The sintering is represented as the ratio of the metal surface area for the sintered catalyst, measured by hydrogen chemisorption, divided by the metal surface area of the corresponding unsintered material. Nickel sinters rapidly on the silica support, but is reasonably well stabilized by the alumina support. These data indicate a stabilizing interaction between the nickel and the alumina, but not with silica.

The exact chemical environment of the metal or supported phase particles is also important in determining the state of dispersion. Redispersion is the phenomenon whereby sintered catalysts can regain some of their lost surface area. This is possible for platinum particles where there is some chloride present in the immediate environment. For example, oxidation treatments are known to redisperse platinum catalysts following sintering. Lieske *et al.* (1983) point to $Pt(OH)_xCl_y$ and $PtO_xCl_y$ as critical intermediates in the redispersion of platinum catalysts. A chloride free catalyst was prepared by these workers and found to have a $Pt_{surf} : Pt_{total}$ ratio, measured by hydrogen chemisorption, of less than 0.2. This sample was subjected to a number of thermal treatments shown in Figure 7.11, none of which changed the dispersion. Treatment in HCl followed by oxidation at 873 K increased the dispersion. The authors point to a role for the oxychloride and hydroxychloride of Pt in the redispersion process (Lieske *et al.*, 1983).

There is also a relationship between the oxidation state of supported platinum and dispersion (Lieske *et al.*, 1983). Figure 7.12 shows the oxidation state of platinum and the dispersion as a function of oxidation temperature, for a catalyst prepared from hexachloroplatinic acid (A second supported platinum catalyst prepared from a chloride free salt showed almost identical behaviour.) For these catalysts the oxidation state of Pt approached 4+ following oxidation at 873 K. Up to this temperature Pt was well dispersed, but this decreased rapidly, as the Pt metal sintered above 873 K. The oxidation state of the platinum mimicked the dispersion and fell to zero as the dispersion decreased. It appears that the 4+ oxidation state for platinum is possible only when the metal dispersion is very high. The predominant oxidation state of larger platinum particles is zero.

### 7.4.3 Stability of support materials

Alumina is a standard catalyst support often intended to present a high surface area over which a precious metal or another oxide may be dispersed. To achieve a high surface area the gamma, delta, kappa or theta form is formed by heating Gibbsite, Boehmite or Bayerite, all forms of aluminium hydroxide $Al(OH)_3$ or oxyhydroxide $AlO(OH)$. The aluminas which form are all referred to as transition aluminas, because they are converted into the thermodynamically stable $\alpha$-$Al_2O_3$ form during treatment at sufficiently high temperatures. The temperatures required for the various transitions tend to be sensitive to preparation method, but only $\alpha$-$Al_2O_3$ is thermodynamically stable above 1500 K, as shown in Figure 7.13 (Oberlander, 1984).

For the most part standard catalytic supports are not sufficiently stable for combustion applications and large changes can occur in the support and the supported phase when used in a combustion environment. For example, Figure 7.14 presents the evolution of the surface area of a sample of $\gamma$-$Al_2O_3$ above 773 K; a pronounced decrease in surface area is observed due to support sintering (Garten *et al.*, 1997).

*Mechanism of sintering of aluminas*

A general feature of all known forms of alumina is that they contain aluminium ions in regular trivalent sites and oxygen anions in regular lattice positions. In addition, there are cation vacancies which could accommodate a divalent cation and anion vacancies some of which contain hydroxyl groups. It appears that all the transition aluminas contain greater or lesser amounts of hydroxyls substituting for oxygen in regular anion sites. All aluminas and their oxyhydroxide and hydroxide precursors may be represented by the formula:

$$(Al_{Al})_2V''_MO_{O(3\text{-}\nu/2)}(OH_O)_\nu V'_{O(1\text{-}\nu/2)} \tag{7.9}$$

where $Al_{Al}$ represents an aluminium ion in a trivalent site, $O_O$ represents an oxygen anion in a regular lattice position, $V''_M$ is a cation vacancy corresponding to a divalent site, $OH_O$ is a hydroxyl group in an anion vacancy and $V'_O$ is an anion vacancy (Aria and Machida, 1996). With $\nu = 2$ the general formula reduces to AlO(OH) (Boehmite) and when $\nu = 0$ the formula becomes $Al_2O_3$. Accordingly, the phase transformation of transition aluminas may be represented by a dehydroxylation, as follows:

$$2(OH_O) \longrightarrow H_2O + O_O + V'_O \tag{7.10}$$

Dehydroxylation is accompanied by the formation of oxygen anion vacancies and the final transformation to alpha-alumina is accompanied by the annihilation of cation and anion vacancies:

$$V''_M + V'_O \longrightarrow \text{null} \tag{7.11}$$

This mechanism implies that cation and anion vacancies are the medium through which movement of material is possible during the solid state transformation of the transition aluminas into the alpha form.

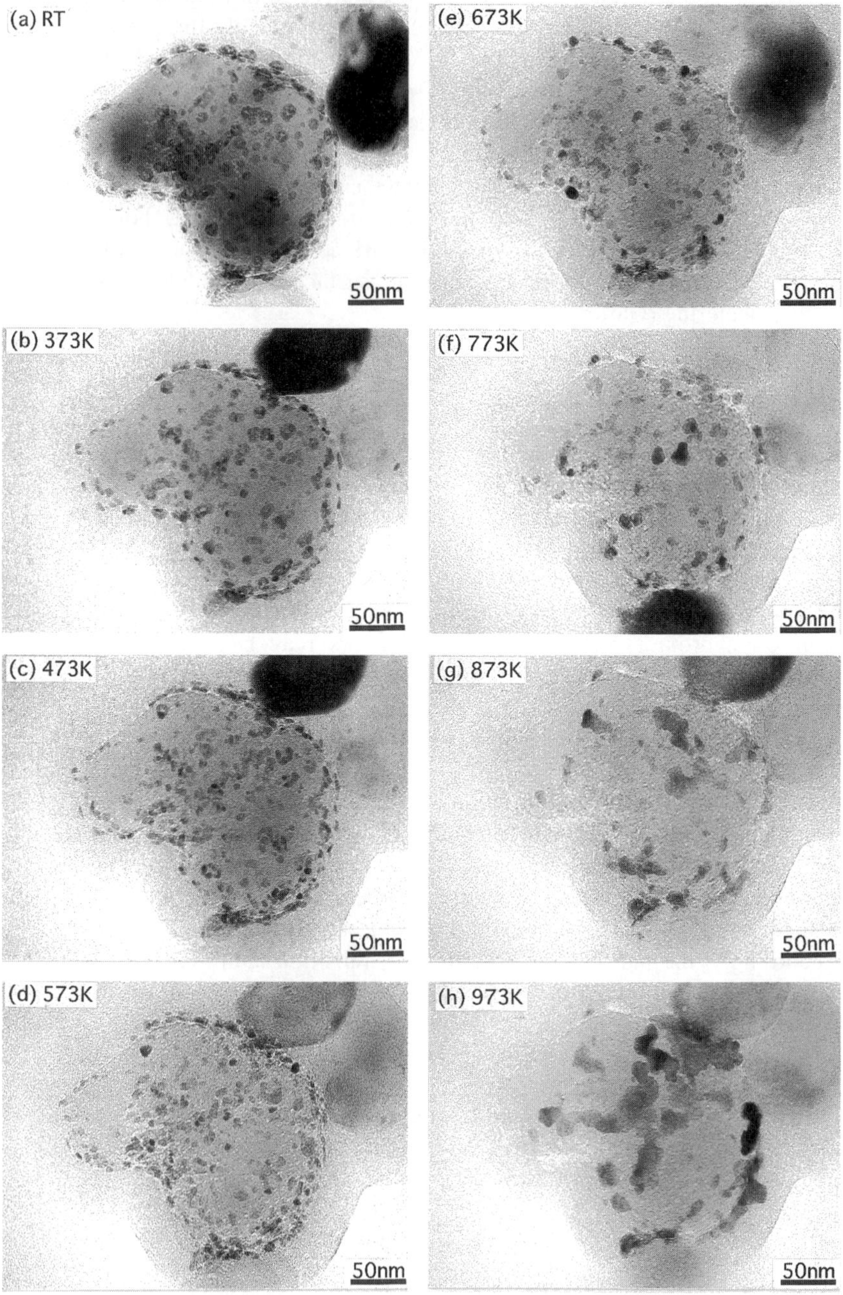
(a) RT
50nm
(e) 673K
50nm
(b) 373K
50nm
(f) 773K
50nm
(c) 473K
50nm
(g) 873K
50nm
(d) 573K
50nm
(h) 973K
50nm

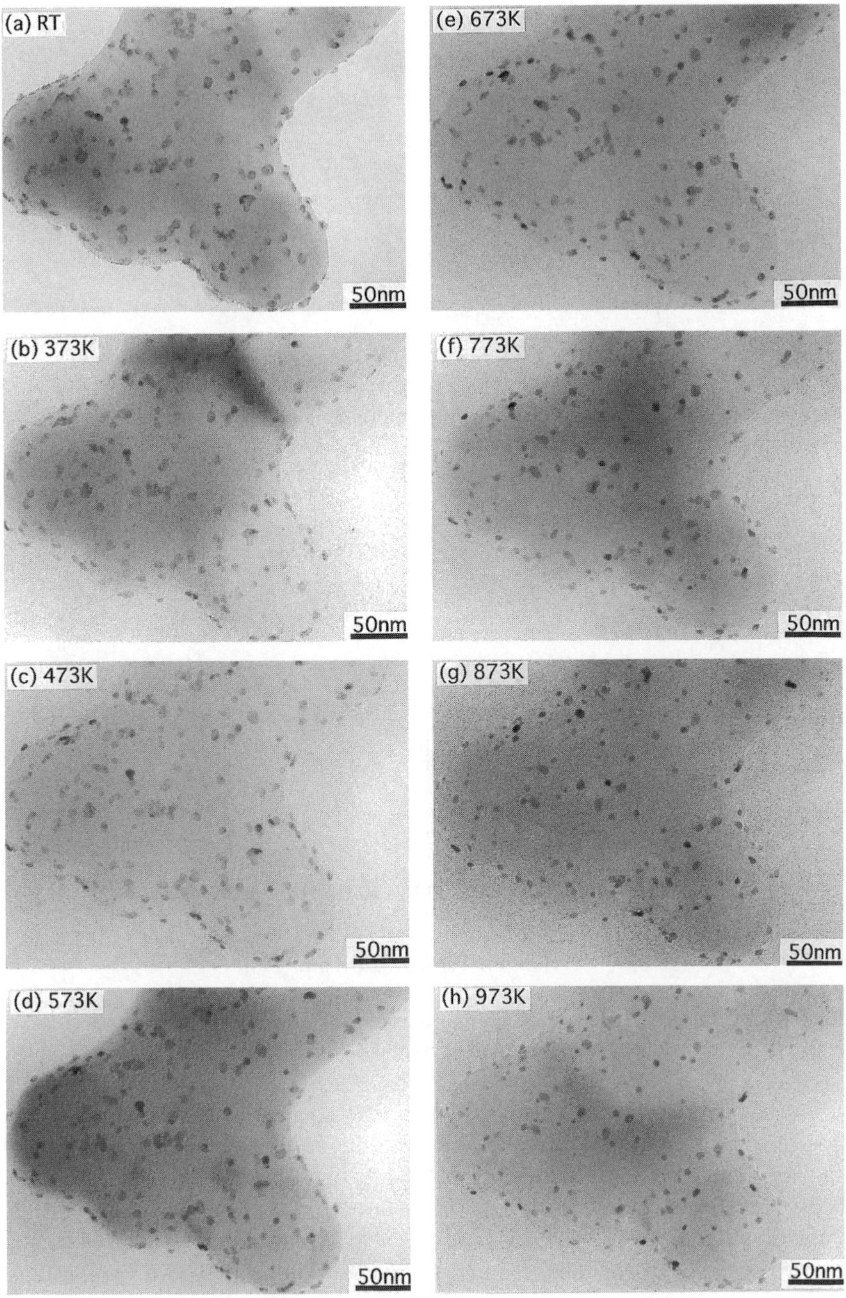


**Figure 7.9** TEM photographs of (left) $PdO/\alpha\text{-}Al_2O_3$ and (above) $Pt\text{-}PdO/\alpha\text{-}Al_2O_3$ during heating to 973 K (Narui *et al.*, 1999) (reproduced by permission of Elsevier Science).

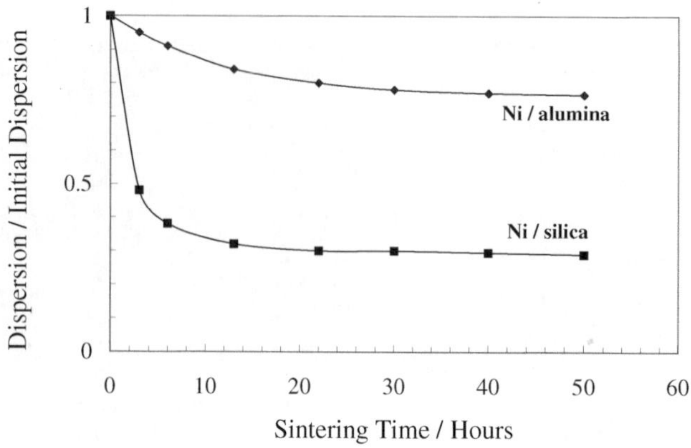


**Figure 7.10** Effect of the support on the sintering of nickel in hydrogen at 1023 K (Bartholomew, 1994) (reproduced by permission of Elsevier Science).

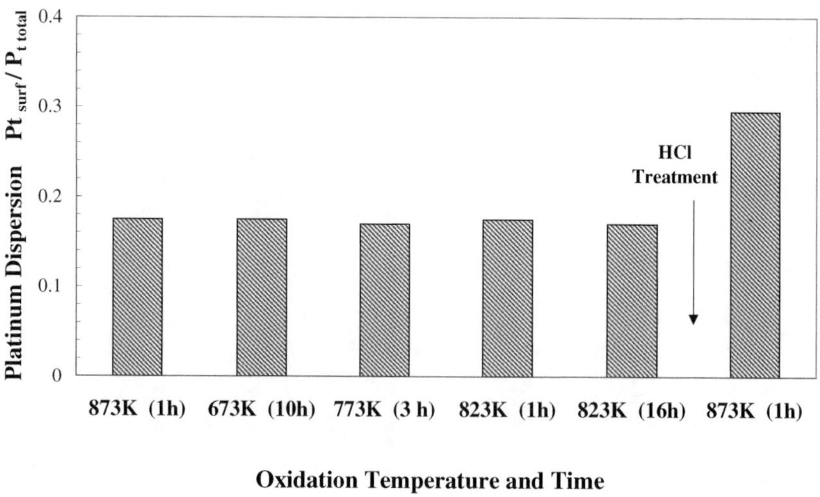


**Figure 7.11** Influence of chloride on the redispersion of a $Pt/Al_2O_3$ catalysts (Lieske *et al.*, 1983) (reproduced by permission of Academic Press).

Surface hydroxyl groups also play a role in particle growth, wherein elimination of water from two hydroxyl groups residing on adjacent particles occurs as shown in Scheme 7.4. This results in the creation of a new Al—O—Al bond between adjacent particles. Phase transformation and surface area loss in transition aluminas are known to be strongly affected by the presence of steam. The presence of steam enhances the surface concentration of hydroxyl groups which favours the process of neck formation shown in Scheme 7.4.

In general, the transition aluminas present surface areas between 150 and 300 $m^2\ g^{-1}$ and contain large numbers of meso- and micropores. When this

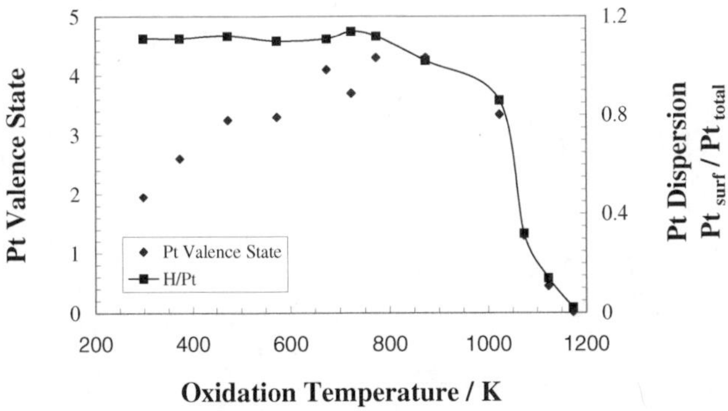


**Figure 7.12** Platinum dispersion and oxidation state as a function of oxidation temperature for 0.5 % Pt/$\gamma$-$Al_2O_3$ prepared from hexachloroplatinic acid (Lieske *et al.*, 1983) (reproduced by permission of Academic Press).

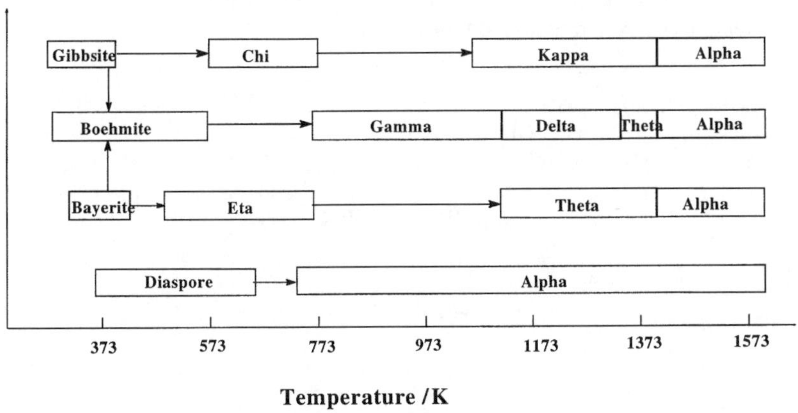


**Figure 7.13** Thermal stability of aluminium hydroxides and oxides (Oberlander, 1984) (reproduced by permission of Academic Press).

material is used as a catalyst support much of the supported phase (a precious metal for example) may be deposited within these pores. Part of the sintering process involves closing of micro- and meso-pores and any supported phase that is present in these pores tends to become trapped inside as shown in Scheme 7.5. The final product of sintering of transition aluminas, namely $\alpha$-$Al_2O_3$, is generally free of micro- and mesopores and presents a very low surface area ($< 1\ m^2\ g^{-1}$).

*Strategies for the stabilization of aluminas*

In general, two approaches have been taken to the question of stabilization of the textural properties of aluminas for applications at high temperatures. The

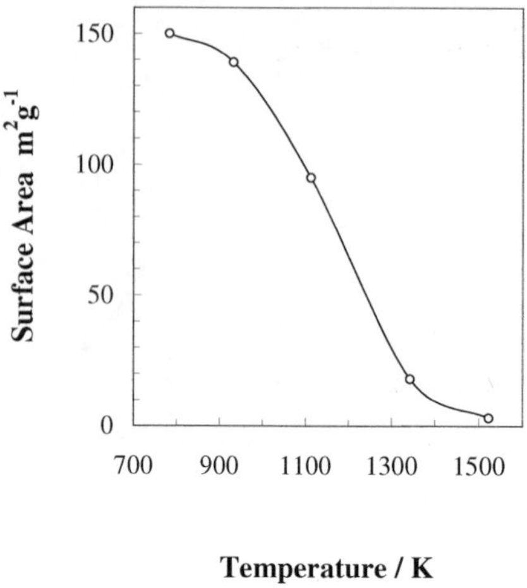


**Figure 7.14** Evolution of the surface area of $\gamma$-$Al_2O_3$ above 773 K (Garten *et al.*, 1997) (reproduced by permission of Wiley–VCH).

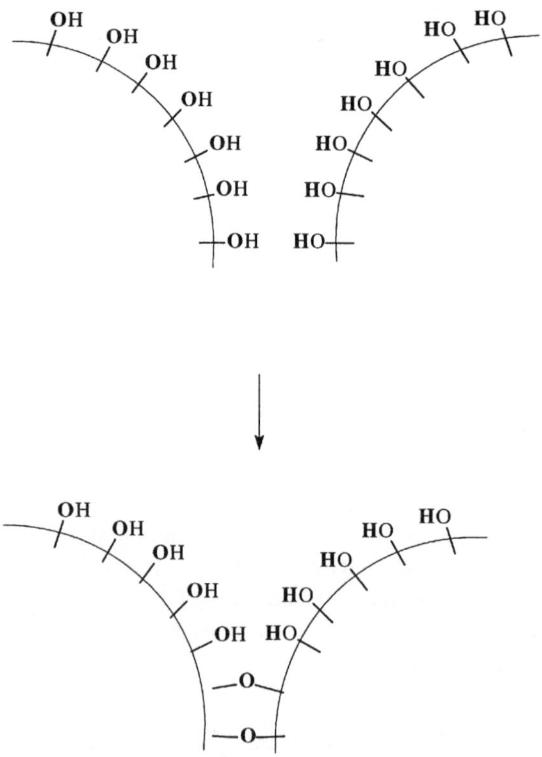


**Scheme 7.4** Surface dehydroxylation from a contact region between two adjacent particles (Aria and Machida, 1996) (reproduced by permission of Elsevier Science).

first is to modify existing support materials and the second is to seek out new supports.

Since sintering of transition aluminas comes about through dehydroxylation and the removal of defects, additives can be designed to occupy sites normally occupied by surface hydroxyl groups and otherwise unoccupied bulk vacancies. This process slows down bulk diffusion because material transport through the network of cationic and anionic defects is no longer possible and lessens the probability of neck formation because there are fewer surface hydroxyl groups present.

Lanthanum is the most widely investigated additive for the stabilization of aluminas. In a typical study a support comprising 5 mol% $La_2O_3/Al_2O_3$ retained a surface area of 50 $m^2\ g^{-1}$ following calcination at 1400 K, whereas the surface area of the unmodified $Al_2O_3$ was just 10 $m^2\ g^{-1}$ after the same treatment. There are several theories to explain the action of lanthanum. These include the formation of a surface layer of $LaAlO_3$ as shown in Figure 7.15. $Al_2O_3$ and $LaAlO_3$ possess similar crystal structures and a coherent interface forms between the two owing to a similarity of oxygen packing in their structures (Aria and Machida, 1996).

Hexaaluminates, of which $\beta$-alumina is one example, are promising new supports for high temperature applications. They are composed of hexagonal layers separated by a mirror plane. The mirror plane is perpendicular to the *c*-axis of the crystal and can accommodate a large cation, thus creating a separation between the layers. This separation means that crystal growth and material transfer along the *c*-axis is very slow, stabilizing the structure. The crystal structure of a hexaaluminate is shown in Figure 7.16 and a TEM image of a catalyst comprising $Mn_3O_4$ supported on hexaaluminate is shown in Figure 7.17. The hexagonal layers can be clearly seen as can the surface layer of the supported phase, $Mn_3O_4$ (Machida *et al.*, 1995).

The extent of thermal stabilization of a range of hexaaluminates to which have been added 8 at% of additive is illustrated in Table 7.5. In this table the stabilization effect is presented as the ratio of surface areas for the additive-stabilized hexaaluminates to the surface area of the additive- free hexaaluminate prepared by the same method (Arai and Machida, 1996).

The effect of restricted material transport along certain directions in the hexaaluminate structure is illustrated by the data in Figure 7.18. A single

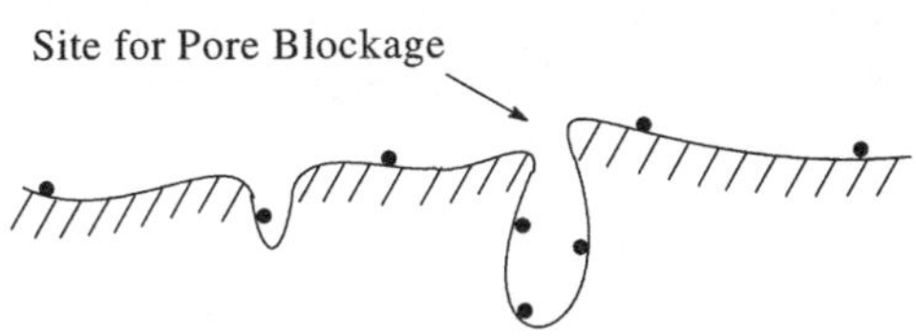


**Scheme 7.5** Schematic representation of metal encapsulation caused by support sintering.

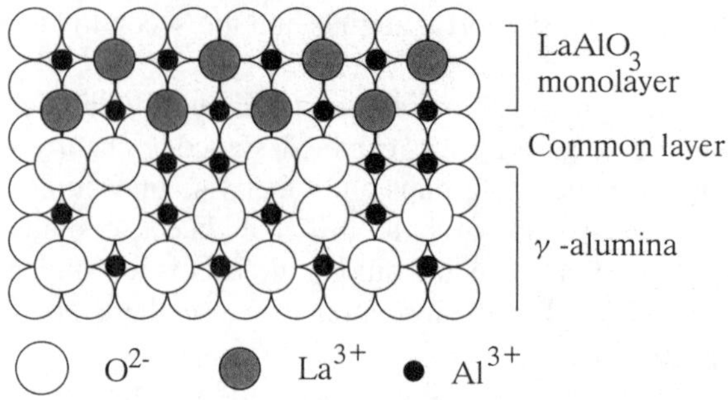


**Figure 7.15** A model of the $LaAlO_3/Al_2O_3$ coherent interface (Aria and Machida, 1996) (reproduced by permission of Elsevier Science).

crystal of Ba hexaaluminate was exposed to $^{18}O_2$ at 1673 K for 120 min and the depth profile of $^{18}O$ and $^{16}O$ was established by SIMS analysis for different directions along the hexaaluminate crystal. When probed along the *c*-axis the exchange between $^{18}O_2$ and $^{16}O_L$ (the original lattice oxygen of the structure) was found not to have proceeded further than 250 nm from the surface. By contrast $^{18}O$ had penetrated more than 1500 nm into the hexaaluminate structure in the direction normal to the *c*-axis. The results show that diffusivities of oxide ions normal to the *c*-axis were 10 times more than diffusivities along the *c*-axis. These data indicate that transport of oxygen anions within the hexagonal layers is possible, but with severely restricted transport across the layers (Machida *et al.*, 1995).

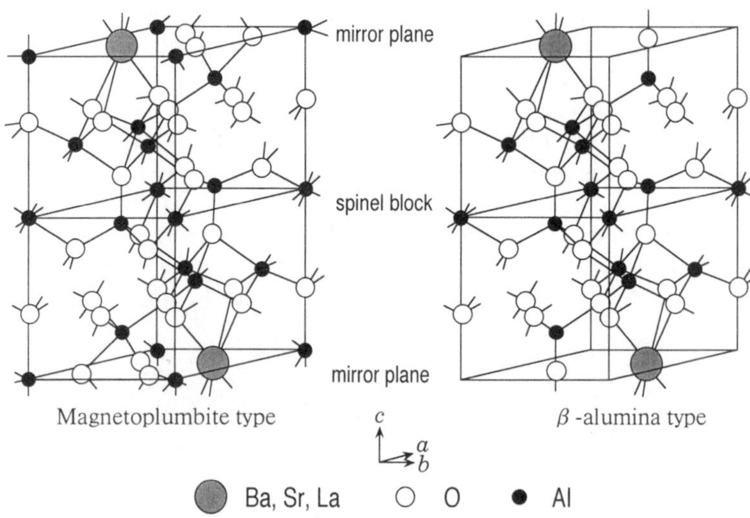


**Figure 7.16** Crystal structure of hexaaluminate (Aria and Machida, 1996) (reproduced by permission of Elsevier Science).

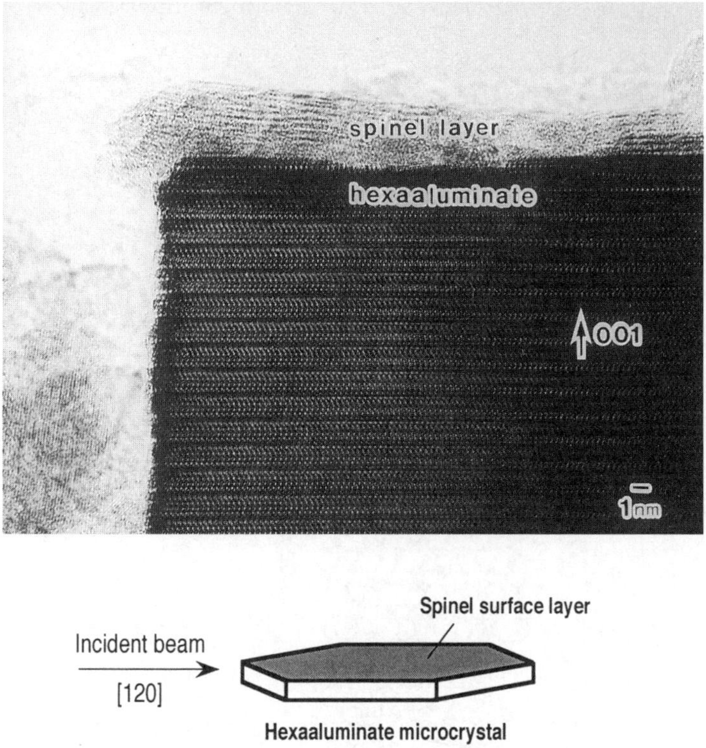


**Figure 7.17** TEM image of 10 wt% $Mn_3O_4$/hexaaluminate taken with the incident beam normal to the (120) plane of hexaaluminate (Machida *et al.*, 1995) (reproduced by permission of Elsevier Science).

**Table 7.5** Influence of additives on the thermal stabilization of the surface area of hexaaluminates (Arai and Machida, 1996)

| *Additive* | *Calcination temperature/K* | *Surface area stabilization*[a] | *Product phase* |
|---|---|---|---|
| $K_2O$ | 1723 | 5.0 | $KAl_{11}O_{17}$ |
| BaO | 1723 | 4.3 | $BaAl_{12}O_{19}$ |
| SrO | 1723 | 3.0 | $SrAl_{12}O_{19}$ |
| CaO | 1723 | 3.6 | $CaAl_{12}O_{19}$ |
| $La_2O_3$ | 1473 | 8.0 | $LaAl_{11}O_{18}$ |
| $Pr_2O_3$ | 1473 | 8.0 | $PrAl_{11}O_{18}$ |
| $Nd_2O_3$ | 1473 | 8.0 | $NdAl_{11}O_{18}$ |

[a](surface area of additive modified hexaaluminate/surface area of additive free hexaaluminate following calcination).

Reproduced by permission of Elsevier Science.

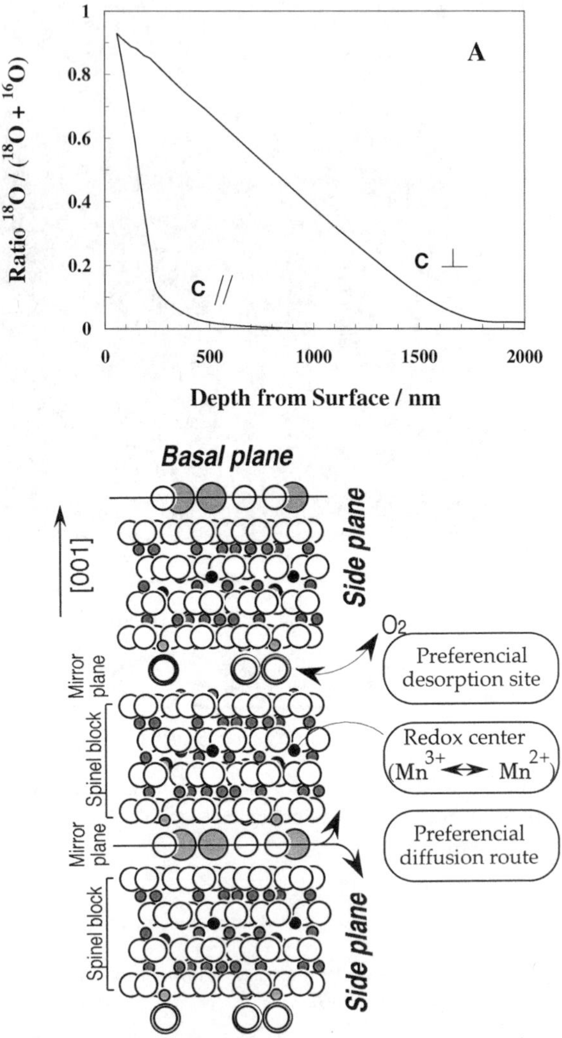


**Figure 7.18** (A) Depth profiles of $^{18}O$ in single crystal of Ba hexaaluminate after isotope exchange at 1673 K and (B) schematic diagram of diffusional routes in Mn-substituted hexaaluminate (Machida *et al.*, 1995) (reproduced by permission of Elsevier Science).

### 7.4.4 Choice of support acid–base properties

Another major characteristic of the support is its acidity. Support acidity seems to be a factor in catalytic oxidation, although not yet understood on a molecular level. Certainly those supports which are classified as acidic tend to give a better performance than neutral or basic supports. Hua and Gao (1996) compared the oxidation activity of platinum supported on alumina, normally classified as an acidic support, and the same amount of platinum supported on sulphated zirconium oxide, a super acidic solid. The acidity of the supports was assessed by testing their responses to a number of dye indicators to generate the Hammett scale of acidity (Ho) where low or negative values indicate a high acidity. Acidity was also assessed from the rate at which these supports aided the isomerization of *n*-butane, a reaction that requires surface acid sites. Both acidity tests, shown in Table 7.6, with other characterization data, indicate that the sulphated zirconium oxide was the more acidic support.

These catalysts were tested in the total oxidation of a range of VOCs including heptane, cyclohexane, toluene, benzene, acetone, ethyl acetate and hexane. The results for the oxidation of heptane over the two catalysts are shown in Figure 7.19. In general a lower temperature was required for total

**Table 7.6** Properties of $Pt/SO_4^{2-}/ZrO_2$ and $Pt/Al_2O_3$ catalysts (Hua and Gao, 1996)

| *Catalyst* | *Support acid strength* | | *Surface area $m^2\,g^{-1}$* | *Pt dispersion/%* |
|---|---|---|---|---|
| | *Ho* | *$n$-$C_4H_{10}$ isomerization ($k \times 10^3/h^{-1}$)* | | |
| 0.5 % $Pt/SO_4^{2-}/ZrO_2$ | −16 | 10.8 | 114 | 12 |
| 0.5 % $Pt/Al_2O_3$ | +4.8 | 0 | 314 | 63 |

Reproduced by permission of Baltzer Scientific Publishers.

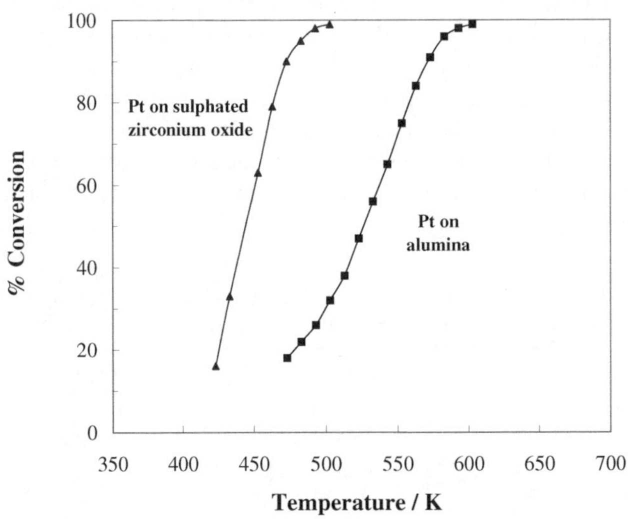


**Figure 7.19** Oxidation of heptane over 0.5 % $Pt/SO_4^{2-}/ZrO_2$ and 0.5 % $Pt/Al_2O_3$. The feed gas was 29.5 $g\,m^{-3}$ of heptane in air (Hua and Gao, 1996) (reproduced by permission of Baltzer Scientific Publishers).

oxidation over the more acidic sulfated zirconium oxide support for all the VOCs tested. The advantage of more acidic supports appears more pronounced at the highest conversion levels, as indicated by the more gradual approach to 100 % conversion for the lower acidity Pt/alumina catalyst in Figure 7.19.

## 7.5 POISONING OF COMBUSTION CATALYSTS

In general, combustion catalysts are required to operate in conditions where the feed gas composition is not always tightly controlled in the way it would be in a normal chemical plant. As a consequence minor components of the feed gas, such as sulfur-bearing compounds and vaporized metals, can interfere with efficient operation and eventually lead to deactivation. The data in Table 7.7 are fairly typical for a platinum-honeycomb catalyst that had operated to treat the exhaust VOCs from a metal strip coating operation. In this case a simple acid wash followed by an alkali wash was sufficient to remove most of the contaminants, restore the surface area and regenerate the activity close to its original level (Heck and Farrauto, 1995).

Two other contaminants that are commonly encountered are sulfur dioxide and hydrogen sulfide. The influence of small amounts of these impurities on the catalytic activity for methane oxidation of Pt, Pd and Rh, the three main metals used in modern three way auto-exhaust catalysts, each supported on alumina, is shown in Figure 7.20 (Meeyoo *et al.*, 1998). These data indicate that Pd is the most reactive of the three metals for methane oxidation in the absence of poisons. Pd and Rh are both poisoned by small amounts of $SO_2$ or $H_2S$. The performance of the Pt based catalyst was improved by exposure to these compounds (Burch *et al.*, 1998, 1999). The reason cited by the authors is that Pd and Rh operate in a partially oxidized state, and that oxides react with $SO_2$ to give the corresponding inactive metal sulfates. The surface of Pt metal is less prone to oxidation but any PtO present can oxidize $H_2S$ into $SO_3$, which in turn can react with the alumina support, making it more acidic. The

**Table 7.7** Build-up of contamination on a platinum-honeycomb catalyst following operation in a metal strip coating operation (Chen *et al.*, 1992)

| *Contaminant* | *Before operation/wt%* | *After operation/wt%* |
|---|---|---|
| Pb | 0.0003 | 0.007 |
| Sn | 0.001 | 0.3 |
| Na | 0.3 | 2 |
| P | 0.03 | 0.3 |
| Cu | 0.003 | 0.02 |
| BET surface area/$m^2\ g^{-1}$ | 15 | 3.6 |

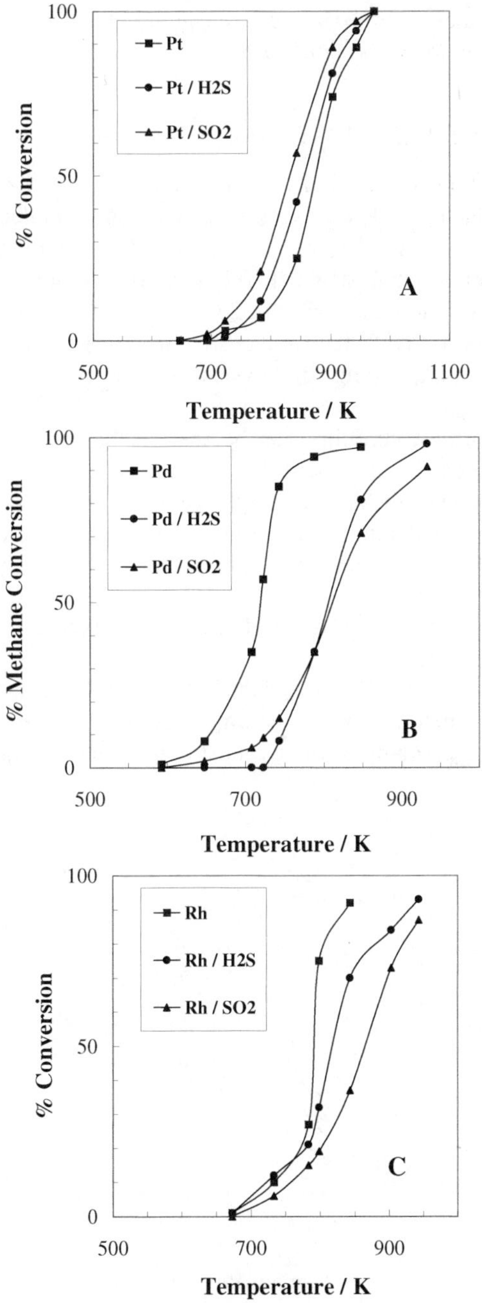


**Figure 7.20** Influence of $H_2S$ and $SO_2$ (20 ppm) on the performance of (A) Pt, (B) Pd and (C) Rh catalysts supported on alumina (1.8 % $CH_4$, 21 % $O_2$) (Meeyoo *et al.*, 1998) (reproduced by permission of Elsevier Science).

beneficial effect is then seen as increasing the acidity of the support, hence improving the activity (Meeyoo *et al.*, 1998).

## 7.6 REACTIVITY OF THE SUBSTRATE

Perusal of the literature on catalytic combustion indicates that the substrate structure is a vital factor in determining reactivity. Some molecules are more easily oxidized than others. Figure 7.21 shows the operation of a typical $Pd/Al_2O_3$ combustion catalyst used for the total oxidation of C1–C4 alkanes. Reaction commences at about 523 K and 100 % conversion of each alkane is possible close to 773 K (Yao, 1980). However, there is a clear order of destructibility, butane being the easiest to oxidize and methane the hardest.

A similar trend in destructibility can be seen in the data shown in Figure 7.22, which compares the destructibility of propan-2-ol, toluene, acetone, acetic acid and methane (O'Malley, 1998). Again a clear order of destructibility emerges. In fact a number of studies have established an empirical order of destructibility which runs as follows (Hermia and Vigneron, 1993):

$$\text{alcohols} > \text{aldehydes} > \text{aromatics} > \text{ketones} > \text{alkenes} > \text{alkanes}$$

Several workers have looked for the reason behind the differing reactivities of different substrates. There is now an accumulation of evidence to indicate that the bond dissociation enthalpy of the weakest bond in the reacting substrate is

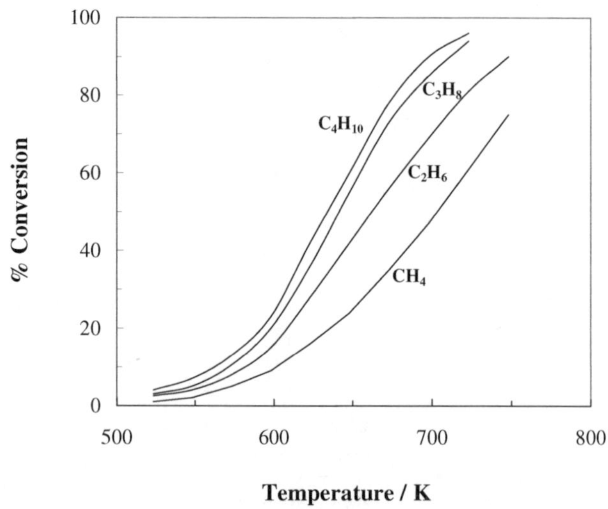


**Figure 7.21** Oxidation of C1–C4 alkanes over $Pd/Al_2O_3$ ($P_{O2} = 5\,\%$; $O_2/HC = 2 \times$ stoichiometric requirement; (Yao, 1980) (reproduced by permission of the American Chemical Society).

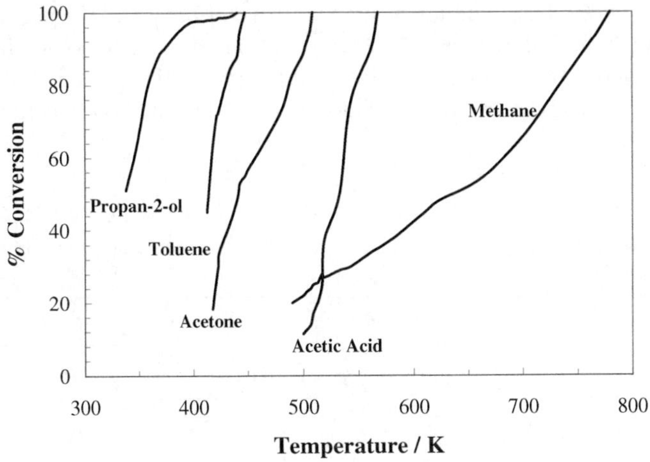


**Figure 7.22** Oxidation of propan-2-ol, toluene, acetone, acetic acid and methane over 0.5 wt% Pt/beta-zeolite (O'Malley, 1998).

one of the primary factors in determining reactivity. Sokolovskii shows an impressive correlation between the rate of oxidation of *n*- alkanes on cuprous oxide and the bond dissociation enthalpy of the weakest C—H bond in the alkane. This correlation is shown in Figure 7.23 and indicates that the weaker the C—H bond strength in an alkane, the more readily it is oxidized over a cuprous oxide catalyst (Sokolovskii, 1990).

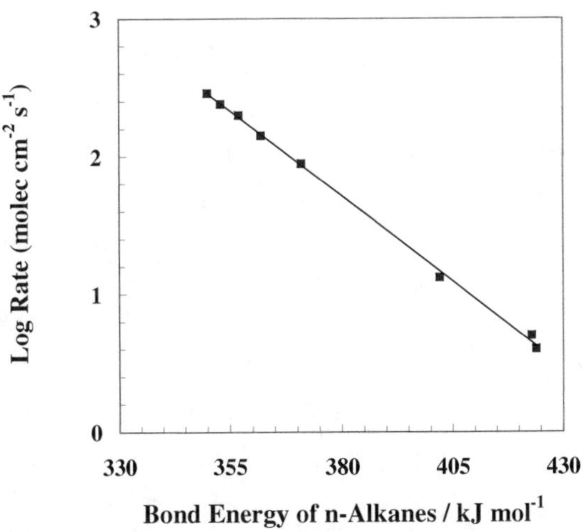


**Figure 7.23** Rate of oxidation of *n*-alkanes on cuprous oxide at 573 K as a function of C—H bond energy (Sokolovskii, 1990) (reproduced by permission of Marcel Dekker).

A similar study was undertaken by Schwartz *et al.* (1971) using a wider range of substrates. They measured the activation energies for oxidation of a number of substrates over Pt or Pd (Table 7.8) and concluded that the reactivity of a substrate towards oxidation is related to the bond energy of the

**Table 7.8** Activation energies for oxidation of organic compounds over Pt and Pd (Schwartz *et al.* 1971)

| *Substrate* | $E_a/kJ\ mol^{-1}$ | | *Substrate* | $E_a/kJ\ mol^{-1}$ | |
|---|---|---|---|---|---|
| | *Pt* | *Pd* | | *Pt* | *Pd* |
| $C_2H_6$ | 114 | | 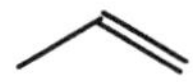 | 92 | 103 |
|  | 74 | 102 | 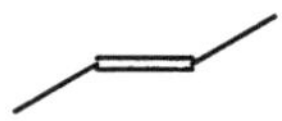 | 69 | |
|  | 71 | | $CH_3OH$ | 56 | 152 |
|  | 43 | | $C_2H_5OH$ | 55 | |
|  | 74 | 109 | 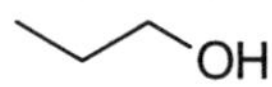 | 55 | |
| 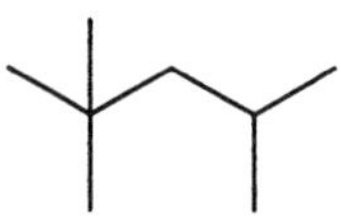 | 58 | | 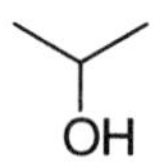 | 34 | 119 |
| 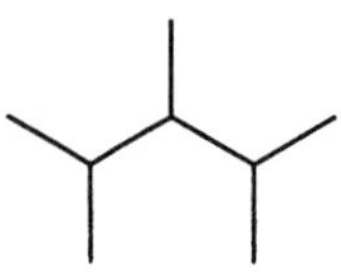 | 57 | | 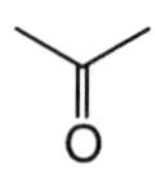 | 66 | |
|  | | 141 | 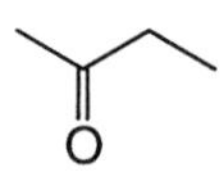 | 62 | |
| $C_2H_4$ | | 66 | 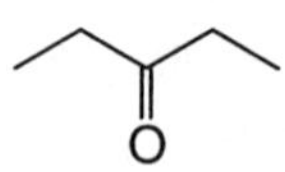 | 61 | |

Reproduced by permission of Academic Press.

weakest bond in the substrate, although no data on bond dissociation energies were presented in that work. This led to a mechanism whereby the substrate was activated on the metal surface through rupture of the weakest C—H bond. Weakest bond energy was cited as the primary factor determining reactivity despite the presence of a wide variety of functional groups in the substrates tested. The species which formed following cleavage of the weakest C—H bond in the substrate then reacted rapidly with surface oxygen species to yield total oxidation products.

The total oxidation of organic compounds involves many reactions and intermediates, and may involve adsorption and readsorption steps on several surface sites in the course of the overall reaction. Typically a reaction sequence for the total oxidation of an alkane may involve the formation of intermediate alkenes, dienes, ethers, alcohols, ketones, aldehydes, carboxylic acids and carbon monoxide. Radical species may also be involved. To date, no study has appeared which seeks to elucidate the full reaction sequence and to define the overall reaction mechanism on the atomic scale. A number of individual studies have focused on the final steps in the sequence, namely the steps whereby carboxylic acids are converted into carbon dioxide.

Many of these studies have involved the use of infrared spectroscopy which is a useful technique to identify adsorbed reaction intermediates. However, a problem which has hindered the use of infrared spectroscopy for the detection of surface reaction intermediates has been the difficulty in distinguishing between reactive and spectator species. The former are defined as the true reaction intermediates, whereas the latter are defined as unreactive adsorbed species. Standard infrared spectroscopy, even when used *in situ*, identifies all infrared active surface species. Some attempts have been made to combine *in situ* infrared techniques with perturbation methods in order to distinguish spectator species from true reaction intermediates. One such method is the application of the temperature jump technique whereby a sharp increase in temperature is applied to the catalyst surface. Reactive intermediates should be transformed during this perturbation, whereas spectator species may remain unaffected.

In one such study Bandara *et al.* (1998) studied the decomposition of formate species on NiO(111) and Ni(111) single crystal surfaces, combining a temperature perturbation with a variant of *in situ* infrared spectroscopy called time-resolved sum-frequency generation (SFG) spectroscopy. These workers used a laser beam to generate a temperature jump at the surface of the catalysts. The laser energy, and hence the temperature jump, was applied to a circular area of the surface 2–5 mm in diameter and was delivered to the surface as 35 ps pulses. The temperature jump and relaxation back to the original temperature lasted no more than 200 ps. The temperature perturbation was applied to the surface about 10 times per second.

In their study of formate species on the NiO(111) and Ni(111) surfaces Bandara *et al.* (1998) directed a flow of deuterated formic acid over their surface at a pressure of $3 \times 10^{-5}$ Pa. In these conditions they identified a formate species on the surface through the occurrence of a C—D stretching vibration at 2160 $cm^{-1}$. This was assigned to a bidentate formate, the structure of which is shown as an inset in Figure 7.24. When the laser pulse was applied, the intensity of this peak diminished and relaxed back to its original value as the temperature re-equilibrated within about 200 ps. At the same time, a new peak appeared in the SFG spectrum at 2190 $cm^{-1}$. This peak mirrored that due to bidentate formate, so that it increased in intensity and relaxed back to zero intensity in the course of the temperature jump. This second species was assigned to a monodentate formate species and this assignement was further supported by the detection of a carbonyl peak at 1718 $cm^{-1}$ over the Ni(111) surface.

Bandara *et al.* (1998) propose a mechanism whereby bidentate formate is first transformed into unidentate formate, allowing rotation of this surface species and approach of the C—D bond towards the catalyst surface where it is cleaved, releasing hydrogen and carbon dioxide. This mechanism is shown in Scheme 7.6.

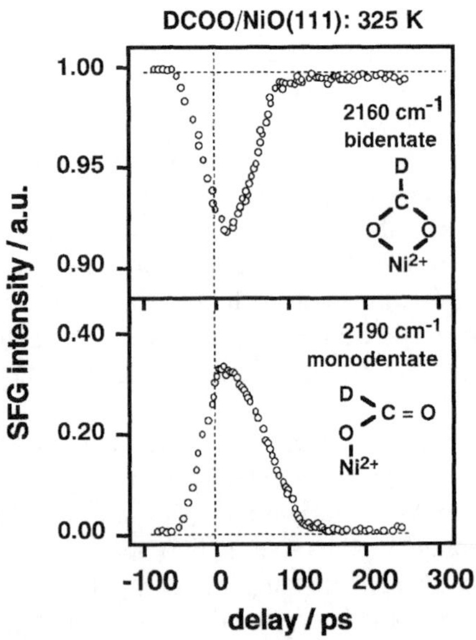


**Figure 7.24** Temporal recovery profiles of the transient SFG signals at 2160 and 2190 $cm^{-1}$ over a NiO(111) surface exposed to a flow of deuterated formic acid. The equilibrium temperature was 325 K prior to and after the application of the laser-induced temperature jump (Bandara *et al.*, 1998) (reproduced by permission of the American Chemical Society).

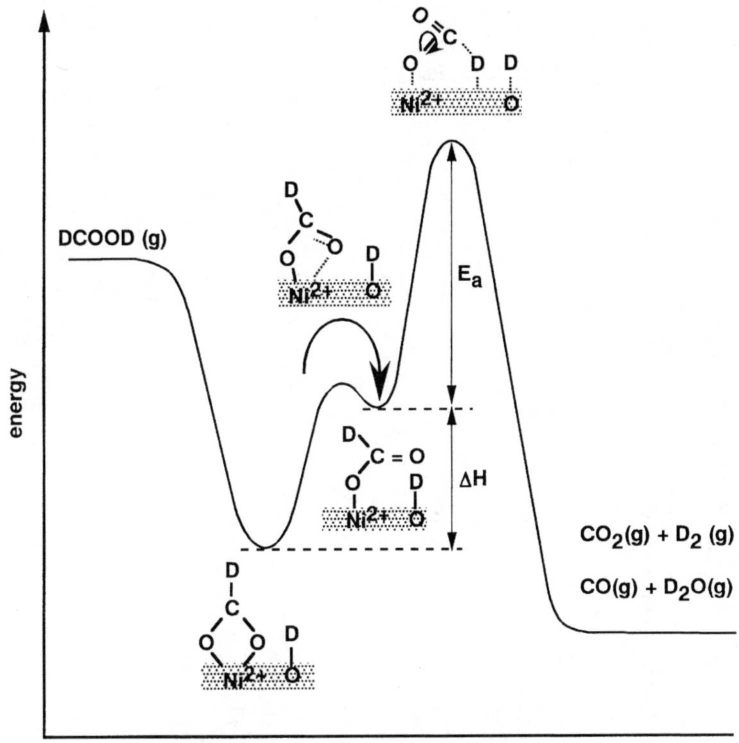


**Scheme 7.6** Energy scheme postulated for the transformation of bidentate formate into monodentate formate and carbon dioxide (Bandara *et al.*, 1998) (reproduced by permission of the American Chemical Society).

## 7.7 KINETICS OF CATALYTIC COMBUSTION REACTIONS

A large number of kinetic studies of total oxidation reactions have been undertaken over an extended time scale. The data are most often presented and analysed in terms of power rate laws and sometimes in terms of surface coverage dependent kinetic models. A selection of power rate laws for the oxidation of a number of substrates is presented in Table 7.9 for carbon monoxide and a range of unsaturated organic substrates over a number of precious metals and base metal oxides. No clear cut trends emerge from this work, except to note that fractional orders are quite common which mostly occur in the range 0–0.75. Fractional orders in substrate are also common, but here negative orders are more common, indicating a degree of inhibition of the reaction as the partial pressure of the substrate increases. Orders in water, where reported, are always negative, pointing to a mechanism involving

**Table 7.9** Summary of power law kinetic expressions of the form $V = k(p_{O_2})^m \, (p_{\text{Substrate}})^n \, (p_{H_2O})^l \exp(-E_a/RT)$ for total oxidation over a range of combustion catalysts (Mezaki and Inoue, 1991)

| *Catalyst* | *Substrate* | *Temperature range/K* | *Order in* $O_2$ (*m*) | *Order in substrate* (*n*) | *Order in* $H_2O$ (*l*) | *Activation energy/kJ mol*$^{-1}$ |
|---|---|---|---|---|---|---|
| $Pt/SiO_2$ | CO | 390–437 | 0.64 | −0.38 | — | 66 |
| Pt | $C_2H_2$ | 411–450 | 1.37 | −0.61 | — | 52 |
| Pt | $C_2H_4$ | 493–533 | −0.09 | 0.83 | — | 71 |
| Pt | i-$C_4H_8$ | 365–391 | 1.05 | 0.41 | — | 95 |
| Pd | CO | 423–433 | 0.93 | −0.53 | — | 48 |
| Pd | $C_2H_2$ | 374–414 | 0.79 | −0.68 | — | 118 |
| Pd | $C_2H_4$ | 610–641 | −1.6 | 1.3 | — | 152 |
| Pd | i-$C_4H_8$ | 405–426 | 0.78 | −0.22 | — | 140 |
| $Cr_2O_3$ | CO | 573–673 | 0.9–0.7 | −0.3–0.4 | −0.3 | 50–63 |
| $Cr_2O_3$ | $C_2H_2$ | 478–518 | 0.09 | 0.58 | — | 117 |
| $Cr_2O_3$ | $C_2H_4$ | 556–614 | 0.17 | 0.78 | — | 92 |
| $Cr_2O_3$ | i-$C_4H_8$ | 515–543 | 0.47 | 0.25 | — | 129 |
| $Co_3O_4$ | CO | 423–473 | 0.5 | 0.5 | −0.3 | 84 |
| $Co_3O_4$ | $C_2H_2$ | 459–480 | 0.71 | −0.28 | — | 173 |
| $Co_3O_4$ | $C_2H_4$ | 523–723 | 0.3 | 0.5 | −0.35 | 96 |
| $Co_3O_4$ | i-$C_4H_8$ | 493–535 | 0.75 | −0.09 | — | 161 |

competition between water and one of the reactants for adsorption sites (Mezaki and Inoue, 1991).

A more coherent picture of power law kinetics relating to the oxidation of alkanes over platinum or palladium catalysts has been presented by Yao (1980) and is reproduced in Table 7.10 This study of oxidation reactions on Pt catalysts consistently shows a negative order in oxygen partial pressure and a positive order in organic substrate. The situation for Pd is different; it usually shows a near zero order dependence in oxygen and a slightly positive order in the organic substrate. Both precious metal catalysts appear to be rather more sensitive to an overexposure to oxygen as indicated by the negative or close to zero dependencies on oxygen partial pressure when the substrate being oxidized is an alkane. With unsaturated alkanes these same metals were less sensitive to overexposure to oxygen, possibly because of differing adsorption characteristics of the alkanes and the unsaturated hydrocarbons.

Other attempts at finding kinetic expressions for total oxidation reactions are summarized in Table 7.11. These are largely of the Langmuir–Hinshelwood type and are generally more successful in that they readily accommodate fractional orders.

**Table 7.10** Kinetic parameters for alkane oxidation over Pt and Pd catalysts (Yao, 1980)

| *Alkane* | *Catalyst* | *Order in $O_2$* | *Order in alkane* | *Activation energy/* kJ $mol^{-1}$ | TOF/$s^{-1}$ [a] (Temp./K) |
|---|---|---|---|---|---|
| $CH_4$ | Pt wire | −0.6 | 1.0 | 88 | 0.13 (773) |
| $CH_4$ | 0.22 % $Pt/Al_2O_3$ | −0.5 | 1.1 | 100 | 0.0098 (773) |
| $C_2H_6$ | 0.22 % $Pt/Al_2O_3$ | −2.7 | 3 | 79 | 0.31 (623) |
| $C_3H_8$ | 0.22 % $Pt/Al_2O_3$ | −1 | 2 | 96 | 1.46 (523) |
| $C_4H_{10}$ | 0.22 % $Pt/Al_2O_3$ | −0.5 | 0.9 | 75 | 5.18 (498) |
| $CH_4$ | Pd wire | 0.1 | 0.8 | 71 | 5.4 (673) |
| $CH_4$ | 1 % $Pd/Al_2O_3$ | 0.1 | 0.7 | 84 | 0.31 (673) |
| $C_2H_6$ | 1 % $Pd/Al_2O_3$ | 0.1 | 0.5 | 96 | 0.37 (673) |
| $C_3H_8$ | 1 % $Pd/Al_2O_3$ | 0.1 | 0.5 | 92 | 0.21 (623) |
| $C_4H_{10}$ | 1 % $Pd/Al_2O_3$ | 0.0 | 0.6 | 88 | 0.3 (623) |

[a]Molecules $CO_2$ produced per surface metal atom per second.

Reproduced by permission of the American Chemical Society.

**Table 7.11** Summary of Langmuir–Hinshelwood kinetic expressions for the total oxidation of carbon monoxide, methane, ethylene and benzene on a range of combustion catalysts (Mezaki and Inoue 1991)

| *Catalyst* | *Substrate* | *Temperature range/K* | *Rate expression* | *Notes* |
|---|---|---|---|---|
| $V_2O_5/SiO_2$ | CO | 733 | $V = Ak_5 \frac{K_{CO} p_{CO}}{1 + K_{CO} p_{CO}}$ | $K_{CO} = 9.7\ l\ mol^{-1}$<br>$Ak_5 = 2 \times 10^{-3}\ l\ (g\ cat)^{-1}\ min^{-1}$ |
| $MnO_2$ | CO | 294 | $V = \frac{kp_{CO}}{[1 + K_1 p_{CO} + K_2 p_{CO_2}]^2}$ | $k = 0.00421$ kg-mol $(kg\ cat)^{-1}\ s^{-1}\ atm^{-1}$<br>$K_1 = 5.30$ atm<br>$K_2 = 2.90$ atm |
| $Pt/Al_2O_3$ | $CH_4$ | 853 | $V = \frac{k_9 p_{CH_4} p_{O_2}}{1 + k_{10} p_{CH_4} + k_{11} p_{O_2}^2}$ | $k_9 = 9.36 \times 10^{-8}$ k mol $(kg\ cat)^{-1}\ s^{-1}$<br>$k_{10} = 7.94 \times 10^{-4}$ k mol $(kg\ cat)^{-1}\ s^{-1}$<br>$k_{11} = 4.20 \times 10^{-3}$ k mol $(kg\ cat)^{-1}\ s^{-1}$ |
| $Pt/\gamma\text{-}Al_2O_3$ | $C_2H_4$ | 373–483 | $V = \frac{kp_{C_2H_4}}{1 + Kp_{C_2H_4}}$ | $k = 1.393 \times 10^8 \exp(-7450/RT)\ cm^3\ (g\ Pt)^{-1}\ s^{-1}$<br>$K = 373.4 \exp(+7150/RT)\ cm^3\ mol^{-1}$ |
| $Pt/N_i/\gamma\text{-}Al_2O_3$ | $C_6H_6$ | 413–443 | $V = \frac{k_{oi} k_i p_{O_2} p_{C_6H_6}}{k_{oi} p_{O_2} + \gamma_i k_i p_{C_6H_6}}$ | $k_{oi} = 3.4 \times 10^{10} \exp(-97939/RT)\ cm^3\ (g\ cat)^{-1}\ s^{-1}$<br>$k_i = 5.347 \times 10^{13} \exp(-96351/RT)\ cm^3\ (g\ cat)^{-1}\ s^{-1}$ |
| Pt/silica gel | $C_6H_6$ | 540 | $V = \frac{kK_{O_2} K_{C_6H_6} p_{O_2} p_{C_6H_6}}{[1 + K_{O_2} p_{O_2} + K_{C_6H_6} p_{C_6H_6}]^2}$ | $k = 5.78 \times 10^{-5}\ mol\ min^{-1}$<br>$K_{O_2} = 5.347\ l\ atm^{-1}$<br>$K_{C_6H_6} = 1.65 \times 10^2\ l\ atm^{-1}$ |

## 7.8 NATURE OF THE ACTIVE SPECIES ON WORKING PRECIOUS METAL CATALYSTS

### 7.8.1 Reactivity of surface species in combustion catalysis

Extensive literature is available covering the surface species which are important in working of precious metal catalysts, but a good point of reference is the thermodynamic data shown in Figure 7.25 (Barin *et al.*, 1989). These data point to zones of stability for precious metal oxides in the temperature range normally encountered in combustion catalysis. There is an important difference in the stability of platinum and palladium oxides; the former is never stable in the temperature range of interest here. However, it should be stressed that the data in Figure 7.25 refer to the bulk state whereas surface properties and the properties of small metal clusters can be significantly different.

In fact, a feature to emerge from single crystal work with precious metals is the propensity of these metals to undergo restructuring, sintering and redispersion, depending on the conditions applied. These effects point to a fluid system whose working state is difficult to define. A typical example concerns the (100) face of platinum. In ultrahigh vacuum conditions this face is stable, but on exposure to small amounts of oxygen, the uppermost layer of platinum atoms rearranges into a buckled hexagonal structure. This process is called restructuring. Oxygen uptake on the restructured surface is very slow at ambient temperatures, but it does increase as the temperature is increased (Lauterbach *et al.*, 1993). In general, adsorption properties can change

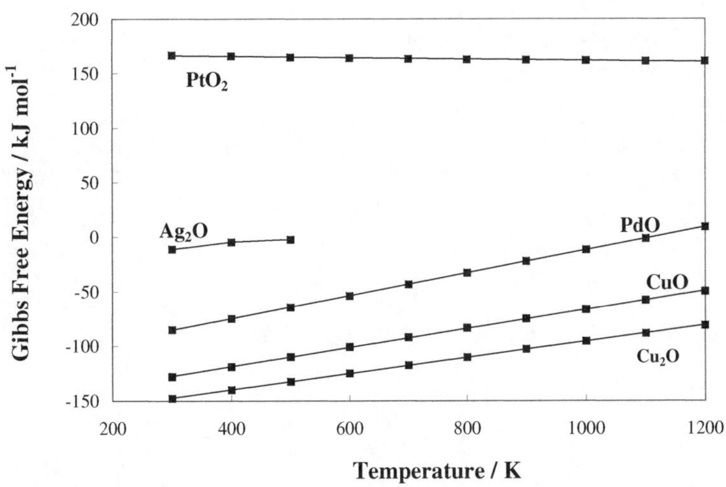


**Figure 7.25** Gibbs free energy of formation of the metal oxides indicated (Barin *et al.*, 1989).

dramatically depending on the crystal surface on which adsorption is attempted.

Sticking coefficients have been measured for oxygen on a number of the different exposed faces of platinum and are presented in Table 7.12. The sticking coefficient in itself gives no indication of the nature of the adsorbed oxygen species created.

The influence of metal dispersion on activity has been addressed by Yao (1980). In a series of tests of alumina supported Pt or Pd catalysts for the combustion of C1–C4 alkanes, metal dispersions were measured which allowed activities to be compared on a turnover frequency (TOF) basis. In addition to measuring the TOF for each supported catalyst, Pt and Pd wires were also examined. The wires exhibited poor overall catalytic activity, but very high TOFs. When the catalysts were compared with the wires (TOF catalyst/TOF wire) a general finding was that the TOF increased as the metal

**Table 7.12** Sticking coefficients for oxygen on some Pt surfaces at 530 K
(Lauterbach *et al.*, 1993)

| *Surface* | *Sticking coefficient* |
|---|---|
| Pt (110) | 0.4–0.6 |
| Pt(111) | 0.046 |
| hex-Pt(100) | $10^{-4}$ |
| Pt(100)[a] | 0.1 |

[a]The Pt(100) surface can be stabilized against restructuring by the adsorption of a small amount of CO.

**Table 7.13** Influence of precious metal dispersion on relative turnover frequency for the oxidation of $C_1$–$C_4$ alkanes
(Yao, 1980)

| *Metal/support* | *Dispersion %* | *TOF (catalyst)/TOF (wire)* | | | |
|---|---|---|---|---|---|
| | | $CH_4$ | $C_2H_6$ | $C_3H_8$ | $C_4H_{10}$ |
| $Pt/Al_2O_3$ | 87 | | | 0.016 | 0.12 |
| | 36 | | | 0.075 | 0.24 |
| | 19 | 0.58 | 0.18 | 0.13 | 0.22 |
| | 6 | 0.75 | 0.33 | | 0.51 |
| | 4 | | | 0.254 | 0.70 |
| $Pd/Al_2O_3$ | 64 | 0.002 | 0.006 | 0.007 | 0.0023 |
| | 51 | 0.004 | 0.008 | 0.015 | 0.004 |
| | 16 | 0.057 | 0.027 | 0.024 | 0.008 |
| | 5 | | | | 0.018 |

dispersion decreased (Table 7.13). There data were interpreted to mean that a minimum metal particle size is necessary to arrive at a good working catalyst (Yao, 1980).

More recent work by Briot *et al.* (1990) supports this hypothesis. These workers noted that the rate of transformation of methane over 1.95 wt% $Pt/Al_2O_3$, expressed in terms of moles of methane oxidized per unit mass of catalyst, was similar for a fresh catalyst with average platinum particle size of 2 nm and the same catalyst after it had been aged, by which time the average platinum particle size had increased to 12 nm. As a consequence, the TOF was much higher for the larger platinum particles. Microcalorimetric measurement of oxygen adsorption showed a decreased heat of chemisorption over the larger platinum particles. For the fresh catalyst with platinum particle size of 2 nm, the initial heat of chemisorption was $-280\ \text{kJ}\,(\text{mol } O_2)^{-1}$, and remained constant until close to monolayer coverage. The aged sample, featuring average Pt particle size of 12 nm, exhibited the same trend in heat of adorption as a function of oxygen coverage, but the heat of chemisorption was never less than $-250\ \text{kJ}\,(\text{mol } O_2)^{-1}$. The authors conclude that the strength of the Pt—O bond depends to some extent on the platinum particle size and the weaker this bond the more reactive the active site. In addition these authors showed that, when hydrogen was contacted with these catalysts at low temperatures, the oxygen species on the larger platinum particles reacted at 195 K, whereas the temperature had to be increased to 293 K before the oxygen species on the smaller particles reacted. These data are consistent with a greater reactivity associated with the more loosely bound chemisorbed oxygen species present on the larger platinum particles.

Ho *et al.* (1996) prepared a series of palladium catalysts supported on $Al_2O_3$, with loadings in the range 0.5–10 wt%. Palladium average particle sizes were in the range 1.2–7.9 nm as measured by hydrogen chemisorption. The heats of oxygen chemisorption at 293 K presented in Table 7.14 are integral values and were measured for the O : Pd ratios shown in column 3. These ratios were reasonably close to the H : Pd ratios used for the measurement of particle size, hence we must conclude that oxygen chemisorption on the palladium surface was very extensive. There is a clear relationship between integral heats of adsorption and palladium particle size, wherein the heat of adsorption is higher for the smaller particles. In fact these authors report an empirical relationship of the form

$$-\Delta H_{\text{chem}}[\text{kJ}\,(\text{mol } O_2)^{-1}] = 373\text{-}14d \tag{7.12}$$

where $d$ is the particle size in nm.

By contrast the heat of formation of palladium oxide at 773 K, also shown in Table 7.14, was somewhat less than the corresponding heats of chemisorption and was independent of the palladium particle size.

Similar results have been observed for palladium catalysts. Particles in the

**Table 7.14** Heats of oxygen chemisorption and PdO formation on $Pd/Al_2O_3$ (Ho *et al.*, 1996)

| *Catalyst wt% Pd* | *Oxygen chemisorption at 298 K* | | | *PdO formation at 773 K* | |
|---|---|---|---|---|---|
| | *Pd particle diameter/nm* | *O : Pd ratio (H : Pd ratio)* | *Heat of adsorption/kJ (mol $O_2$)$^{-1}$* | *O : Pd ratio* | *Heat of formation/kJ (mol $O_2$)$^{-1}$* |
| 10.1 | 7.9 | 0.10 (0.14) | −253 | 0.93 | −188 |
| 4.72 | 5.2 | 0.14 (0.21) | −293 | 1.05 | −184 |
| 2.8 | 3.7 | 0.21 (0.30) | −330 | 1.04 | −188 |
| 2.04 | 4.4 | 0.19 (0.25) | −326 | 0.96 | −184 |
| 1.07 | 2.1 | 0.33 (0.52) | −334 | 1.10 | −188 |
| 1.08 | 1.4 | 0.48 (0.80) | −346 | 1.24 | −185 |
| 0.55 | 1.2 | 0.35 (0.91) | −359 | 1.32 | −173 |



range 3–1000 nm gave integral heats of oxygen chemisorption close to −210 kJ (mol $O_2$)$^{-1}$ and this value decreased sharply to −333 kJ (mol $O_2$)$^{-1}$ for particles less than 3 nm. These effects were associated with changing electronic structure of the small palladium particles differing from those of bulk palladium (Chou and Vannice, 1987).

Hicks and coworkers compared TOFs of small and large palladium particles for methane combustion. Values of 0.02 $s^{-1}$ were recorded for small Pd particles on alumina and 1.3 $s^{-1}$ for large particles. In each case the activation energy for methane combustion was close to 120 kJ $mol^{-1}$. These workers claim that methane oxidation on Pd is a structure sensitive reaction and that different TOFs result from differing reactivities of adsorbed oxygen species on different cleavage planes of palladium (Hicks *et al.*, 1990a and b).

In general, there seems to be an accumulation of evidence that larger precious metal particles lead to lower heats of oxygen chemisorption and that this loosely held surface oxygen species is more reactive in combustion reactions. These data can be interpreted in terms of the influence of heat of oxygen chemisorption on the overall reaction profile as shown in Scheme 7.7. High heats of adsorption, predominating on small metal particles, lead to deeper thermodynamic states (corresponding to chemisorbed oxygen intermediates) than the corresponding states on larger particles. The activation energy typically measured in heterogeneous catalysis is in fact the apparent activation energy shown in Scheme 7.7. The deeper thermodynamic state associated with oxygen chemisorption on smaller metal particles leads to a higher apparent activation energy, hence a lower reaction rate or TOF at a given temperature. Conversely the more shallow thermodynamic state associated with oxygen chemisorption on larger particles will lead to a lower apparent activation energy and a faster reaction rate at a given set of reaction conditions.

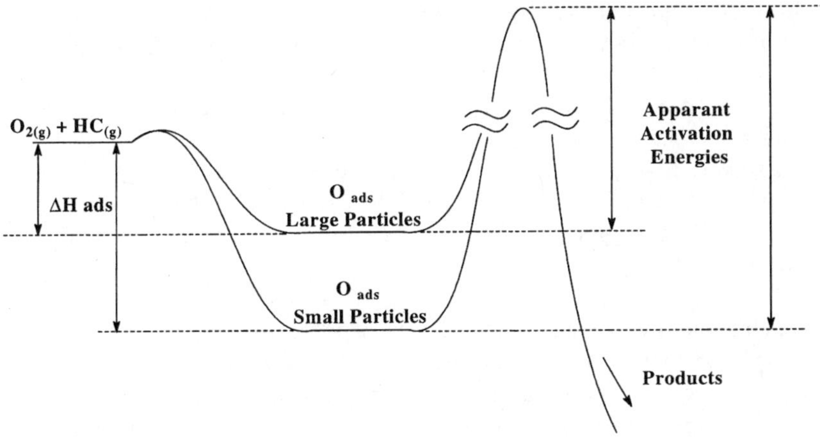


**Scheme 7.7** Reaction energy profile for the activation of oxygen on small and large metal particles.

### 7.8.2 Nature of oxygen species in combustion catalysis

Using the TAP reactor technique, Mallens *et al.* (1995) examined the reactivity of different forms of oxygen on a platinum gauze catalyst in an attempt to determine the reactivity of the different forms of oxygen on platinum. The type of catalyst used was composed of large platinum crystallites, compacted and worked into the form of a metal gauze. The reaction investigated was the partial oxidation of methane for the production of synthesis gas:

$$CH_4 + 0.5O_2 \longrightarrow CO + 2H_2 \quad (7.13)$$

The desired reaction products are carbon monoxide and hydrogen in a ratio 1 : 2 but overoxidation can occur via two routes, namely the direct oxidation of methane to form carbon dioxide and water, or through the consecutive oxidation of carbon monoxide and hydrogen.

These workers started from the basis that, on platinum, adsorbed atomic oxygen tends to predominate between 150 and 400 K. Above 773 K a small amount of oxygen becomes incorporated into the platinum structure and two different forms of oxygen have been proposed. The first is dissolved or subsurface atomic oxygen; the second is an oxide of platinum, PtO or $PtO_2$. These oxides can develop to several layers below the surface, representing significant diffusion of oxygen into the bulk of platinum. Because of the very large platinum particle size, the predominant oxidation state of the platinum remained close to zero even with oxygen incorporation, consistent with the bulk thermodynamic data (see Figure 7.25).

In their study, carried out between 973 and 1073 K, Mallens *et al.* determined the lifetimes of the three types of catalyst bound oxygen species:

- Platinum oxide ($PtO_x$) was stable at the reaction temperature for several hours.
- Dissolved oxygen, which formed to the extent of 0.7–2.5 theoretical monolayers, could be removed in less than 10 min by vacuum pumping.
- Chemisorbed oxygen species were stable for a few seconds at the reaction temperature.

When the platinum gauze catalyst was exposed to oxygen at 1023 K and pumped for more than 10 min chemisorbed and dissolved oxygen species were removed. A pulse of methane was then admitted and the reaction between $PtO_x$ and methane was monitored. The TAP profile indicated in Figure 7.26 was recorded. Several features are important. Hydrogen emerges from the reactor before water, and carbon monoxide emerges before carbon dioxide. These results indicate that hydrogen and carbon monoxide are formed initially by reaction with $PtO_x$ and are then transformed into water and carbon dioxide respectively by further reaction with $PtO_x$.

When the delay between the oxygen treatment of the catalyst and the introduction of the methane pulse was less than 10 min, thus allowing both dissolved oxygen and $PtO_x$ species to react with the methane pulse, the TAP profile shown in Figure 7.27 was recorded. There are some significant changes in the order of appearance of the reaction products. Now carbon dioxide

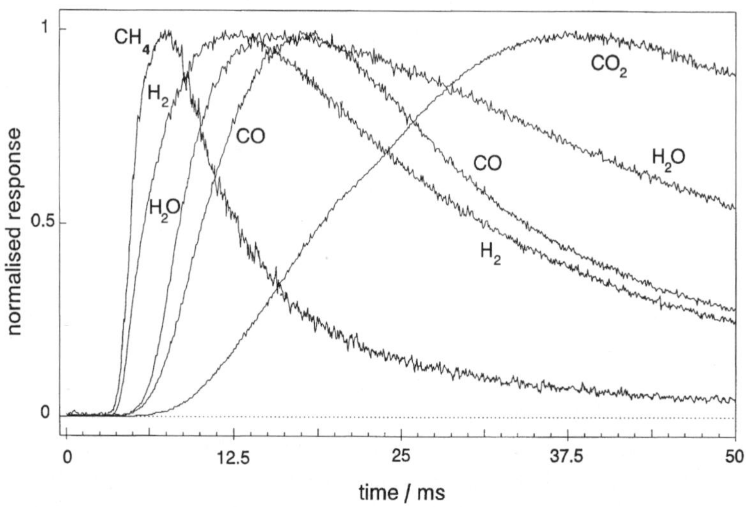


**Figure 7.26** TAP profile from the reaction of methane with $PtO_x$ on a Pt gauze catalyst at 1023 K (Mallens *et al.*, 1995) (reproduced by permission of Baltzer Scientific Publishers).

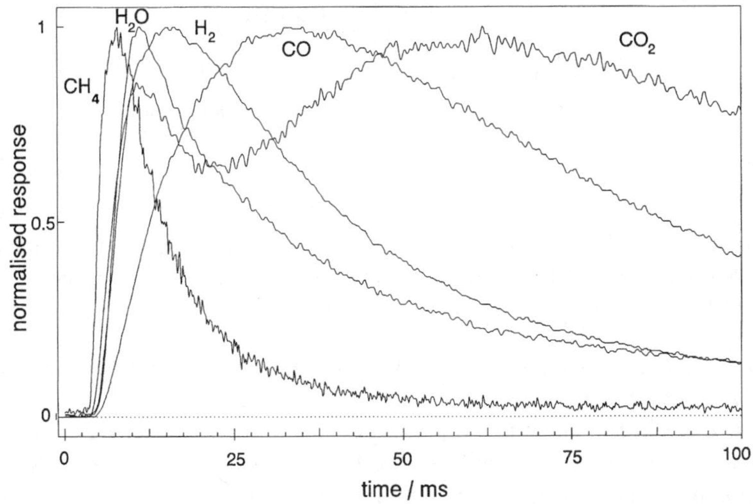


**Figure 7.27** TAP profile from the reaction of methane with $PtO_x$ and dissolved oxygen on a Pt gauze catalyst at 1023 K, (Mallens *et al.*, 1995) (reproduced by permission of Baltzer Scientific Publishers).

appears as two peaks, the first preceding the carbon monoxide and the water peak appears before the hydrogen peak. The overall selectivity to carbon monoxide and hydrogen was less than that in Figure 7.26. The additional carbon dioxide and water, appearing earlier than carbon monoxide and hydrogen respectively, point to a direct non-selective reaction between methane and dissolved oxygen.

When pulses of oxygen and methane were passed over the surface in quick succession, so that chemisorbed, dissolved oxygen and $PtO_x$ were available to react with the methane pulse, the selectivity to carbon monoxide and hydrogen declined further, consistent with a role for chemisorbed species on platinum in total oxidation reactions.

Simultaneous pulsing of $CH_4$ and $^{18}O_2$ over a platinum catalyst that had previously been treated with $^{16}O_2$, yielded $C^{16}O$, $C^{16}O^{18}O$, $C^{16}O_2$, $H_2{}^{16}O$ and $H_2{}^{18}O$. Neither $C^{18}O$ nor $C^{18}O_2$ was detected. In fact, the $^{18}O$ containing reaction products were only 3 % of the total. These results indicate that carbon monoxide is produced only via $^{16}O$ present in the form of $PtO_x$. Chemisorbed oxygen species are incorporated into carbon dioxide in the form of $C^{16}O^{18}O$, which confirms their role in the consecutive oxidation of carbon monoxide.

The roles of the individual forms of oxygen on and in platinum in the partial oxidation of methane are summarized in Scheme 7.8. Methane can dissociate on the bare platinum surface and the methane fragments react selectively with $PtO_x$ to give the desired partial oxidation products, carbon monoxide and hydrogen. Thereafter, all forms of oxygen on and in platinum contribute to total oxidation reactions. Dissolved oxygen directly oxidizes

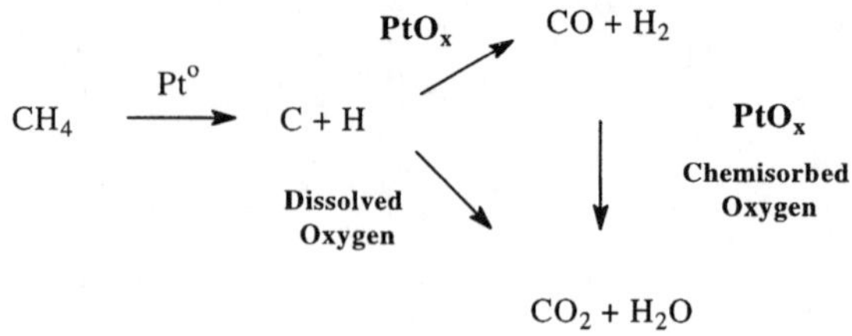


**Scheme 7.8** Role of platinum oxide, dissolved oxygen and chemisorbed oxygen in the partial oxidation of methane on Pt gauze between 973 and 1073 K (Mallens *et al.*, 1995) (reproduced by permission of Elsevier Science).

methane fragments whereas consecutive oxidation of CO and $H_2$ can occur via $PtO_x$ or chemisorbed oxygen species.

These experiments point to a role for quasi-platinum oxide species, unlikely to be stable on the basis of bulk thermodynamic data alone, in the selective oxidation of methane to carbon monoxide and hydrogen. This situation is very similar to that observed in conventional selective oxidation reactions over metal oxide catalysts, with the operation of a Mars and van Krevelen mechanism. By contrast, for total oxidation reactions on platinum surfaces, the more loosely held oxygen species seem to be more important.

Farrauto *et al.* (1995) studied a number of supported Pd catalysts. When supported PdO was heated in air it decomposed to $Pd^0$ at a temperature that depended on the support material. Upon cooling in air a part of the palladium oxide, referred to as Pd : O, reformed. The process is shown in Figure 7.28. These workers observed that the values for Td, Tr and Td − Tr depended strongly on the support used. Their data are reproduced in Table 7.15. The important point to come from this work is that these catalysts were active in methane oxidation only when they were in the partially oxidized state. More recent work with Auger electron spectroscopy of Pd foil before and after reaction indicate that only Pd metal was present before the reaction, whereas post-reaction analysis showed significant formation of a partial oxide of palladium labeled as $PdO_x$ (Descorme *et al.*, 1998).

Burch and Urbano (1995) also examined the relationship between the oxidation state of supported palladium and the activity of palladium catalysts for methane oxidation. Their catalyst comprised 4% $Pd/Al_2O_3$. Following reduction in hydrogen to the metallic state, oxygen uptake at 573 K was monitored. Typical results are presented in Figure 7.29 in terms of the evolving $O : Pd_{surf}$ ratio and the $O : Pd_{total}$ ratio. These data show rapid uptake of oxygen. A monolayer of oxygen was formed on Pd in a few seconds ($O : Pd_{surface} = 1 : 1$). In the following 600 s oxygen continued to be taken up more slowly until the $O : Pd_{total}$ ratio was close to 1 : 1, consistent with the formation of a material close in composition to bulk PdO.

Testing of the 4% $Pd/Al_2O_3$ catalyst for methane oxidation is shown in Figure 7.30. Initially, the catalyst was fully in the metallic state, and exhibited

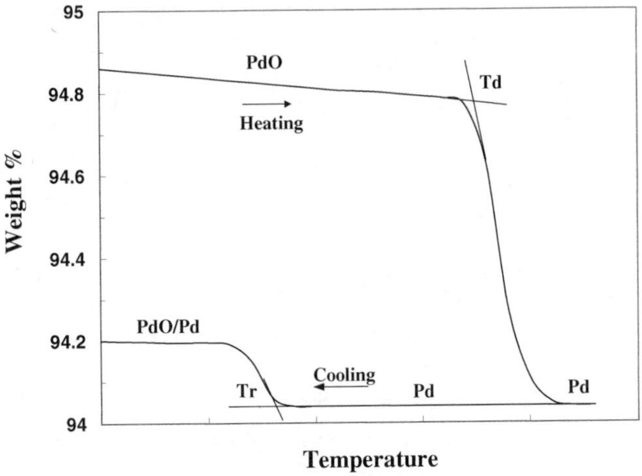


**Figure 7.28** Thermogravimetric cycle in air for the decomposition of supported PdO to Pd metal during heating in air and its partial reformation ($Pd^0 \longrightarrow Pd:O$) on cooling in air. The decomposition temperature is indicated by Td and the reformation temperature is indicated by Tr (Farrauto *et al.*, 1995) (reproduced by permission of Elsevier Science).

**Table 7.15** Decomposition of PdO on various supports (Farrauto *et al.*, 1995)

| *Catalyst* | *Temperature/K* | | |
|---|---|---|---|
| | *Td* | *Tr* | *Td − Tr* |
| 4% $PdO/Al_2O_3$ | 1083 | 873 | 210 |
| 4% $PdO/Ta_2O_3$ | 1083 | 923 | 160 |
| 4% $PdO/TiO_2$ | 1088 | 1008 | 80 |
| 4% $PdO/CeO_2$ | 1048 | 1003 | 45 |
| 4% $PdO/ZrO_2$ | 955 | 743 | 212 |

Reproduced by permission of Elsevier Science.

no activity. As the methane–air feed was passed over the catalyst the activity improved on a time scale similar to that observed for the uptake of oxygen in Figure 7.29. It is important to realize here that the first monolayer of oxygen forms on Pd metal in these conditions within a few seconds of exposure to the oxygen source. However, the catalytic activity improved on a much slower time scale, indicating that neither metallic palladium nor chemisorbed oxygen were particularly active for methane oxidation. It appears that the species responsible for methane oxidation has formed by the time the $O : Pd_{surf}$ ratio has reached a value of 3–4 : 1. Thereafter, the formation of a bulk PdO does not inhibit in any way the oxidation of methane.

Haack and Otto (1995) have taken these studies a step further by probing

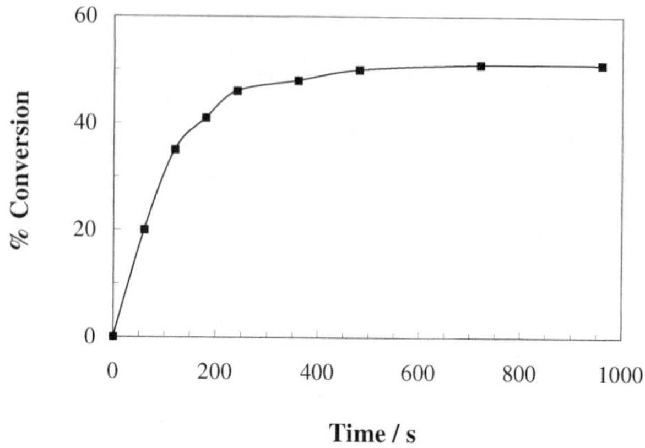


Figure 7.29 Oxygen uptake on 4% $Pd/Al_2O_3$ at 573 K. Prior to the first run the catalyst was reduced at 773 K in hydrogen for 16 h and at 573 K in hydrogen prior to the second run (Burch and Urbano, 1995) (reproduced by permission of Elsevier Science).

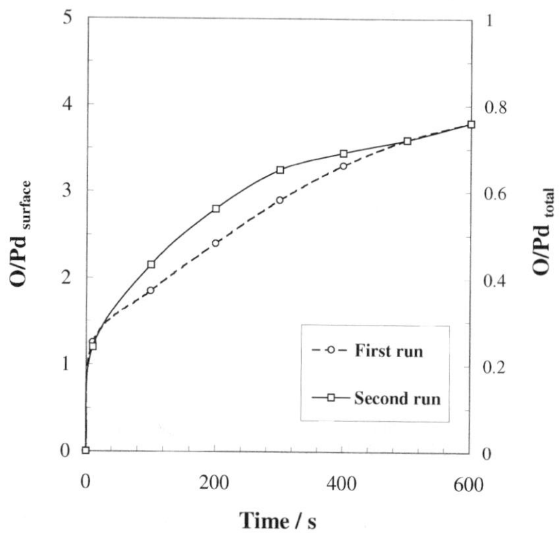


Figure 7.30 Methane combustion experiments performed in pulse mode over reduced 4% $Pd/Al_2O_3$ at 573 K. Prior to the experiment the sample had been reduced in hydrogen at 773 K for 16 h (Burch and Urbano, 1995) (reproduced by permission of Elsevier Science).

the working state of a palladium catalyst. The catalysts in question were 1.5 and 10 wt% palladium supported on alumina and a simple palladium foil. These catalysts were tested in the methane–oxygen mixtures defined in Table 7.16, which were designed to operate the various catalysts in net oxidizing and reducing conditions. Each catalyst was tested in the specified

**Table 7.16** Percentage of palladium in the $Pd^{2+}$ state by XPS analysis following exposure to a flow of methane and oxygen of the compositions shown (Haack and Otto, 1995)

| *$CH_4 : O_2$ molar feed ratio* | *% $Pd^{2+}$* | | |
|---|---|---|---|
| | *1.5 wt% Pd* | *10 wt% Pd* | *Pd foil* |
| 1 : 16 | 9 | 100 | 100 |
| 1 : 8 | 0 | 89 | 100 |
| 1 : 4 | 0 | 0 | 100 |
| 1 : 2 | — | — | 62 |

Reproduced by permission of Baltzer Scientific Publishers.

reaction conditions at 823 K for 4 h. Thereafter it was cooled to 373 K in argon, evacuated and transferred, without exposure to the ambient atmosphere, to the analysis chamber of an X-ray photoelectron spectrometer. The worked catalysts were then analysed for the palladium oxidation state and the results are summarized in Table 7.16.

A clear result to emerge from this work is that the oxidation state of the uppermost layer of palladium is a complex function of the palladium particle size and the operating conditions. With small palladium particles, corresponding to the lowest palladium loading, the surface of the palladium particle is present largely as palladium metal, except for the most oxidizing of conditions

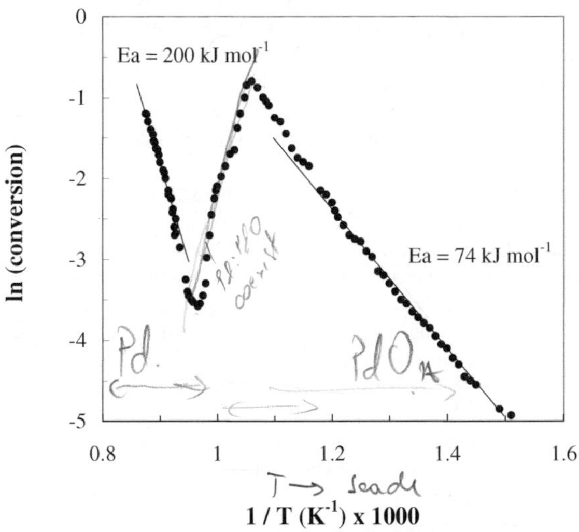


**Figure 7.31** Arrhenius plot for the combustion of methane over a palladium catalyst supported on alumina. The catalyst was first heated to 1073 K, and data were collected as the temperature was decreased in stages (Lyubovsky and Pfefferle, 1998) (reproduced by permission of Elsevier Science).

(the lowest $CH_4 : O_2$ ratio), where a small amount of palladium oxide is present. By contrast, the surface of palladium foil, with large palladium particles, was almost exclusively palladium oxide even for the richest $CH_4 : O_2$ ratio used.

The complex nature of the $PdO_x$–Pd equilibrium has been demonstrated in the work of Lyubovsky and Pfefferle (1998) from which a typical Arrhenius plot for methane combustion is presented in Figure 7.31. In this work a palladium catalyst supported on $\alpha$-alumina was pretreated above 1073 K so that all the metal was in the $Pd^0$ oxidation state. On cooling, the rate of methane oxidation decreased rapidly, but then increased as the experimental condition favoured $PdO_x$ formation. Once formed the $PdO_x$ compound far outperfomed metallic Pd as a combustion catalyst.

## REFERENCES TO CHAPTER 7

Arai H and Machida M (1996) Thermal stability of catalyst supports and their application to high-temperature catalytic combustion. *Appl. Catal. A General* **138**: 161–176.

Bandara A, Kubota J, Onda K, Wada A, Kano SS, Domen K and Hirose C (1998) Short-lived reactive intermediate in the decomposition of formate on NiO (111) surface observed by picosecond temperature jump. *J. Phys. Chem. B* **102**: 5951–5954.

Bartholomew CH (1994) Sintering kinetics of supported metals: Perspectives from a generalised power law approach. In *Catalysts Deactivation*, Delmon B and Froment GF (Eds). *Stud. Surf. Sci. Catal.* **88**: 1–18.

Barin I, Sauert F, Schultze-Rhonhof E and Sheng WS (1989) *Thermochemical Data of Pure Substances*. VCH, Berlin.

Briot P, Auroux A, Jones D and Primet M (1990) Effect of particle size on the reactivity of oxygen-adsorbed platinum supported on alumina. *Appl. Catal.* **59**: 141–152.

Burch R, Crittle DJ and Hayes MJ (1999) C—H bond activation in hydrocarbon oxidation on heterogeneous catalysts. *Catal. Today* **47**: 229–234.

Burch R, Halpin E, Hayes M, Ruth K and Sullivan JA (1998) The nature of activity enhancement for propane oxidation over supported Pt catalysts exposed to sulphur dioxide. *Appl. Catal. A General* **19**: 199–207.

Burch R and Urbano FJ (1995) Investigation of the active state of supported palladium catalysts in the combustion of methane. *Appl. Catal. A General* **124**: 121–138.

Chen J, Heck RM and Farrauto RJ (1992) Deactivation, regeneration and poison-resistant catalysts: Commercial experience in stationary pollution abatement. *Catal. Today* **11**: 517–545.

Chou P and Vannice MA (1987) Calorimetric heats of adsorption measurements on palladium: III Influence of crystalline size and support on oxygen chemisorption. *J. Catal.* **105**: 342–351.

Crump DR (1995) Volatile organic compounds in indoor air. In *Volatile Organic Compounds in the Atmosphere*, Hester RE and Harrison RM (Eds). The Royal Society of Chemistry, Cambridge, pp 109–124.

Descorme C, Jacobs PW and Somorjai GA (1998) Catalytic combustion of ethane over palladium foil in the 300–450°C range: Kinetics and surface composition studies. *J. Catal.* **178**: 668-678.

Eurostat Yearbook (1997) *A Statistical Eye on Europe 1986– 1996*. Office of Official Publications of the European Communities, Luxembourg.

Farrauto RJ, Lampert JK, Hobson MC and Waterman EM (1995) Thermal decomposition and reformation of PdO catalysts; support effects. *Appl. Catal. B Environmental* **6**: 263–270.

Garten RL, Dalla Betta RA and Schlatter JC (1997) Catalytic combustion. In *Handbook of Heterogeneous Catalysis*, Vol. 2, Ertl G, Knozinger H and Weitkamp J (Eds). Wiley–VCH, New York, pp 1668–1677.

Haack LP and Otto K (1995) X-ray photoelectron spectroscopy of Pd/$\gamma$-alumina and Pd foil after catalytic methane oxidation. *Catal. Letts.* **34**: 31–40.

Heck RM and Farrauto RJ (1995) *Catalytic Air Pollution Control: Commercial Technology*. Van Nostrand Rheinhold, New York.

Hermia J and Vigneron S (1993) Catalytic incineration for odour abatement and VOC destruction. *Catal. Today* **17**: 349–358.

Hicks RF, Qi H, Young ML and Lee RG (1990a) Structure sensitivity of methane oxidation over platinum and palladium. *J. Catal.* **122**: 280–294.

Hicks RF, Qi H, Young ML and Lee RG (1990b) Effect of catalyst structure on methane oxidation over palladium on alumina. *J. Catal.* **122**: 295–306.

Ho Y-S, Wang C-B and Yeh C-T (1996) Calorimetric study on interaction of dioxygen with alumina supported palladium. *J. Mol. Catal. A General* **112**: 287–294.

Hua W and Gao Z (1996) Low-temperature catalytic combustion of $Pt/SO_4^{2-}/ZrO_2$ and $Pd/SO_4^{2-}/ZrO_2$ catalysts. *Catal. Lett.* **42**: 209–212.

Knozinger H and Taglauer E (1997) Spreading and wetting. In *Handbook of Heterogeneous Catalysis*, Vol. 1, Ertl G, Knozinger H and Weitkamp J (Eds). Wiley–VCH, New York, pp 216–231.

Lieske H, Lietz G, Spindler H and Volter J (1983) Reactions of platinum in oxygen- and hydrogen-treated $Pt/\gamma\text{-}Al_2O_3$ catalysts. *J. Catal.* **81**: 8–16.

Lauterbach J, Haas G, Rotermund HH and Ertl G (1993) Spatio- temporal pattern formation on polycrystalline platinum surfaces during catalytic CO oxidation. *Surf. Sci.* **294**: 116–130.

Lyubovsky M and Pfefferle L (1998) Methane combustion over the $\alpha$-alumina supported Pd catalyst: Activity of the mixed Pd/PdO state. *Appl. Catal. A General* **173**: 107–119.

Mallens EPJ, Hoebink JHBJ and Marin GB (1995) An investigation of the reaction mechanism of the partial oxidation of methane to synthesis gas over platinum. *Catal. Lett.* **33**: 291–304.

Machida M, Sato A, Kijima T, Inoue H, Eguchi K and Arai H (1995) Catalysis properties and surface modification of hexaaluminate microcrystals for combustion catalysts. *Catal. Today* **26**: 239–245.

Meeyoo V, Trimm DL and Cant NW (1998) The effect of sulfur containing pollutants on the oxidation activity of precious metals used in vehicle exhaust catalysts. *Appl. Catal. B Environmental* **16**: L101–L104.

Mezaki R and Inoue H (1991) *Rates Equations of Solid-Catalysed Reactions*. University of Tokyo Press, Tokyo.

Morooka Y and Ozaki A (1966) Regularities in catalytic properties of metal oxides in propylene oxidation. *J. Catal.* **5**: 116–124.

Narui K, Yata H, Furuta K, Nishida A, Kohtoku Y and Matsuzaki T (1999) Effects of addition of Pt to $PdO/Al_2O_3$ catalysts on catalytic activity for methane combustion and TEM observations of supported particles. *Appl. Catal. A General* **179**: 165–173.

*National Air Quality and Emissions Trends Report* (1996). Office of the Air Quality and Planning Standards of the United States Environmental Protection Agency.

Oberlander RK (1984) Aluminas for catalysis—their preparation and properties. In *Applied Industrial Catalysis*, Vol. 3, Leach BE (Ed.). Academic Press, New York, pp 64–112.

O'Malley A (1998) Thesis, University of Limerick.

Schwartz A, Holbrook LL and Wise H (1971) Catalytic oxidation studies with platinum and palladium. *J. Catal.* **21**: 199–207.

Sokolovskii VD (1990) Principles of oxidative catalysis on solid oxides. *Catal. Rev.-Sci. Eng.* **32**: 1–49.

Warnatz J, Maas U and Dibble RW (1996) *Combustion: Physical and Chemical Fundamentals, Modelling and Simulation, Experiments, Pollutant Formation*. Springer, Berlin.

Yao Y-F Y (1980) Oxidation of alkanes over noble metal catalysts. *Ind. Eng. Chem. Prod. Res. Dev.* **19**: 293–298.

Zwinkels MFM, Jaras SG, Menon PG and Griffin TA (1993) Catalytic materials for high-temperature combustion. *Catal. Rev.- Sci. Eng.* **35**: 319–358.

# 8

# Activation and Insertion of Nitrogen into Organic Compounds


8.1 Introduction 241
8.2 Technology for Propene Ammoxidation 243
8.2.1 Fluidizatiom 247
8.3 Catalysts Used in Ammoxidation 249
8.4 The Ammoxidation Reaction Network 250
8.5 Mechanism of Nitrogen Insertion in Ammoxidation 254
8.6 Nitrogen Insertion in Ammoximation 260
References to Chapter 8 262



Summary

Acrylonitrile and caprolactam are the two major nitrogen containing bulk organic chemicals.

Acrylonitrile is manufactured by the ammoxidation of propene. In this process propene is mixed with ammonia and air and converted to acrylonitrile over a multi-component bismuth molybdate catalyst operating in a fluidized bed reactor. Propene oxidation and ammoxidation have many features in common. Similar catalysts are employed and the same $\pi$-allyl species is a common intermediate in both reactions. Commercial advances in ammoxidation are being attempted in propane and toluene ammoxidation and many of the new mechanistic insights into ammoxidation reactions in general have come from investigations of these reactions.

Commercial advances have also been made in the area of ammoximation with the development of the titanium silicalite catalysts for cyclohexanone oxime production, a key intermediate in the production of caprolactam.

This chapter describes

- The technology for propene ammoxidation,
- Catalysts used in ammoxidation,
- The ammoxidation reaction network,

- The mechanism of nitrogen insertion in ammoxidation,
- Nitrogen insertion in ammoximation.

## 8.1 INTRODUCTION

Acrylonitrile is by far the most important bulk chemical manufactured by ammoxidation. Its principal uses are in the manufacture of acrylic fibres, acrylonitrile–butadiene–styrene rubber and styrene–acrylonitrile rubber (see Table 8.1). World production is dominated by the BP (America), previously SOHIO, propene ammoxidation process (ACS Office of Public Outreach, 1996), which uses multicomponent bismuth molybdate catalysts in fluidized bed reactors (Brazdil, 1991). The basic catalyst is rendered attrition resistant for use in these reactors by coprecipitation with colloidal silica. Otherwise, the catalyst is similar to that used for propene oxidation to acrolein (Grasselli, 1997) .

Table 8.2 presents pertinent summary data for the ammoxidation of propene to acrylonitrile and literature reports of the performance of ammoxidation catalysts for this same reaction are presented in Figure 8.1 as a multiple selectivity–conversion plot.

Pilot scale plants for the production of acrylonitrile from propane have been operated successfully and there are plans for commercial scale units in the near future (Fairley, 1998). A summary of literature selectivity–conversion data for this reaction is presented in Figure 8.2. More importantly from the perspective of this chapter, studies of propane (equation (8.1)) and toluene (equation (8.2)) ammoxidation in recent years have improved our understanding of the mechanism of this type of reaction (Rizayev *et al.*, 1992; Martin and Lucke, 1996):

$$C_3H_8 + NH_3 + 2O_2 \longrightarrow CH_2{=}CH{-}CN + 4H_2O$$
$$\Delta H_{298\,K} = -632\ \text{KJ mol}^{-1} \tag{8.1}$$

**Table 8.1** Annual production and principal uses of acrylonitrile (*Chemical Week*, 1997)

| *Location* | *Production ktonne yr$^{-1}$* | *Principal Uses* | |
|---|---|---|---|
| America | 1600 | Acrylic Fibres | 53 % |
| Europe | 1300 | ABS/SAN | 30 % |
| Pacific | 1700 | Others | 17 % |
| Total | 4600 | | |

**Table 8.2** Summary information for propene ammoxidation (Zabetakis, 1965; Cox and Pitcher, 1970)

*Selective ammoxidation*

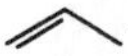 $+1.5O_2 + NH_3 \longrightarrow$ 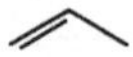 $+ 3H_2O$ $\Delta G_{298\,K} = -578\ kJ\,mol^{-1}$

*Non-selective*

$C_3H_6 + 3O_2 + 3NH_3 \longrightarrow 3HCN + 6H_2O$ $\Delta G_{298\,K} = -1050\ kJ\,mol^{-1}$

| Catalyst | Temp/K | Pressure/atm |
|---|---|---|
| $50\%Bi_9PMo_{12}O_{52} - 50\%SiO_2$ | 723–773 | 1–3 |

| | $LFL_{298}$ [a] | $UFL_{298}$ [a] | AIT[a]/K |
|---|---|---|---|
| $C_3H_6$ | 2.4 | 11 | 733 |
| $C_3H_3N$ | 3.0 | – | – |
| $NH_3$ | 15 | 28 | – |

[a] Lower and upper flammability limits at 298 K and the auto ignition temperature.

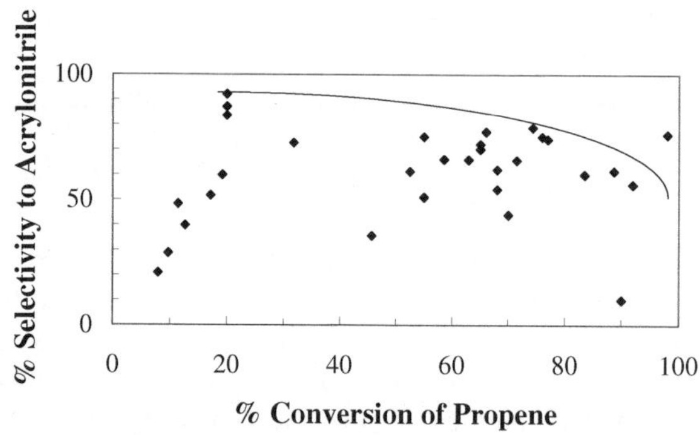


**Figure 8.1** Multiple selectivity–conversion plot for propene ammoxidation to acrylonitrile (Cassidy, 1999).

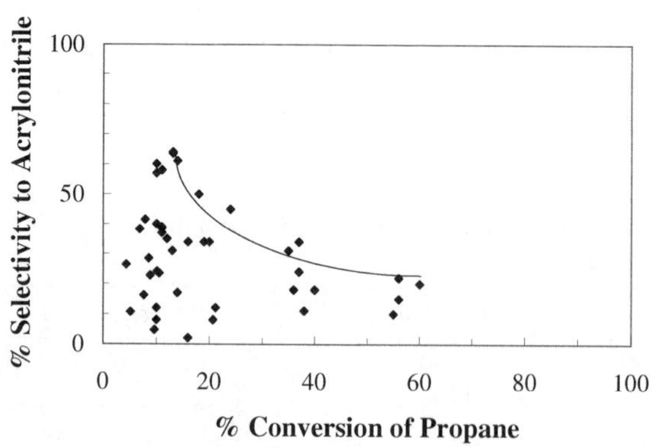


**Figure 8.2** Multiple selectivity–conversion plot for propane ammoxidation to acrylonitrile (Cassidy, 1999).

**Table 8.3** C—H bond energies in propene, propane, acrolein and acrylonitrile/ kJ mol$^{-1}$
(Cassidy, 1999)

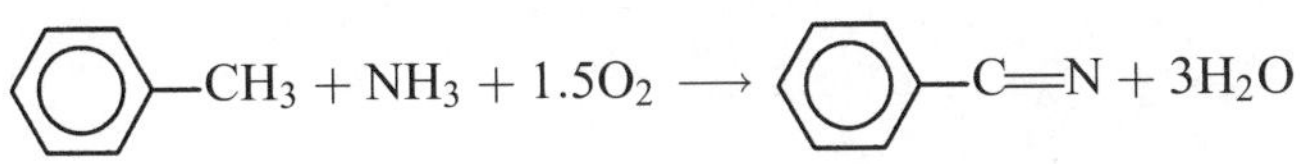

$$\Delta H_{298\,K} = -513\ \text{KJ mol}^{-1} \qquad (8.2)$$

The BP (America) commercial propene ammoxidation process dominates the market and is reported to operate at 98 % propene conversion and an acrylonitrile selectivity of 79 % (Brazdil, 1991). The very high conversion and selectivity is possible because of the extreme stability of the acrylonitrile by comparison with propene as indicated by the bond energies shown in Table 8.3. The absence of any weak bonds in the nitrile product, makes it very resistant to oxidation. These data also demonstrate why acrylonitrile can be produced commercially from propane specifically because of the high stability of the nitrile vis a vis the starting alkane. A similar situation pertains with *n*-butane selective oxidation to maleic anhydride (see Chapter 5). By contrast attempts to make acrolein from propane have not been successful because of the ease of activation of the aldehydic C—H bond in this material (see Table 8.3).

## 8.2 TECHNOLOGY FOR PROPENE AMMOXIDATION

The BP (America) fluidized bed technology uses reactor diameters of up to 10 m containing up to 75 tonne of a silica supported multicomponent bismuth molybdate catalyst (Grasselli, 1997). A typical plant flow-sheet is presented in Figure 8.3. The process achieves 98 % conversion of propene in a single pass through the reactor. Typical reactor inlet and outlet compositions are indicated in Table 8.4. In commercial operation the reactor effluent is quenched with water in a countercurrent absorber. Unreacted ammonia is neutralized with sulfuric acid and removed as ammonium sulfate, and off-gas, comprising nitrogen, carbon monoxide, carbon dioxide and unreacted propene, is sent for incineration. The acrylonitrile solution is passed to a recovery

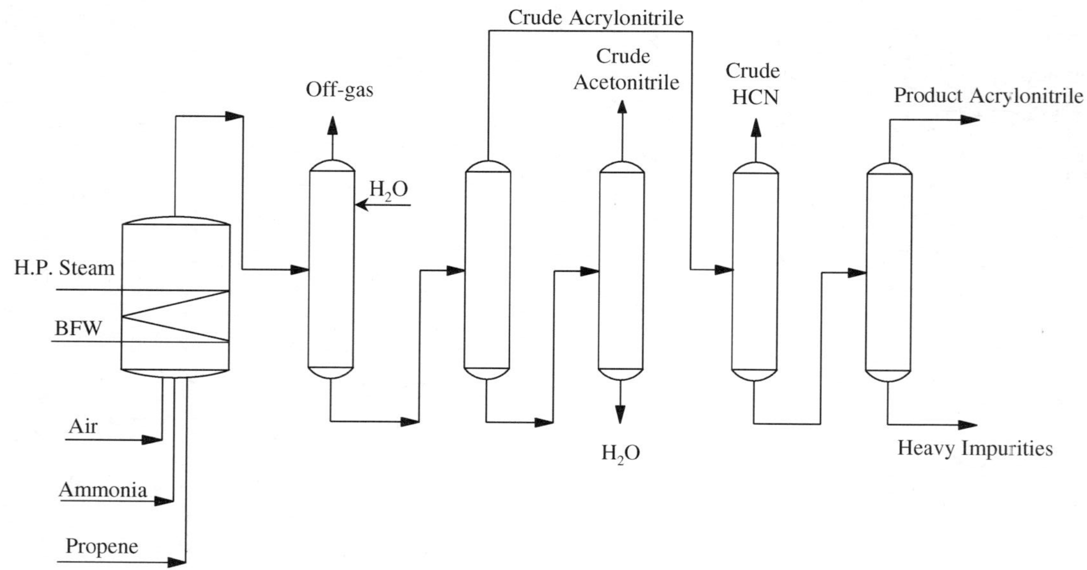


**Figure 8.3** Flow diagram for the production of acrylonitrile from propene (Brazdil, 1991) (reproduced by permission of John Wiley and Sons).

**Table 8.4** Typical inlet and outlet compositions for propene ammoxidation over 50 %$Bi_9PMo_{12}O_{52}$–50 %$SiO_2$ in a fluidized bed reactor at 743 K; contact time, 9.1 s (Grasselli, 1997; Zabetakis, 1965; Callaghan *et al.*, 1970)

| *Component* | *Inlet composition mol%* | *Outlet composition mol%*[a] | *$LFL_{298}$/Vol%* |
|---|---|---|---|
| $CH_2{=}CHCH_3$ | 6.5 | 0 | 2.4 |
| Air ($O_2$) | 86.2 (17.2) | 13.7 (2.7) | — |
| $NH_3$ | 7.2 | 0.6 | 15 |
| $CH_2{=}CHCN$ | 0 | 4.2 | 3 |
| $CH_3CN$ | 0 | 0.4 | 3 (est) |
| HCN | 0 | 0.8 | 5.6 |
| CO | 0 | 1.1 | 12.5 |
| $CO_2$ | 0 | 3.3 | — |
| $H_2O$ | 0 | 19.4 | — |
| $N_2$ | 0 | 58 | — |
| Totals | 99.9 | 101.5 | |
| Total inerts[b] ($N_2 + CO_2 + H_2O$) | 0 | 80.7 | |
| Total combustibles | 13.7 % | 7.1 % | |
| $LFL_{298}$[c] mixture | 4.3 % | 3.7 % | |
| $UFL_{298}$ mixture | 13.5 % | 12.5% | |

[a]Based on the inlet composition.
[b]Nitrogen associated with inlet and outlet air is not included in this value.
[c]Lower flammability limit in air at 298 K. For mixtures the LFL was calculated using equation (2.6).
Reproduced with permission of the American Chemical Society.

unit. Crude acrylonitrile is recovered overhead; this stream also contains HCN. The water phase is then passed to another recovery unit where water is removed and acetonitrile is produced. Acrylonitrile is produced at greater than 99.2 % purity by fractional distillation of the crude acrylonitrile steam to remove HCN, water and heavy impurities. Byproduct HCN, which is recovered is used primarily in the manufacture of methyl methacrylate (Brazdil, 1991).

Propene ammoxidation differs from oxidation to acrolein in one important way, namely the need to add ammonia to the reactor feed. Because ammonia is itself combustible it contributes to the flammability limit of the propene–ammonia mixture. Figure 8.4 shows the upper and lower flammability limits (UFL and LFL) for various mixtures of propene and ammonia in air at 298 K. These limits, indicated as the combined vol% of propene and ammonia, increase as the ammonia : propene ratio increases, reflecting the higher flammability limit associated with pure ammonia (see Table 2.2). However,

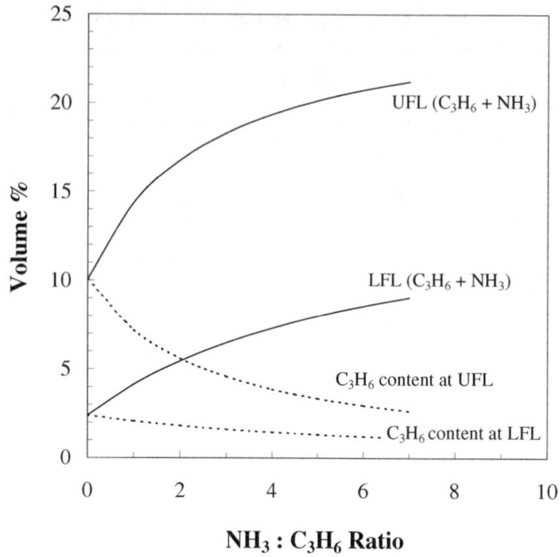


**Figure 8.4** Influence of $NH_3 : C_3H_6$ ratio on the flammability limits of the mixture at 298 K (Zabetakis, 1965). The vol% of propene in each mixture is indictated by the dashed lines.

the underlying propene concentration that can be tolerated for this mixture to remain, for example, below the LFL decreases from a value of 2.4 vol% without added ammonia to 1.6 vol% for an ammonia : propene ratio of 3 : 1.

Commercial practice operates with ammonia : propene ratios very close to 1 : 1 (see Table 8.4). This maximizes the amount of propene that can be included in the reactor feed, cuts down on non-selective utilization of ammonia and reduces the amount of treatment needed for the effluent gases.

A much bigger boost in terms of allowed propene concentrations is achieved by operating with a fluidized bed reactor. Because of the extreme turbulence in these beds, the inlet and exit concentrations become instantly equalized, in the manner shown in Figure 8.5. A feature of reactor design is the separate injection of combustibles (propene and ammonia) and oxidant into the reactor.

Table 8.4 presents the inlet and outlet gas compositions for a typical propene ammoxidation run in a fluidized bed reactor (Grasselli, 1997). The reactor operates very close to 100 % conversion of propene and about 90 % conversion for ammonia. The air or oxygen conversion is nearly 85 %. The product slate contains four important combustible gases, namely, acrylonitrile, acetonitrile, hydrogen cyanide and carbon monoxide and some unreacted ammonia. These constitute about 7.1 % of the outlet stream. This composition would be well above the LFL if mixed with air (Zabetakis, 1965). The operation is safe because the effluent gases contain very little air. The

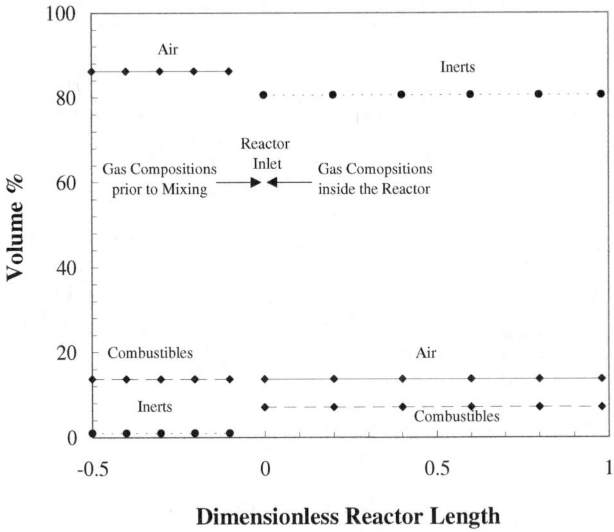


**Figure 8.5** Composition profile for ammoxidation of propene in a fluidized bed reactor, based on the data in Table 8.5.

combined effects of oxygen depletion and carbon dioxide and water formation in the reactor effectively means that the concentration of inerts at the reactor exit is close to 80 % (nitrogen associated with the unreacted air is not included in this figure).

Figure 8.6 presents the % combustibles (unreacted propene and ammonia and the combustible products) as a function of % conversion. Superimposed on this plot are the flammability limits for the propene–air–inert system and the (propene + ammonia)–air–inert system. The % inerts, normally plotted on the *x*-axis in this type of diagram, has been replaced with % conversion. In propene ammoxidation each 20 % increment in conversion adds about 15 % inerts to the system. Presented in this way Figure 8.6 clearly shows that when operated at high conversions (close to 100 % oxygen consumption in commercial operation) this system offers the security that a flammable mixture can never be created irrespective of the % combustible in the mixture.

### 8.2.1 Fluidization

Fluidized bed reactors are being applied to an increasing number of catalytic processes. All catalytic cracking (fluid catalytic cracking) now operates in fluidized bed reactors as does propene ammoxidation and this technology is now in fairly widespread use for the oxidation of *n*-butane to maleic anhydride. In terms of oxidation the main advantages of this type of reactor

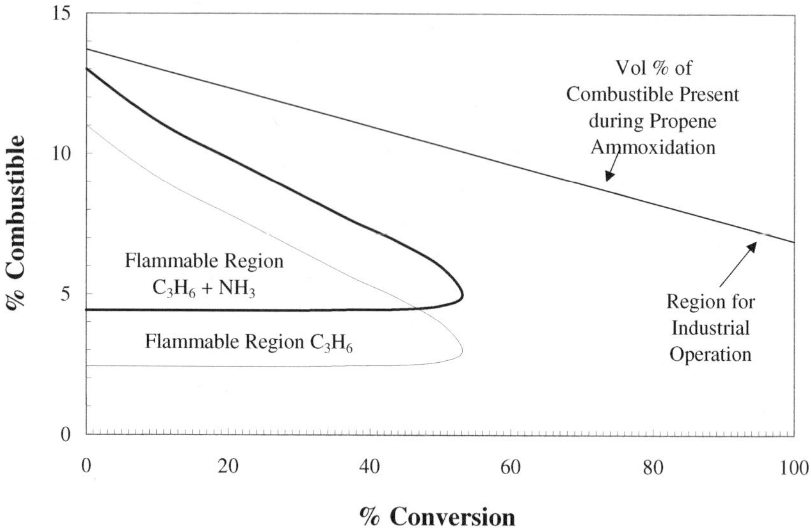


**Figure 8.6** Flammability limits and % combustible as a function of the % conversion for the propene ammoxidation related to the data in Table 8.5.

are that it allows operation with richer hydrocarbon-oxidant mixtures in an essentially temperature-gradient free environment within the reactor.

Fluidization converts a bed of suitable solid particles into an expanded, suspended mass that exhibits many liquid like properties. Common problems encountered in attempting to use catalysts in fluidized bed operation are agglomeration of particles which can occur when a component of the catalyst becomes soft and sticky at its reaction temperature and the converse, the breakdown or attrition of particles through the enormous mechanical forces generated at high superficial gas velocities.

Uniform particle size and attrition resistance are two key properties in catalysts suitable for operation in fluidized beds. Particle size is usually in the range $10^{-5}$–$10^{-6}$ m, but should be as uniform as possible to achieve good fluid properties. The upward or superficial gas velocity is usually between 0.15 and 6 m s$^{-1}$.

An essential part of the fluidized bed reactor is the gas distributor, which is designed as a multipoint gas inlet system and is placed at the bottom of the bed. The distributor is intended to introduce the gas as a uniform flow and in the special case of oxidation, where it is essential that the hydrocarbon and oxidant are not mixed outside the bed, the distributor design must be such as to mix these components inside the reactor.

When gas passes upward through a bed of particles a pressure drop is generated, which increases as the superficial gas velocity is increased. At the velocity where the pressure drop is equal to the mass of the bed divided by its cross-sectional area:

$$\Delta p = m/A \tag{8.3}$$

the particles become fluidized and this is the minimum fluidization velocity. As the superficial velocity is further increased the particles will expand uniformly until a velocity is reached where gas bubbles will form (minimum bubbling velocity). As the gas velocity is further increased, turbulence increases and bed density decreases. At still higher gas velocities bubbles disappear and streamers of solids and gas appear. Beyond this velocity pneumatic transport of particles is possible (Froment and Bischoff, 1990)

## 8.3 CATALYSTS USED IN AMMOXIDATION

Many of the multicomponent bismuth molybdate catalysts used for propene oxidation to acrolein can also be successfully used at the laboratory scale in fixed bed reactors, at comparatively higher temperatures, for the ammoxidation of propene. Otherwise, many of the factors relating to catalyst composition, lattice oxygen mobility, anion vacancies and solid solution formation are identical to those discussed in Chapter 4. This point is amply demonstrated in Figure 8.7, which shows the same $Pb_{1-x}Bi_{2x}\phi_x(MoO_4)$ series of solid solutions with the Scheelite structure used for propene oxidation to acrolein (see Figure 4.10) used here for propene ammoxidation to acrylonitrile

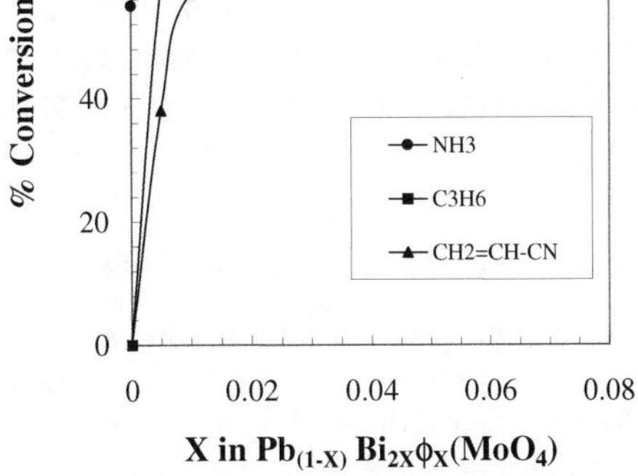


**Figure 8.7** Influence of cation vacancy concentration on the % conversion of propene and ammonia and the yield of acrylonitrile during ammoxidation at 713 K over $Pb_{1-x}Bi_{2x}\phi_x(MoO_4)$ catalysts. Contact time 6 s, 4 % $C_3H_6$, 4.8 % $NH_3$, 47.7 % air, 43.55 % $N_2$ (Sleight, 1977) (reproduced by permission of Academic Press).

(Sleight, 1977). Again the % conversion of ammonia and propene as well as the acrylonitrile yield increase as the cation vacancy concentration increases.

Many catalysts are still being actively researched for propane ammoxidation, but an increasing number of studies and patents have focused on the use of vanadium antimony oxides (Centi *et al.*, 1996). These oxides are characterized by the presence of a non-stoichiometric vanadium antimonate phase, $VSbO_4$, with the rutile structure, and when the Sb : V ratio exceeds 1 : 1 $\alpha$-$Sb_2O_4$ appears as a separate phase in the X-ray diffraction pattern. Attempts to prepare pure $VSbO_4$ have produced a material that contains amorphous vanadium oxide and $\alpha$-$Sb_2O_4$ on the surface. Increasing the Sb : V ratio above 1 : 1 decreases the amount of amorphous vanadium oxide that is present, but crystalline $\alpha$-$Sb_2O_4$ then appears. The catalytic activity improves as the Sb : V ratio is increased, because the amorphous vanadium oxides tend to oxidize ammonia to $N_2$, reducing the amount available for the desired ammoxidation reaction.

Typical propane ammoxidation catalysts prepared for commercial operation are $VSb_5W_{0.5}Te_{0.5}Sn_{0.5}O_x$ supported on $SiO_2/Al_2O_3$ (reported conversion 69 %, selectivity 57 %, yield 39 %) and $VSb_{3.5}Bi_2Fe_2O_x/Al_2O_3$ (conversion 21 %, selectivity 90 %, yield 19 %) (Grasselli *et al.*, 1997).

A much bigger variety of catalysts have been used for the ammoxidation of toluene. Probably the system that has received most attention is the $V_2O_5/TiO_2$ system, close in composition and structure to that used for the oxidation of *o*-xylene to phthalic anhydride (*Centi*, 1996). This catalyst is often described as a monolayer catalyst with a few layers of amorphous vanadium oxide, not necessarily $V_2O_5$, supported on a $TiO_2$ anatase. A transmission electron micrograph of this type of catalyst is shown in Figure 3.12 and clearly shows surface layers of vanadium oxide on anatase. Note also that there is no epitaxial relationship between the vanadium oxide layer and the underyling support. When the rutile form of $TiO_2$ is used the layer is somewhat thinner because some of the vanadium oxide can migrate into the support to form a solid solution therein. When the vanadium oxide composition exceeds a threshold value crystalline $V_2O_5$ starts to appear (Sanati *et al.*, 1991)

## 8.4 THE AMMOXIDATION REACTION NETWORK

Figure 8.8 presents the evolution of reaction products at various conversions for propane ammoxidation over a V–Sb–O catalyst supported on $Al_2O_3$. In this figure various propane conversions at 773 K were generated by varying the contact time, with the highest conversion achieved at the highest contact time. Propene predominates at short contact times and is then converted into acrylonitrile. At longer contact times carbon oxides tend to dominate (Centi *et al.*, 1992).

A typical TAP reactor run for propane ammoxidation over V–Sb–O

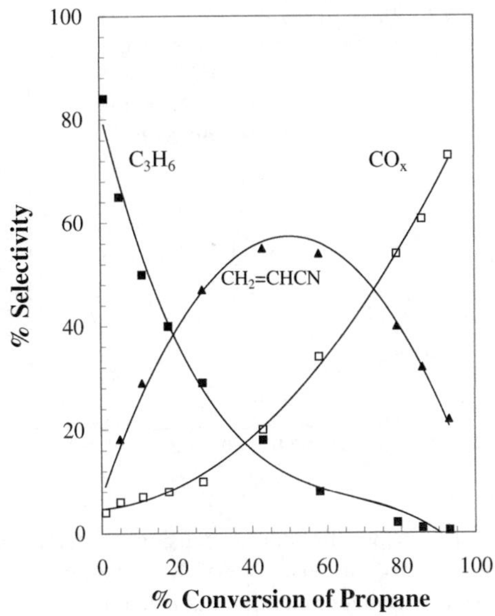


**Figure 8.8** Selectivity to propene, acrylonitrile and carbon oxides for propane ammoxidation over V–Sb–O catalysts at increasing conversions. Temperature = 773 K; $C_3H_8$ = 1.6 vol%; $O_2$ = 8 vol%; $NH_3$ = 8 vol%; mass of catalyst = 7 g (Centi *et al.*, 1992) (reproduced by permission of Elsevier Science).

catalyst is shown in Figure 8.9. This technique allows much shorter contact times to be explored in comparison with conventional flow reactors and in these conditions propene and acrolein were observed as intermediates in the route to acrylonitrile (Zanthoff and Buchholz, 1997). It should be recalled also that small amounts of acrolein are usually reported to emerge even from commercial propene ammoxidation reactors (Grasselli, 1997). Figure 8.10 shows the results of sequentially passing a pulse of ammonia followed by a propane pulse with pulse intervals from 0.1 to 1.4 s. As the pulse interval increased, the amount of acrolein detected also increased (Zanthoff and Buchholz, 1996), indicating that the nitrogen species generated on V–Sb–O when the ammonia pulse passed over the catalyst was short lived because the maximum amount of acrylonitrile forms only when the pulse interval is less than 0.1 s. These data are not consistent with the incorporation of any appreciable amounts of nitrogen species into the catalyst structure, and point instead to a role for short lived adsorbed species. These data are consistent with the reaction network shown in Scheme 8.1 for propane ammoxidation.

This network has been further refined by Zhang and co-workers (Zhang *et al.*, 1997) for toluene ammoxidation over $(VO)_2P_2O_7$ catalysts. They exposed this catalyst to toluene, benzaldehyde and benzylamine in ammoxidation reaction conditions at 673 K. *In situ* infrared spectroscopy revealed the

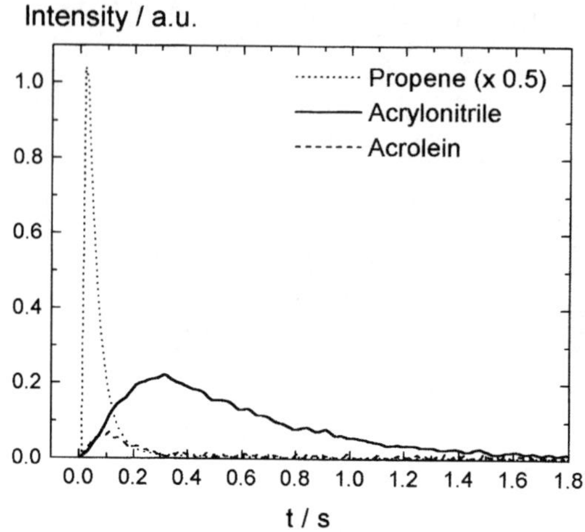


**Figure 8.9** Formation of acrylonitrile during pulsing of $C_3H_6$/Ne over $\alpha$-$(NH_4)_2[(VO)_3(P_2O_7)_2]$ in the absence of $NH_3$ at 653 K (Zanthoff and Buchholz, 1997) (reproduced by permission of Baltzer Scientific Publishers).

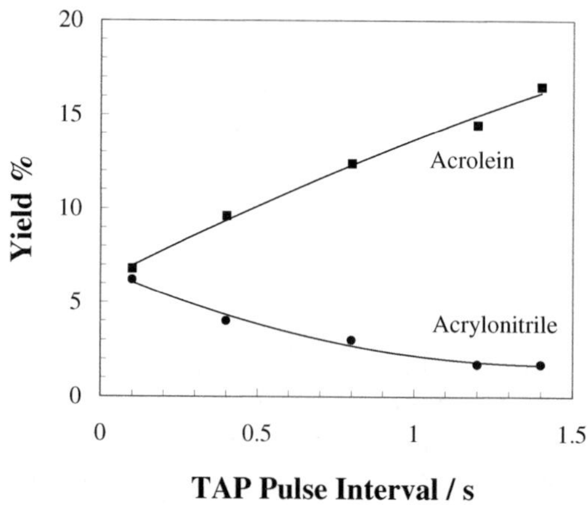


**Figure 8.10** Acrolein and acrylonitrile formation during sequential pulsing of ammonia followed by propane over $VSb_5O_x$, reaction temperature 697 K (Zanthoff and Buchholz, 1996) (reproduced by permission of Baltzer Scientific Publishers).

presence of surface-bound benzylimine species during toluene ammoxidation. When benzaldehyde was fed to the same catalyst in identical reaction conditions, bands due to benzylimine were again observed and benzonitrile was observed in the reaction products. When benzylamine was fed to the catalyst in ammoxidation conditions a weak peak due to benzylimine was observed, but benzonitrile was not observed in the reaction products. These

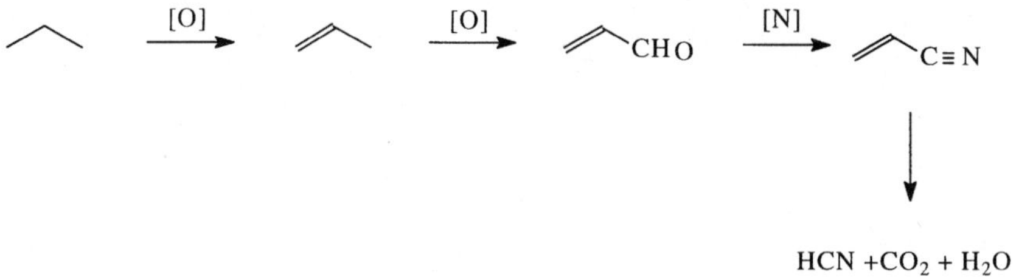


**Scheme 8.1** Reaction network for the ammoxidation of propane.

data are summarized in Table 8.5. They are cited as evidence that benzaldehyde and benzylimine are reaction intermediates in the ammoxidation of toluene, as shown in Scheme 8.2. This proposal is consistent with the classical mechanism for the addition of ammonia to aldehydes (Scheme 8.3)

**Table 8.5** Characteristic infrared bands ($cm^{-1}$) observed during ammoxidation of toluene, benzaldehyde and benzylamine over $(VO)_2P_2O_7$ at 673 K (Zhang *et al.*, 1997)

| *Reactant* | | | |
|---|---|---|---|
| *Toluene* | *Benzaldehyde* | *Benzylamine* | *Assignment* |
| 2235 (w) | 2235 (w) | | $\nu C{\equiv}N$ |
| 1675 (sh) | 1675 (sh) | 1675 (w) | $\nu HC{=}NH$ |
| 1610 (s) | 1610 (s) | 1610 (m) | $\delta NH_3 + \nu C{=}C$ |
| 1560 (sh) | 1560 (sh) | 1560 (w) | $\nu_{as}$ $COO^-$ |
| 1425 (vs) | 1425 (vs) | 1425 (vs) | $\delta NH_4^+ + \nu_S COO$ |

vs = very strong; s = strong; m = medium; w = weak; sh = shoulder.
Reprinted by permission from Elsevier Science.

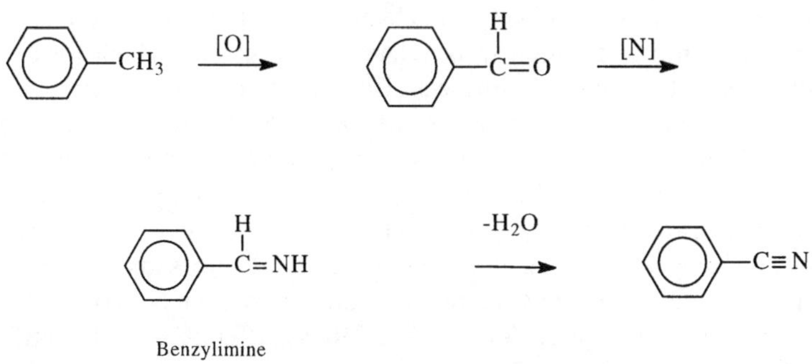


**Scheme 8.2** Reaction network of the ammoxidation of toluene (Zhang *et al.*, 1997)

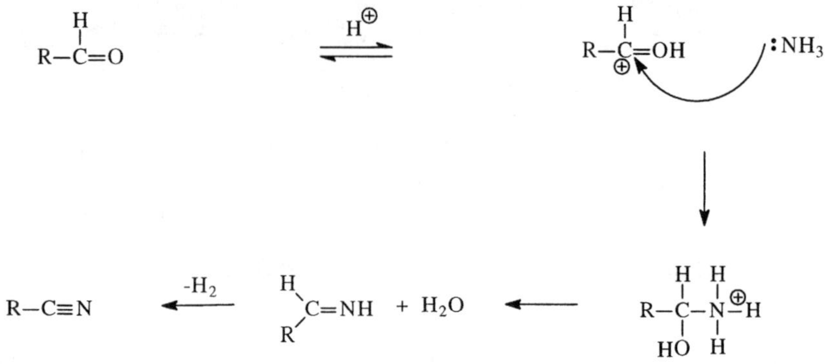

**Scheme 8.3** Mechanism for propene ammoxidation (Burrington *et al.*, 1983) (reproduced by permission of Academic Press).

## 8.5 MECHANISM OF NITROGEN INSERTION IN AMMOXIDATION

A kinetic isotope effect study has been carried out for propane ammoxidation over V–Sb–O supported on alumina and the results are summarized in Table 3.2. The results are consistent with cleavage of the weakest C—H bond in the propane as the rate determining step (Brazdil *et al.*, 1996). In addition, it is generally accepted that the kinetic isotope effect work for propene oxidation to acrolein also applies to ammoxidation (Kranzke and Keulks, 1980). In this case also the slow step is the cleavage of the weakest C—H bond in the reactant and the same $\pi$-allyl species has been associated with the reaction mechanism (Grasselli, 1997).

Hence, it appears that, in ammoxidation of propane and propene, activation of the hydrocarbon is the slow step. For propane ammoxidation the reaction conditions are more severe than for propene ammoxidation so that activation of the latter proceeds easily once the first C—H bond is activated. A remaining point of discussion is the mechanism of nitrogen insertion.

There is a wealth of infrared work related to the species which form when ammonia is adsorbed on oxide surfaces (Busca, 1993: Cavani *et al.*, 1987). A point of some importance in using the infrared technique is that these measurement are not usually performed in the same experimental conditions used for testing catalysts. This point is amply demonstrated in Figure 8.11 which shows the relative abundance of $NH_4^+$ and $NH_3$ species on a V–Sb–O catalyst as a function of evacuation temperature. These same catalysts are normally active and selective in propane ammoxidation between 673 and 773 K, but in this temperature range, as the data in Figure 8.11 demonstrate, nearly all the chemisorbed ammonia species have desorbed from the catalyst surface.

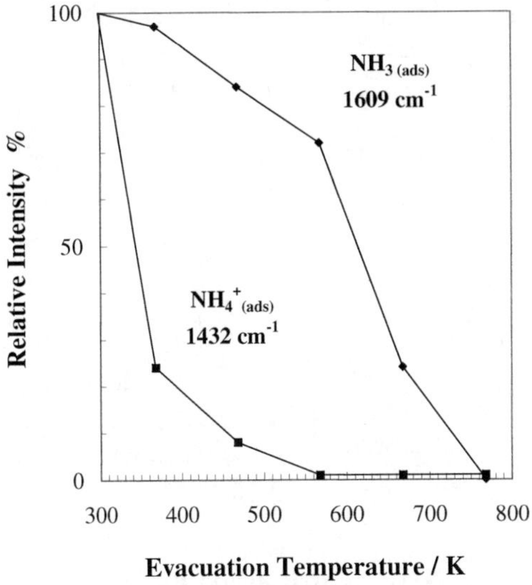


**Figure 8.11** Intensity of $NH_4^+$ and $NH_3$ species on V–Sb–O. Ammonia was contacted with the catalyst at room temperature and evacuated at successively higher temperatures (Centi *et al.*, 1997) (reproduced by permission of Elsevier Science).

There is ample evidence that added ammonia does compete with the other reactants for adsorption sites on ammoxidation catalysts. Figure 8.12 shows the influence of added ammonia on the conversion of propene and propane over bismuth molybdate catalysts and V–Sb–O catalysts respectively (Grasselli *et al.*, 1981; Centi *et al.*, 1997). At 593 K propene conversion is inhibited by added ammonia; this inhibition disappears at 703 K and was probably due to excessive adsorption of ammonia on the catalyst surface at the lower temperature thereby denying access for propene to the active sites. For propane ammoxidation over V–Sb–O at 753 K there is also inhibition, again probably due to competition between propane and ammonia for adsorption on the active sites.

A more complete study of toluene ammoxidation was undertaken by Sanati and Andersson (1991) who investigated the kinetics of toluene oxidation over a vanadium oxide catalyst supported on titanium oxide. The vanadium loading corresponded to a theoretical monolayer. The partial pressures of ammonia, oxygen and toluene were varied and rates of formation for benzonitrile measured in the range 613–673 K. Benzaldehyde was a significant second product detected in all reaction conditions.

The rate of benzonitrile formation as a function of ammonia partial pressure is presented in Figure 8.13. For low pressures the rate rapidly increases as the ammonia pressure is increased, then passes through a maximum and gradually falls to lower values. The dependence of rate of benzonitrile formation on the oxygen partial pressure is shown in Figure 8.14

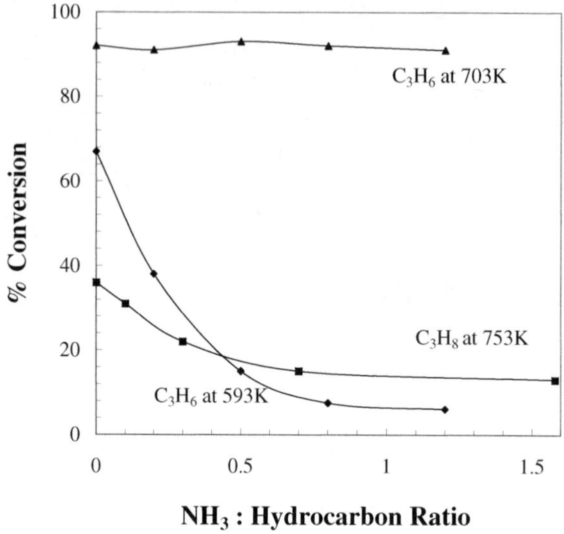


**Figure 8.12** Influence of $NH_3$ : hydrocarbon ratio on propene conversion at 593 and 703 K and on propane conversion at 753 K (Grasselli *et al.*, 1981; Centi *et al.*, 1997).

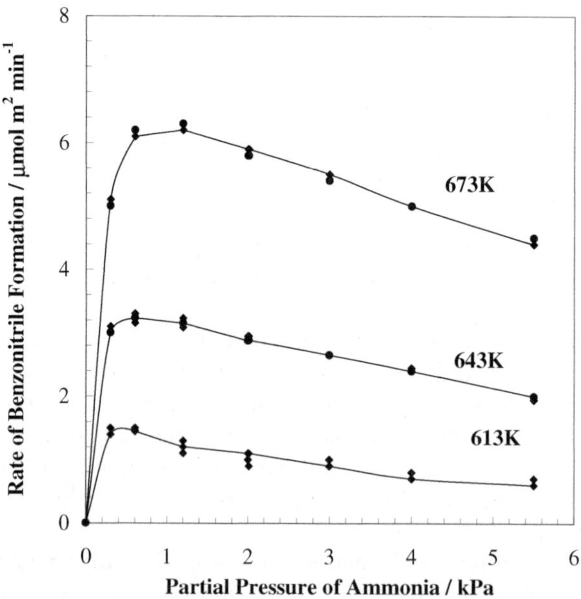


**Figure 8.13** Rate of formation of benzonitrile as a function of ammonia partial pressure at the temperatures indicated: $P_{O_2} = 11.4$ KPa; $P_{toluene} = 0.77$ KPa (Sanati and Andersson, 1991) (reproduced by permission of the American Chemical Society).

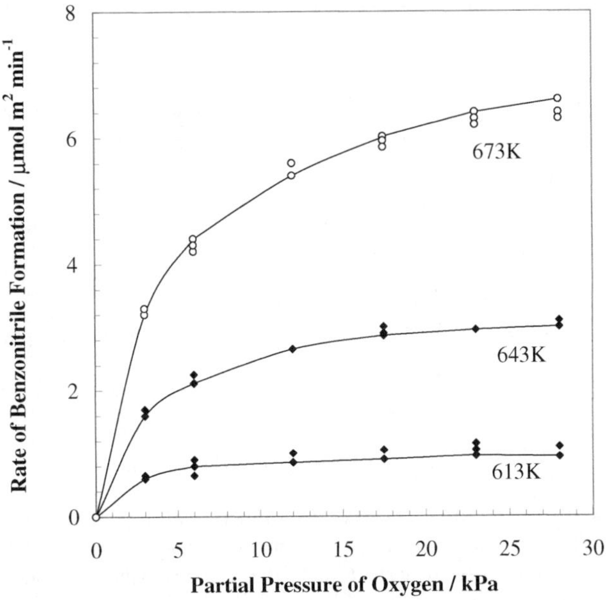


**Figure 8.14** Rate of formation of benzonitrile as a function of oxygen partial pressure at the temperatures indicated: $P_{NH_3} = 2.85$ KPa; $P_{toluene} = 0.77$ KPa (Sanati and Andersson, 1991) (reproduced by permission of the American Chemical Society).

for a range of temperatures. The rate increases initially as the oxygen pressure is increased, and then reaches a saturation level at an oxygen pressure that increases as the reaction temperature increases. A similar trend was observed for the dependence of rate on toluene partial pressure, shown in Figure 8.15 for a number of temperatures and with low and high partial pressures of ammonia. These data conform to many of the features of a Langmuir–Hinshelwood type of reaction mechanism. The data could be fitted to the following rate expression:

$$r_{CN} = \frac{l_1 \dfrac{P_O P_A}{P_T} + l_2 \dfrac{P_O {P_A}^2}{P_T}}{1 + l_3 \dfrac{P_O}{{P_T}^2} + l_4 \dfrac{P_O}{P_T} + l_5 \dfrac{P_O P_A}{P_T} + l_6 \dfrac{P_O {P_A}^2}{P_T} + l_7 \dfrac{{P_A}^2}{P_T} + l_8 \dfrac{P_A}{P_T}} \tag{8.4}$$

where $P_O$, $P_A$ and $P_T$ are the partial pressures of oxygen, ammonia and toluene respectively, and $l_1 - l_8$ are complex functions of individual rate and equilibrium constants in the overall reaction sequence. The results were interpreted in terms of benzonitrile formation over these catalysts via two routes. The first route, predominant at low ammonia partial pressures,

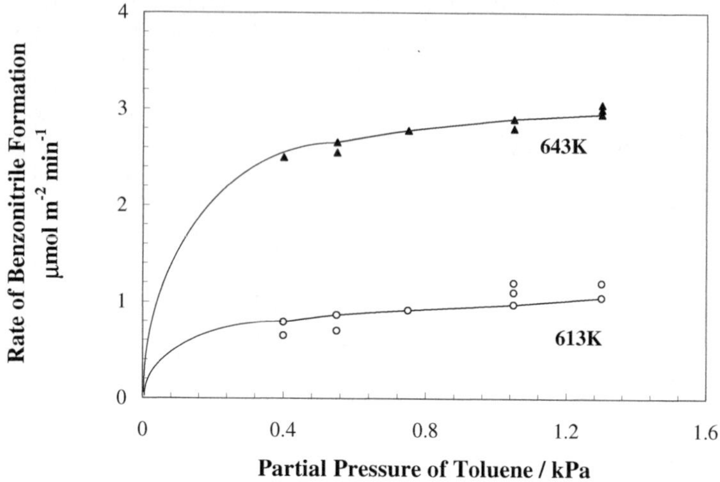


**Figure 8.15** Rate of benzonitrile formation as a function of toluene partial pressure at the ammonia partial pressures and tempertures indicated: $P_{O_2} = 11.4$ KPa (Sanati and Andersson, 1991) (reproduced by permission of the American Chemical Society).

involved reaction at a site where one ammonia molecule was adsorbed. At higher pressures of ammonia a mechanism involving adsorption of two ammonia molecules at the same site became important. In both routes the adsorption of ammonia was regarded as fast and the surface reaction was seen as the slow step in the overall mechanism. A further feature of the proposed mechanism is dissociative adsorption of oxygen on anion vacancies following desorption of the nitrile product.

The first coherent picture of propene ammoxidation was developed by Grasselli and co-workers (Grasselli, 1997), who proposed the mechanism for the insertion of nitrogen, which is shown as Scheme 8.4 (Burrington *et al.*, 1983). The central element here is the proposal that ammonia interacts with the fully oxidized bismuth molybdate catalyst to form an imine structure, namely Mo═NH, which inserts into the same $\pi$-allyl species that forms during propene oxidation. The mechanism shows cleavage of the allylic C—H bond with formation of a surface $\pi$-allyl species as the slow step. Thereafter nitrogen is inserted via an Mo═NH species in a similar fashion to the mode of insertion of oxygen during propene oxidation to acrolein over very similar catalysts. This mechanism is widely quoted for the ammoxidation of propene but there is no direct experimental evidence in support of the formation of the Mo═NH surface species.

The other major attempt to explain the mechanism of ammoxidation has centred around the classical addition reaction of ammonia with aldehydes, in which nitriles are formed through the nucleophilic attack of ammonia on the carbonyl as shown in Scheme 8.3. Subsequent elimination of hydrogen gives

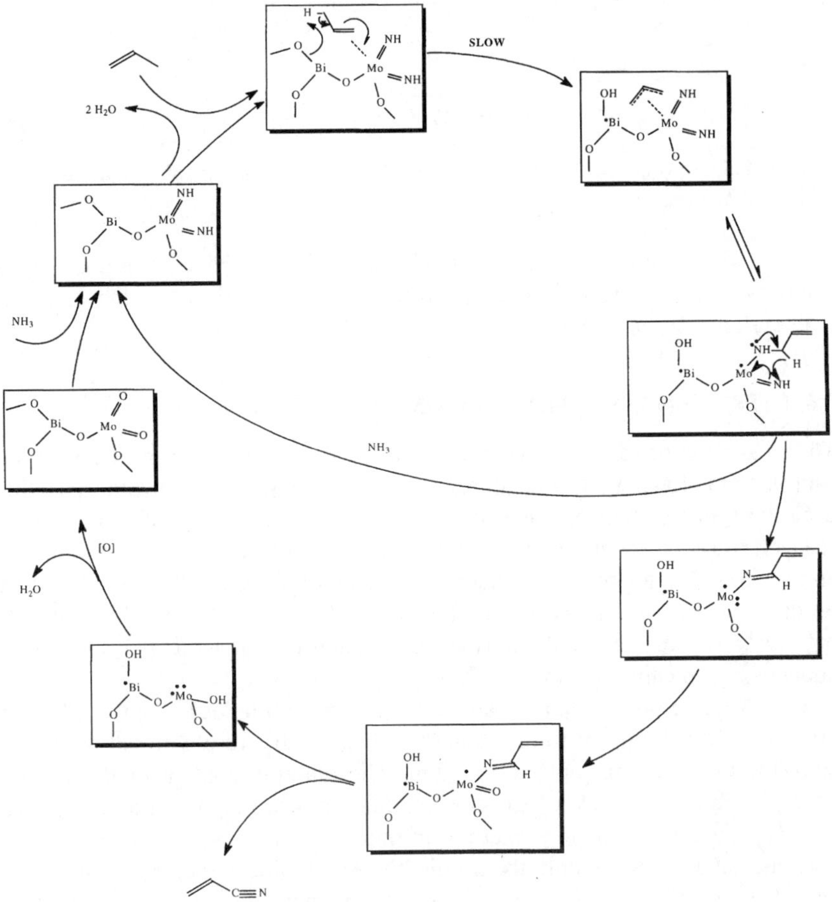


**Scheme 8.4** Classical mechanism for the addition of ammonia to aldehydes.

the product nitrile. The more recent work in ammoxidation, with the detection of propene and acrolein as a reaction intermediate in a TAP study of propane ammoxidation (Zanthoff and Buchholz, 1996) and the detection of surface benzimine species in toluene ammoxidation (Zhang *et al.*, 1997), adds support to this type of mechanism. What is not clear is the extent to which a discrete aldehyde species is formed. Some workers propose a surface-bound aldehyde like species shown in Scheme 8.5, which reacts with a surface ammonia species (Centi *et al.*, 1996). This proposal would be consistent with the small amounts of aldehyde normally detected in ammoxidation reactions.

The infrared studies to date have not been able to identify clearly the nature of the adsorbed nitrogen species which is inserted into the product, because

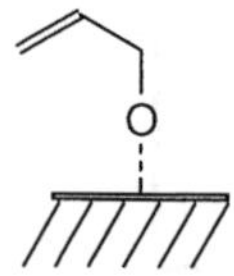

**Scheme 8.5** Proposed surface-bound intermediate for the ammoxidation of propene (Centi *et al.*, 1997).

most of the ammonia-derived nitrogen species detected on ammoxidation catalysts do not appear in the spectrum of these catalysts at the reaction temperature (Busca, 1993).

## 8.6 NITROGEN INSERTION IN AMMOXIMATION

There are about five or six industrial processes for the production of caprolactam. Each in turn features five or six steps in the synthesis, with cyclohexanone or benzene as starting materials. The number of steps can be reduced to two through the ammoximation of cyclohexanone by a mixture of ammonia and hydrogen peroxide over a titanium silicalite (TS-1) catalyst to produce cyclohexanone oxime, as shown in Scheme 8.6. A second step is still necessary, namely the acid-catalysed Beckmann rearrangement of cyclohexanoneoxime to caprolactam.

The structure of the titanium silicalite catalyst will be described in Chapter 10 as a solid solution of titanium in the silicalite structure with isomorphous substitution of silicon ions by titanium. The titanium content of the zeolite never exceeds about 2.5 %, and the titanium sites are regarded as the active sites for the ammoximation reaction. In this regard they are relatively easy to describe since they essentially mimic the crystalline structure of the host silicalite, and their small number is consistent with the normal requirement for dilution of active sites in selective oxidation reactions (see Section 3.4).

Table 8.6 presents data for the ammoximation of a number of oximes with

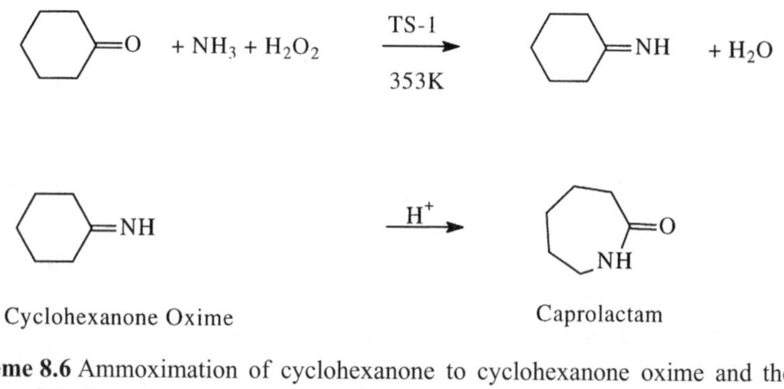


**Scheme 8.6** Ammoximation of cyclohexanone to cyclohexanone oxime and the acid catalysed Beckmann rearrangement to yield caprolactam.

**Table 8.6** Ammoximation of ketones of varying size over titanium silicalite-1 $H_2O_2$: ketone = 1.02–1.08: 1; temperature = 353 K
(Zecchina *et al.*, 1993)

| *Ketone* | *Structure* | *Adsorption/ mmol (100 g of TS-1)$^{-1}$* | *% Conv*[a] | *% Sel* | *% Yield*[b] |
|---|---|---|---|---|---|
| Cyclohexanone | 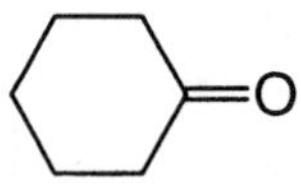 | 13 | 92.4 | 99.9 | 93.8 |
| Cyclpentanone | 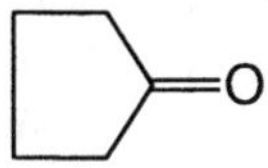 | — | 99.9 | 90.4 | 84.8 |
| 4-*t*-Butyl-cyclohexanone | 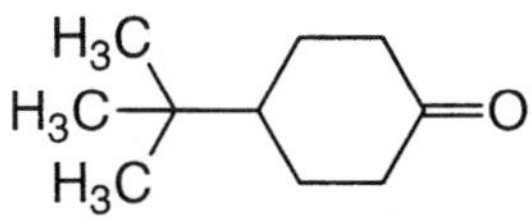 | 0 | 93.6 | 91 | 82.8 |
| Cyclododecanone | 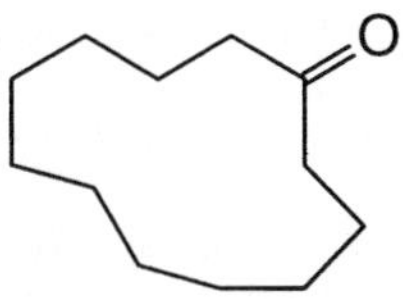 | 0 | 53.0 | 99.9 | 52 |
| Acetophenone | 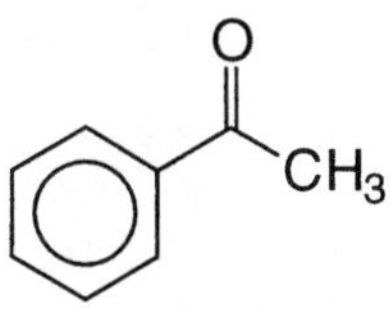 | — | 50.8 | 98.1 | 48.9 |
| 2-*t*-Butyl-cyclohexanone | 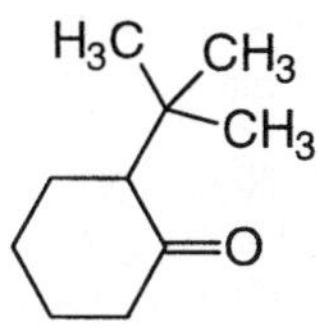 | 0 | 3.8 | 14.4 | 0.5 |

[a]% Conversion of the ketone.
[b]% Yield based on $H_2O_2$ consumed.


differing molecular sizes. Adsorption was not observed for ketones larger than cyclohexanone. The other ketones listed were too large to enter into the zeolite microporosity. However, all the ketones, with the exception of 2-*t*-butyl-cyclohexanone, were readily ammoximized over titanium silicalite. When 2-*t*-butyl-cyclohexanone was oximized in standard solution phase reaction conditions, in which diffusion was not a factor, it did not react at all, indicating that the bulky *t*-butyl group adjacent to the ketone functional group inhibited the reaction. Hence, on the TS-1 even those oximes which could not

**Scheme 8.7** Reaction of hydroxylamine with cyclohexanone.

reasonably be expected to diffuse into the interior of the zeolite are ammoximized as easily as the smaller ketones. These data point to a mechanism in which the smaller reactants, ammonia and hydrogen peroxide, diffuse into the zeolite and react there to form an intermediate which is stable in the absence of the ketone but reacts rapidly when contacted with the ketone at the exit of the pores. The proposed intermediate is hydroxylamine ($NH_2OH$), which is known to react with ketones in mild condition to produce oximes as shown in Scheme 8.7 (Zecchina *et al.*, 1993).

## REFERENCES TO CHAPTER 8

ACS Office of Public Outreach (1996) *The SOHIO Acrylonitrile Process*. American Chemical Society, Division of the History of Chemistry and the Office of Public Outreach.

Brazdil JF (1991) Acrylonitrile, In *Kirk–Othmer Encyclopedia of Chemical Technology*, Vol. 1. Wiley, New York, 4th edition, pp. 352–369.

Brazdil LA, Ebner AM and Brazdil JF (1996) Kinetic isotope effect in the ammoxidation of propane over alumina supported vanadium antimony oxide catalyst. *J. Catal.* **163**: 117–121.

Burrington JD, Kartisek CT and Grasselli RK (1983) Mechanism of nitrogen insertion in ammoxidation catalysis. *J. Catal.* **81**: 489–498.

Busca G (1993) Selective and nonselective pathways in oxidation and ammoxidation of methylaromatic compounds over vanadia–titania catalysts. Fourier transform infrared spectroscopic studies. In *Catalytic Selective Oxidation. ACS symp.* Ser. 523, Oyama ST and Hightower JW (Eds). American Chemical Society, Washington, DC, pp. 168–182.

Callaghan JL, Grasselli RK, Milberger EC and Strecker HA (1970) Oxidation and ammoxidation of propylene over bismuth molybdate catalysts. *Ind. Eng. Chem. Prod. Res. Dev.* **9**: 134–142.

Cassidy FE, (1999) Thesis, University of Limerick.

Cavani FE, Foresti, Trifiro F and Busca G (1987) Nature of active species in the ammoxidation of toluene over $V_2O_5/TiO_2$ catalysts prepared by flash-drying. *J. Catal.* **106**: 251–262.

Centi G (1996) Nature of active layer in vanadium oxide supported in titanium oxide and contriol of its reactivity in the selective oxidation and ammoxidation of alkylaromatics. *Appl. Catal. A General* **147**: 267–298.

Centi G, Grasselli RK and Trifiro F (1992) Propane ammoxidation to acrylonitrile—An overview. *Catal. Today* **13**: 661–666.

Centi G, Marchi F and Perathoner S (1996) Surface chemistry of V–Sb–oxide in relation to the mechanism of acrylonitrile synthesis from propane. *J. Chem. Soc., Faraday Trans.* **92** (24) 5141–5149.

Centi G, Marchi F and Perathoner S (1997) Effect of ammonia chemisorption on the surface reactivity of V–Sb–oxide catalysts for propane ammoxidation. *Appl. Catal. A General* **149**: 225–244.

Chemical Week (1997) (http:\\www.chemweek.com\marketplace\prod_focus.html).

Cox JD and Pitcher G (1970) *Thermochemistry of Organic and Organometallic Compounds*. Academic Press, New York.

Fairley P (1998) Process technology–Piloting propane to acrylonitrile. *Chem. Week* **160** (36) 45–45.

Froment GF and Bischoff KB (1990) *Chemical Reactor Analysis and Design*. 2nd edition, Wiley, New York.

Grasselli RK, (1997) Ammoxidation. In *Handbook of Heterogeneous Catalysis*. VCH Weinheim, pp. 2302–2326.

Grasselli RK, Burrington JD and Brazdil JF (1981) Mechanistic features of selective oxidation and ammoxidation catalysis. *Faraday Discuss. Chem. Soc*. **72**: 203–223.

Kranzke LD and Keulks GW (1980) The catalytic oxidation of propylene:VI. Mechanistic studies utilizing isotopic tracers. *J.Catal*. **61**: 316–325.

Martin A and Lucke B (1996) Ammoxidation of methylaromatics over vanadium phosphate catalysts. I. Effect of size, position and electronic properties of substituents on the catalytic conversion of methylaromartics to the corresponding nitriles. *Catal. Today* **32**: 279–283.

Rizayev RG, Mamedov EA, Vislovskii VP and Sheinen VE (1992) Some fundamental and practical aspects of the ammoxidation of alkylbenzenes. *Appl. Catal. A General* **83**: 103–140.

Sanati M and Andersson A (1991) Kinetics and mechanism in the ammoxidation of toluene over a $TiO_2$ (B)-supported vanadium oxide monolayer catalyst. 1 Selective reactions. *Ind. Eng. Chem. Res*. **30**: 312–320.

Sanati M, Wallenberg LR, Andersson A, Jansen S and Tu Y (1991) Vanadia catalysts on anatase, rutile and $TiO_2$(B) for the ammoxidation of toluene: An ESR and high-resolution electron microscopy characterization. *J. Catal*. **132**: 128–144.

Sleight AW (1977) Crystal chemistry and catalytic properties of oxides with the scheelite structure. In *Advanced Materials in Catalysis. Materials Science and Technology Series* Burton JJ and Garten T (Eds) Academic Press, New York, Chapter 6, pp. 181–208.

Zabetakis MC (1965) *Flammability Characteristics of Combustible Gases and Vapors*. US Dept of the Interior, Bureau of Mines, Washington, DC.

Zanthoff HW and Buchholz SA (1996) In *Heterogeneous Hydrocarbon Oxidation. ACS Symp. Ser*. 638, Warren BK and Oyama ST (Eds). Americal Chemical Society, Washington, DC, Chapter 19.

Zanthoff HW and Buchholz SA (1997) On the mechanism of ammoxidation of propane over vanadium oxide catalysts. *Catal. Lett*. **49**: 213–217.

Zecchina A, Spoto G, Bondiga S, Geobaldo F, Petrini G, Leofanti G, Padovan M, Mantegazza M and Roffia P (1993) Ammoximation of cyclohexanone on titanium silicalite: Investigation of the reaction mechanism. *Stud. Surf. Sci. Catal*. **74**: 719–729.

Zhang Y, Martin A, Berndt H, Lucke B and Meisel M (1997) FTIR investigation of surface intermediates formed during the ammoxidation of toluene over vanadyl pyrophosphate. *J. Mol. Catal. A Chem*. **118**: 205–214.

# 9

# Catalytic Oxidation Reactions Involving Chlorine: Oxychlorination and the Oxidative Destruction of Chlorinated Hydrocarbons

9.1 Introduction 265
9.2 Vinyl chloride by oxychlorination 266
9.3 Technology for vinyl chloride production 268
9.3.1 Direct chlorination of ethylene 268
9.3.2 Oxychlorination of ethylene 270
9.4 Oxychlorination catalyst composition, preparation and characterization 275
9.5 The mechanism of oxychlorination 276
9.6 Destruction of chlorinated hydrocarbons by catalytic oxidation 285
9.7 By-product formation during the catalytic oxidation of chlorinated hydrocarbons 290
9.8 Polychlorinated biphenyls, dioxins and the environment 295
References to Chapter 9 299

Summary
Vinyl chloride monomer is the largest bulk chlorinated organic chemical. It is currently manufactured by a materially and energetically balanced chlorination–oxychlorination–cracking process. The oxychlorination part of the process involves reacting a mixture of ethylene, hydrochloric acid and oxygen over a heterogeneous catalyst. The product finds wide spread application in the polymer industry, as well as serving as a starting point for the manufacture of other chlorinated and fluorinated organic chemicals. In addition, this chapter describes catalytic technologies that are being evaluated for the destruction, through oxidation, of chlorinated hydrocarbons in waste stream, without the production of environmentally harmful

by-products, such as dioxins and polychlorinated biphenyls.

This chapter describes

Technology for vinyl chloride production,
The preparation and composition of oxychlorination catalysts,
The mechanism of oxychlorination,
Catalytic oxidation for the destruction of chlorinated hydrocarbons,
By-product formation during catalytic oxidation,
Polychlorinated biphenyls and dioxins and their impact on the environment.

## 9.1 INTRODUCTION

Interest in catalysts capable of manipulating chlorine began with the Deacon reaction, which is used commercially for the generation of chlorine gas from waste streams of hydrochloric acid. The basic reaction involved is:

$$4HCl + O_2 \rightarrow 2Cl_2 + 2H_2O \tag{9.1}$$

and it operates in the temperature range 673–733 K. Copper chloride catalysts promoted with potassium and rare earth chlorides were found to work well for this reaction and there is evidence that the supported $CuCl_x$–KCl–$LaCl_3$ phase is present as a liquid or melt at the reaction temperature (Naworski and Velez, 1983).

In the 1960s, oxychlorination of ethylene:

$$CH_2{=\!=}CH_2 + 2HCl + 0.5O_2 \rightarrow CH_2Cl{-\!-}CH_2Cl + H_2O \tag{9.2}$$

was developed. Promoted copper chloride catalysts, similar in composition to those used in the Deacon reaction, were found to catalyse this reaction efficiently at lower temperatures (493–503 K). Oxychlorination now forms the basis of all major commercial vinyl chloride monomer production units.

With the emergence of environmental issues associated with the use and widespread dispersion throughout the environment of polychlorinated hydrocarbons, the option of applying oxidation catalysts to their destruction has been intensively researched. These developments have been coupled with concerns that surface catalysed transformations, occurring on particulate matter generated within flames (e.g. copper particles in fly-ash present during combustion of municipal waste), may be involved in the insertion of chlorine into incompletely combusted hydrocarbon molecules, leading to the formation of polychlorinated hydrocarbons, and ultimately toxic chemicals such as polychlorinated biphenyls (PCB), dibenzo-*p*-dioxins (PCDDs), dibenzofurans (PCDFs) (Froese and Hutzinger, 1997).

This chapter is divided into two sections. The first deals with the chemistry and technology involved in the insertion of chlorine into ethylene by oxychlorination to form 1,2-dichloroethane and vinyl chloride monomer. The second section deals with the destruction of chlorinated hydrocarbons by catalytic oxidation.

## 9.2 VINYL CHLORIDE BY OXYCHLORINATION

Vinyl chloride monomer is the largest tonnage organochlorine compound in current use. It is manufactured from ethylene, chlorine and oxygen via a coupled series of chlorination, oxychlorination and cracking reactions. Vinyl chloride is used principally in the manufacture of polyvinyl chlorine. On a smaller scale it is the starting point for the manufacture of other organochlorine and organofluorine compounds. A brief summary of the principal uses of vinyl chloride monomer is presented in Scheme 9.1 (Reed, 1993).

In the 1930s vinyl chloride was manufactured by the addition of hydrogen chloride to acetylene:

$$CH{\equiv}CH + HCl \rightarrow CH_2{=}CHCl \tag{9.3}$$

This process has long been replaced by processes that use the cheaper and more abundant ethylene as feed stock. Modern production methods couple direct ethylene chlorination and cracking with oxychlorination and cracking in a manner that is outlined in Scheme 9.2. The driving force for the

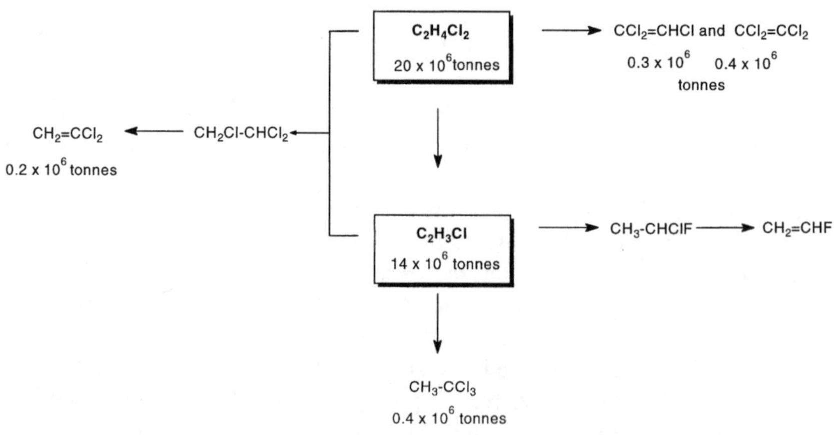


**Scheme 9.1** Organochlorines and fluorines from 1,2-dichloroethane and vinyl chloride monomer (Reed, 1993).

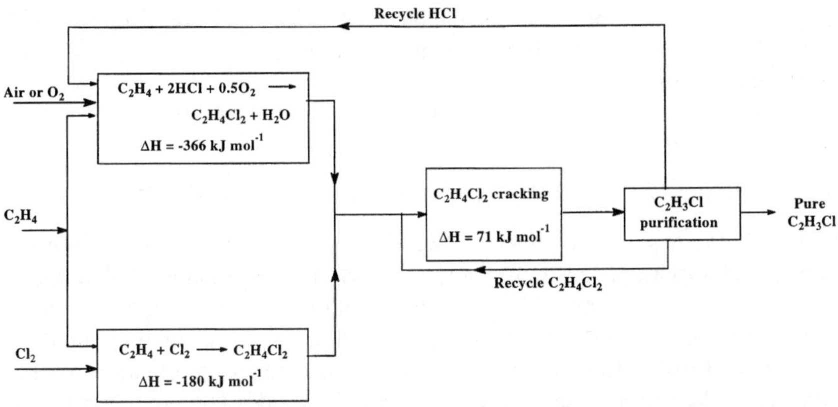


**Scheme 9.2** Balanced chlorination–oxychlorination–cracking route to vinyl chloride monomer from ethylene, chlorine and oxygen.

development of the oxychlorination process was to utilize efficiently the HCl produced in the cracking of 1,2-dichloroethane. The recycled HCl is fed to the oxychlorination reactors (Naworski and Velez, 1983). The overall process is balanced in the sense that just enough HCl is produced to feed the oxychlorination plant and in terms of exothermic and endothermic steps.

The overall reaction can be represented by:

$$2CH_2{=\!=}CH_2 + Cl_2 + 0.5O_2 \rightarrow 2CH_2{=\!=}CHCl \tag{9.4}$$

where 1 mol of vinyl chloride is produced via the direct chlorination route and the second comes from the oxychlorination route. Data pertinent to the oxychlorination of ethylene are presented in Table 9.1.

Because heterogeneous catalysis is involved only in the oxychlorination

**Table 9.1** Summary data for vinyl chloride production by oxychlorination (Cox and Pitcher, 1970; Zabetakis, 1965)

Oxychlorination step

$CH_2{=\!=}CH_2 + 2HCl + 0.5\ O_2 \rightarrow CH_2Cl{-\!-}CH_2Cl + H_2O$ $\quad \Delta G_{298} = -300\ kJ\ mol^{-1}$

Cracking step

$CH_2Cl{-\!-}CH_2Cl \rightarrow CH_2{=\!=}CHCl + HCl$ $\quad \Delta G_{298} = 71\ kJ\ mol^{-1}$

| Catalyst | Support | Additive(s) | Temp/K | Pressure/bar |
|---|---|---|---|---|
| $CuCl_2$ | $\gamma$-$Al_2O_3$ | KCl | 493–503 | 2–3 |
| 3–7 wt% Cu | 150–250 $m^2\ g^{-1}$ | 3 wt% | | |

| | | $LFL_{298}$/% | $UFL_{298}$/% | AIT/K |
|---|---|---|---|---|
| | $C_2H_4$ | 2.7 | 36 | 763 |
| | $C_2H_3Cl$ | 3.6 | 33 | |

part of the balanced process, this chapter will concentrate on this step, with just short descriptions of the direct chlorination, cracking and downstream purification steps.

## 9.3 TECHNOLOGY FOR VINYL CHLORIDE PRODUCTION

Figure 9.1 shows a detailed flow diagram for the balanced direct chlorination–oxychlorination–cracking route to vinyl chloride monomer. The major units in the flow diagram include the direct chlorination reactor, an oxychlorination reactor, which can be the fixed or fluid bed type, the cracking unit and a number of distillation and recycle units. In the direct chlorination reactor, ethylene and chlorine react to give ethylene dichloride. The product ethylene dichloride is mixed with the ethylene dichloride from the oxychlorination reactor and separated from the heavy residues. Both oxychlorination and direct chlorination are exothermic reactions and the heat of reaction is used for the distillation process. The ethylene dichloride thus purified is thermally cracked to vinyl chloride and hydrochloric acid. The effluent from the thermal cracker is quenched and fed to a separation unit. Hydrochloric acid is first taken off overhead and recycled to the oxychlorination reactor. Hydrogen can be added prior to the oxychlorination reactor to convert trace amounts of acetylene into ethylene. This leads to more efficient utilization of chlorine. The light ends from the oxychlorination reactor and the heavy ends from the direct chlorination rector can be treated in a Catoxid reactor. This part of the process recovers chlorine as HCl and generates steam (Cowfer and Gorensek, 1997).

### 9.3.1 Direct chlorination of ethylene

Direct chlorination is generally carried out in liquid 1,2-dichloroethane in a bubble column reactor. Ethylene and chlorine dissolve in the product phase; sometimes a small amount of dissolved ferric chloride is used as catalyst but the reaction rate is determined by the rate of ethylene absorption in the liquid phase. The reaction proceeds by electrophilic addition. It is very efficient and selectivity to 1,2-dichloroethane is greater than 99 %. The main by-product is 1,1,2-trichlorethane formed by the reaction of chlorine radicals with 1,2-dichloroethane. Oxygen present at less than 0.5 % can suppress this reaction. The reaction is exothermic, as shown in Scheme 9.2, and the heat of reaction is used to maintain the reaction close to the boiling point of 1,2-dichloroethane. This allows the product to vaporize as it is formed. At this point the crude 1,2-dichlorethane manufactured by direct chlorination is mixed with the crude product coming from the oxychlorination side for further downstream treatment and cracking (Cowfer and Gorensek, 1997).

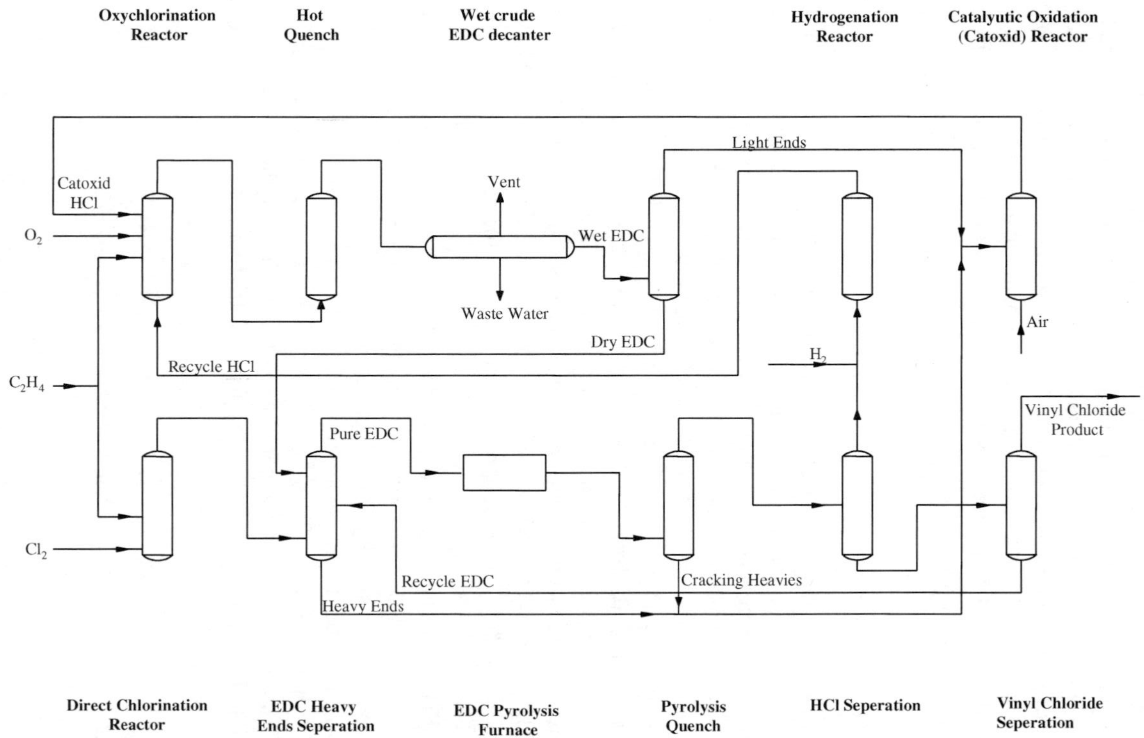


**Figure 9.1** Process flow sheet for a balanced vinyl chloride production facility (Cowfer and Gorensek, 1997) (reproduced by permission of John Wiley and Sons).

### 9.3.2 Oxychlorination of ethylene

There are about ten mature processes for vinyl chloride production by the oxychlorination of ethylene (Naworski and Velez, 1983). Fluidized bed and fixed bed operations are possible and each of these reactor types can be operated with air or oxygen as the oxidant. In each case recycled HCl is fed to the reactor and the crude 1,2-dichlorethene product, after suitable treatment, is mixed with the same product coming from the direct chlorination unit prior to cracking to vinyl chloride monomer and hydrochloric acid. These processes are very efficient, allowing up to 99.9 % HCl conversion and selectivity to 1,2-dichloroethane above 97 %.

*Fixed bed operation*

Fixed bed oxychlorination units are usually of multistage design in which the ethylene and HCl are passed successively through three multi-tubular fixed bed reactors. Air is injected separately into each of the reactors.

The oxychlorination reaction is exothermic (see Scheme 9.2) and discernible hot spots occur in fixed bed oxychlorination reactors, even though reactors and catalysts are designed to minimize these. Typical temperature profiles for the first of three fixed bed reactors operating in series with air or oxygen as oxidants are shown in Figure 9.2(A). The hot spots are less pronounced in the second and third reactors (Figure 9.2(B)) (Markeloff, 1984). (The reasons for two hot spots in the first reactor will be explained below.) The hot spots are most pronounced close to the reactor inlet, where the reaction proceeds vigorously because of the ready availability of all reactants. The hot spot may move somewhat along the reactor bed as the catalyst particles lose their activity owing to deposition of poisons or deactivation through sublimation of copper chloride.

Minimization of hot spots is an important consideration in the operation of oxychlorination reactors. Cracking of 1,2-dichlorethane can start in the oxychlorination unit if the temperature is too high, and the product vinyl chloride can be further oxychlorinated to 1,1,2-trichloroethane, an unwanted by-product.

$$CH_2Cl{-}CH_2Cl \rightarrow CH_2{=}CHCl + HCl \quad (9.5)$$

$$CH_2{=}CHCl + 2HCl + 0.5O_2 \rightarrow CHCl_2{-}CH_2Cl \quad (9.6)$$

Higher temperatures favour complete oxidation of ethylene to carbon oxides, as shown in Figure 9.3. Coke formation and loss of copper by sublimation of copper chloride can also occur (Markeloff, 1984).

In some instances hot spot reduction is achieved by diluting the catalyst with a material, such as graphite. The reactor is divided into zones and the catalyst to diluent ratio varies from zone to zone, as shown in Table 9.2. The

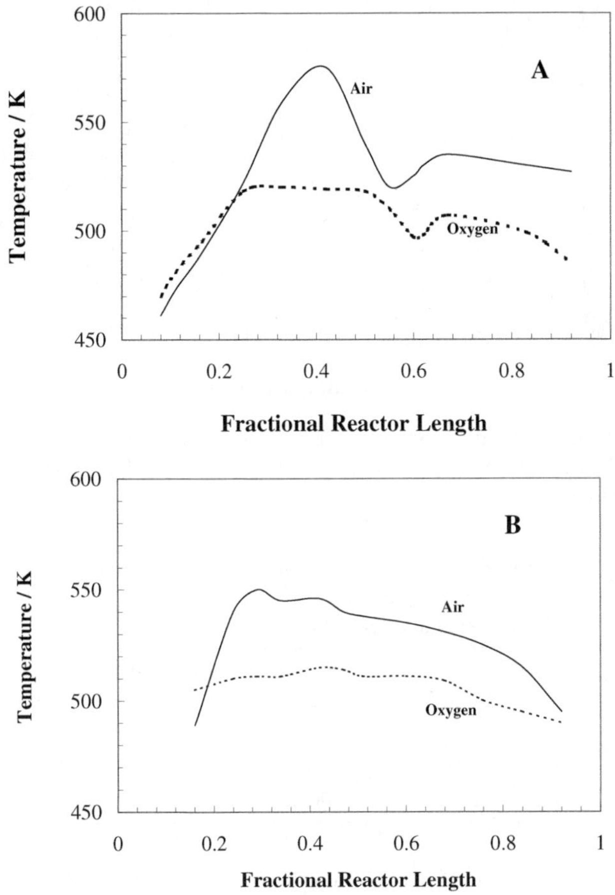


**Figure 9.2** Temperature profiles along the lengths of two oxychlorination reactors operating in series (A followed by B) with air and oxygen as oxidants (Markeloff, 1984) (reproduced by permission of Hydrocarbon Processing © Gulf Publishing Co.).

amount of catalyst and the local temperature determine the amount of reaction which occurs in a particular zone. By restricting the amount of catalyst at the inlet, where most of the reaction would otherwise occur, the reaction can be made to occur more evenly along the entire reactor length, minimizing the tendency for hot spot occurrence (Naworski and Velez, 1983).

A second strategy used with fixed bed reactors operated in series is to use catalysts of differing compositions. A typical strategy is shown in Scheme 9.3. Three catalysts are prepared with varying $CuCl_2$ and KCl loadings. The least active catalyst has the lowest copper loading (6 wt% $CuCl_2$–3 wt% KCl). This catalyst is loaded into the first zone (60 vol%) of reactor no. 1, and the remaining 40 % of the reactor is loaded with the catalyst with the highest

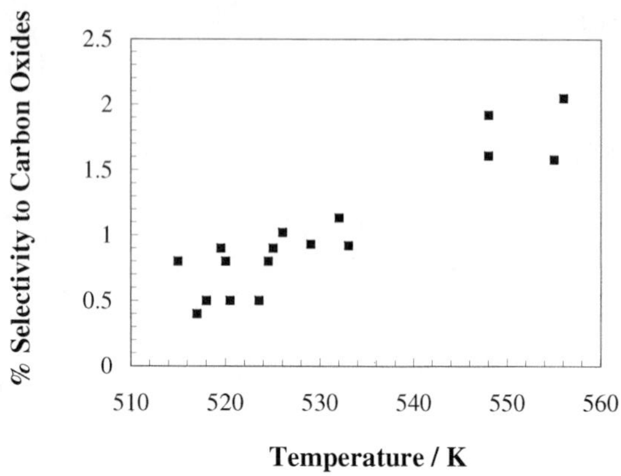


**Figure 9.3** Selectivity to carbon oxides during oxychlorination as a function of hot spot temperature (Markeloff, 1984) (reproduced by permission of Hydrocarbon Processing © Gulf Publishing Co.).

**Table 9.2** Ratio of catalyst to diluent in a fixed bed oxychlorination reactor (Naworski and Velez, 1983)

| *Zone* | *Catalyst/vol %* | *Diluent/vol%* |
|---|---|---|
| 1 | 7 | 93 |
| 2 | 15 | 85 |
| 3 | 40 | 60 |
| 4 | 100 | 0 |

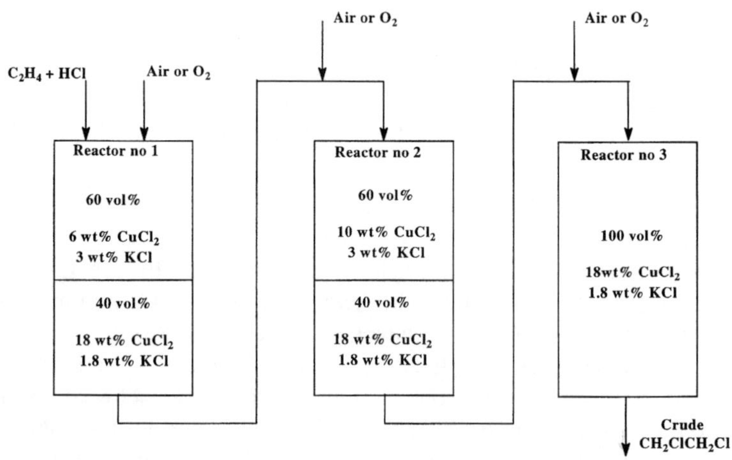


**Scheme 9.3** Strategy for the staged loading of catalysts in fixed bed reactors operated in series for the oxychlorination of ethylene (Naworski and Velez, 1983).

copper loading(18 wt% $CuCl_2$–1.8 wt% KCl). The first 60 % of reactor no. 2 is loaded with a catalyst of intermediate loading (10 wt% $CuCl_2$–3 wt% KCl) and activity and the remainder is filled with the most active catalyst. The last reactor is completely filled with the most active catalyst. In this way the reaction is made to occur as much as possible along the entire length of the reactor (Naworski and Velez, 1983). The temperature profiles shown in Figure 9.2(A) and (B) show the effect of using staged catalyst loading, with the appearance of two hot spots, corresponding to the two sections of a reactor. Hot spots tend to be less pronounced in the second and third reactors.

Staged addition of oxygen or air along the entire reactor length can also help to spread the reaction and further minimize hot spots.

*Fluidized bed operation*

Compressed air or oxygen, HCl recycled from the EDC cracking unit and ethylene are introduced separately at the bottom of a fluidized bed reactor and passed through a $CuCl_2/Al_2O_3$ catalyst in the temperature range 493–503 K. This temperature range tends to be a little lower than that used in fixed bed reactors. The Cu loading is restricted to less than 10 wt%. Above this value the copper chloride phase becomes soft and sticky and the catalyst tends to cake, making fluidization impossible. In addition, the particle size should be as uniform as possible in the range $(20–80) \times 10^{-6}$ m to achieve good fluidization characteristics, the material must be attrition resistant and it should show good adsorptivity of the impregnating salts. Even with good preparation procedures, adverse operating conditions can lead to the growth of dendritic cupric chloride on the surface of individual catalyst particles, increasing particle to particle interactions and agglomeration.

In fluidized bed based oxychlorination operations, the reactor effluent gases pass through a cyclone separator where catalyst fines can be trapped and removed from the system. Ethylene dichloride yields above 98 % based on HCl consumed and above 96 % based on ethylene consumed have been reported (Naworski and Velez, 1983).

*Air or oxygen as oxidant*

Air is a cheap oxidant, but by definition brings with it four times as much inert nitrogen as oxidant. This inert can be an important factor in the safe operation of the plant outside the explosion envelope, but there is a large cost in terms of the energy requirement to pump this material through the plant, and in terms of environmental impact by virtue of the amount of material which must be vented (typically 1 kg of vented gas per kg of 1,2-dichloroethane produced). Oxychlorination plants are operated so as to achieve nearly full conversion of HCl. Air based plants tend to use slight excesses of ethylene (2–20 %) and air (10–80 %) to ensure that all the HCl reacts. Unreacted ethylene is sent to a small chlorination unit or destroyed by incineration (Naworski and Velez, 1983).

Oxygen based plants operate with a large excess of ethylene; unreacted ethylene is recycled. There are costs associated with the production of pure oxygen, but the volume of vented material can be 100 times less than with a similarly sized air based plant. In addition, the throughput of an oxygen based plant tends to be greater. There is some evidence that in fixed bed reactors hot spot temperatures tend to be less than with air based plants (see Figure 9.2).

Table 9.3 compares performance indicators of an oxychlorination plant before and after conversion from air to oxygen.

*Downstream treatment and product purification*

The main unit operations which follow the chlorination and oxychlorination steps are quenching, distillation and cracking. The sequence of steps for a fully balanced process is presented in Figure 9.1.

1,2-Dichloroethane from the direct chlorination route has a typical purity in excess of 99.5 wt%; small amounts of ferric chloride catalyst are separated when the 1,2-dichloroethane is distilled from the bubble column reactor.

1,2-Dichloroethane from the oxychlorination route is less pure and must be hot quenched, washed with water and a caustic solution to remove water soluble impurities, such as chloral. The wet 1,2-dichloroethane effluent is then distilled to remove water, and the purified 1,2-dichloethane is sent for cracking.

The cracking process at 750–800 K and 15–30 atm is a homogeneous free

**Table 9.3** Comparison of operating selectivities of air based and oxygen based oxychlorination plants
(Markeloff 1984)

| | *Air based plant* | *Oxygen based plant* |
|---|---|---|
| Selectivity to | | |
| 1,2-Dichloroethane | 95.1 | 97.3 |
| Ethyl chloride | 1.7 | 1.5 |
| Carbon oxides | 1.8 | 0.7 |
| 1,1,2-Trichloroethane | 0.9 | 0.1 |
| Others | 0.5 | 0.4 |
| %HCl converted | 99.1 | 99.8 |

radical reaction involving initiation, propagation and termination steps, as follows:

$$\textit{Initiation} \quad CH_2Cl{-}CH_2Cl \rightarrow CH_2Cl{-}CH_2^{\bullet} + Cl^{\bullet} \qquad (9.7)$$

$$\textit{Propagation} \quad Cl^{\bullet} + CH_2Cl{-}CH_2Cl \rightarrow CH_2Cl{-}CHCl^{\bullet} + HCl \qquad (9.8)$$

$$CH_2Cl{-}CHCl^{\bullet} \rightarrow CH_2{=}CHCl + Cl^{\bullet} \qquad (9.9)$$

$$\textit{Termination} \quad Cl^{\bullet} + CH_2Cl{-}CH_2^{\bullet} \rightarrow CH_2{=}CHCl + HCl \qquad (9.10)$$

Added propene slows down the cracking reaction by the consumption of some of the radical chain propagators. The allylic C—H bonds in propene are very susceptible to hydrogen abstraction producing an allylic radical which can react with chlorine radicals to form allyl chloride. In this process two radical propagators are consumed.

Per pass conversions tend to be in the range 50–65 % and selectivities to vinyl chloride are in excess of 99 %. Rapid quenching is necessary to avoid coke formation downstream, and this is achieved by rapid contact with cold 1,2-dichloroethane. The final steps involve distillation to recover and recycle HCl, followed by distillation to recover vinyl chloride. Unreacted 1,2-dichloroethane is recycled for cracking (Naworski and Velez, 1983; Cowfer and Gorensek, 1997).

## 9.4 OXYCHLORINATION CATALYST COMPOSITION, PREPARATION AND CHARACTERIZATION

Copper chloride in one form or another seems to be the material of choice for reactions involving the formation and destruction of chlorine or chloride species in the gas phase. Modern oxychlorination catalysts are usually made up of $CuCl_2$, one or more alkali metal chloride and one or more rare earth chloride, often lanthanum chloride, all supported on alumina. The K : Cu atomic ratio is usually about 1 : 1 and the rare earth : Cu ratio is usually about 0.15 : 1 (Villadsen and Livbjerg, 1978).

Volatility of copper chlorides from catalysts used in the Deacon process was a problem prior to the use of KCl as a promoter. KCl is usually seen as a promoter that controls the volatility of the copper chloride phase. In addition $LaCl_3$ and KCl both improve the rate of uptake of oxygen by the catalyst (see below).

Figure 9.4 presents a phase diagram for the CuCl–KCl system. The diagram shows regions for CuCl, KCl and $K_2CuCl_3$ and a liquid region (Fontana *et al.*, 1952). A feature of the diagram is a eutectic, a specific composition which gives the lowest melting point, in this case at 423 K. In fact the phase diagram for the system $CuCl_2$–CuCl–KCl becomes more

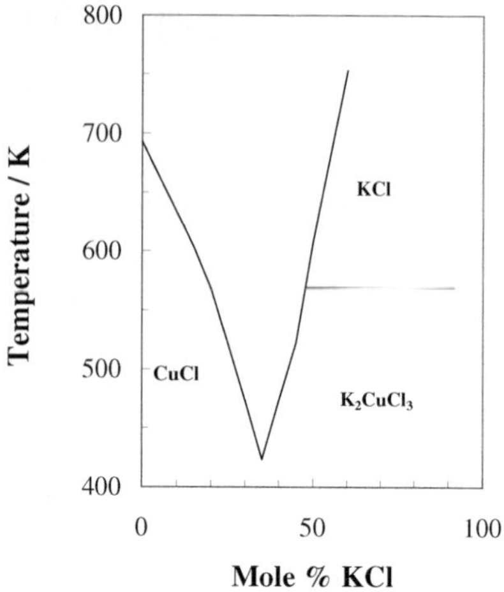


**Figure 9.4** Phase diagram for the system CuCl–KCl (Fontana *et al.*, 1952) (reproduced by permission of the American Chemical Society).

complicated, and the position of the eutectic depends on the $Cu^{2+} : Cu^{1+}$ ratio in the composition. Eutectics occur for all $Cu^{2+} : Cu^{1+}$ ratios and cover the range 423–650 K. Table 9.4 lists the eutectic compositions and temperatures for a range of $CuCl_2$–CuCl–KCl compositions (Fontana *et al.*, 1952).

There is a considerable body of evidence that at the operating conditions some of the supported phase is present on the catalyst as a liquid. Recall that a typical reaction range is 493–503 K. This should be regarded as the nominal temperature range, as hot spots can often elevate the real temperature much above this value.

The exact formulation of modern oxychlorination catalysts varies with the type of reactor in which it is to be used. Copper loadings for fluidized bed operation tend to be somewhat lower, because of the need to accommodate as much as possible of the copper phase inside the porosity, in order to avoid a sticky phase on the external surface where adhesion of particles may occur.

## 9.5 THE MECHANISM OF OXYCHLORINATION

It is tempting to think that oxychlorination proceeds via the generation of chlorine from HCl oxidized in the Deacon reaction. However, the much higher reaction temperature required for the Deacon reaction makes this unlikely. However, there is evidence that olefins can be chlorinated by supported cupric chloride. Arganbright and Yates (1962) passed a number of alkenes through a fluidized bed of $CuCl_2$ supported on pumice and measured the temperature required for the observation of chlorinated products. Their experiments are summarized in Table 9.5. Typically a fluidized bed containing 0.29 mol of cupric chloride reacted to completion with 1,3-butadiene at 513–523 K to

**Table 9.4** Summary of melting points for a range of $CuCl-CuCl_2-KCl$ compositions (Fontana *et al.*, 1952)

| *Solid phase* | *Melt composition/mol%* | | | *Melting point/K* |
|---|---|---|---|---|
| | *KCl* | *CuCl* | *$CuCl_2$* | |
| Compounds | | | | |
| CuCl | 0 | 100 | 0 | 695 |
| $CuCl_2$ | 0 | 0 | 100 | 895 |
| KCl | 100 | 0 | 0 | 1033 |
| $KCuCl_3$ | 50 | 0 | 50 | 637 |
| $K_2CuCl_4$ | 67 | 0 | 33 | 603 |
| $K_2CuCl_3$ | 67 | 33 | 0 | – |
| Binary eutectics | | | | |
| $CuCl-CuCl_2$ | 0 | 87 | 13 | 651 |
| $CuCl-K_2CuCl_3$ | 34 | 66 | 0 | 423 |
| $CuCl_2$-$KCuCl_3$ | 46 | 0 | 54 | 633 |
| $KCuCl_3-K_2CuCl_4$ | 62 | 0 | 38 | 593 |
| $K_2CuCl_4-KCl$ | 70 | 0 | 30 | 598 |
| Binary transition | | | | |
| $K_2CuCl_3-KCl$ | 47 | 53 | 0 | 518 |
| Ternary transitions | | | | |
| $CuCl-CuCl_2-KCuCl_3$ | 24 | 60 | 16 | 523 |
| $K_2CuCl_3-KCl-KCuCl_3$ | 51 | 38 | 11 | 498 |
| $K_2CuCl_4-KCl-KCuCl_3$ | 56 | 22 | 22 | 540 |



yield 0.12 mol of dichlorobutenes. The maximum yield expected from the reaction:

$$C_4H_6 + 2CuCl_2 \rightarrow C_4H_6Cl_2 + Cu_2Cl_2 \tag{9.11}$$

was 0.14 mol. Treatment in a flow of HCl and air for 10 min at 553–573 K regenerated the cupric chloride phase and a second exposure to 1,3-butadiene yielded 0.11 mol of dichlorobutenes. Similar results were reported for $C_2$–$C_4$ alkenes and acetylene.

More recently a similar experiment was performed using a series of metal chlorides and with propene as the probe olefin. A typical set of results is presented in Table 9.6 (Miyake and Hanaya, 1995). These tests were carried out by flowing propene over the individual metal chlorides at 513 K and

**Table 9.5** Reactions of olefins with $CuCl_2$ on pumice (Arganbright and Yates, 1962)

| *Olefin* | *Minimum reaction temp/K* | *mol %* | *Products* |
|---|---|---|---|
| 1,3-Butadiene | 493 | 42 | 3,4-dichlorobut-1-ene |
| | | 15 | *cis*-1,4-dichlorobut-2-ene |
| | | 41 | *trans*-1,4-dichlorobut-2-ene |
| | | 2 | 1-chlorobutadiene |
| Acetylene | 513–523 | 92 | *trans*-1,2-dichloroethylene |
| | | 2 | *cis*-1,2-dichloroethylene |
| | | 6 | 1,1,2-trichloroethylene |
| But-1-ene | 553–563 | 35 | *meso*-2,3-dichlorobutane |
| | | 30 | dl-2,3-dichlorobutane |
| | | 35 | 2,3-dichlorobutane |
| *cis*-But-2-ene | 563 | 99 | dl-2,3-dichlorobutane |
| | | 1 | monochlorobutenes |
| *trans*-But-2-ene | 563 | 76 | meso-2,3dichlorobutane |
| | | 19 | dl-2,3-dichlorobutane |
| Propylene | 583–588 | >90 | 1,2-dichloropropane |
| Ethylene | 598 | >90 | dichloroethane |

**Table 9.6** Products of the reaction of propene with metal chlorides (Miyake and Hanaya, 1995)

| *$MeCl_x$* | *Me—Cl bond energy/kJ $mol^{-1}$*[a] | *% Conversion of $MeCl_x$* | *% Selectivity* | | |
|---|---|---|---|---|---|
| | | | *Allyl chloride* | *i-Propyl chloride* | *Dichloropropane* |
| $LaCl_3$ | 356 | 0 | 0 | 0 | 0 |
| $MnCl_2$ | 240 | 0 | 0 | 0 | 0 |
| $CoCl_2$ | 156 | 0 | 0 | 0 | 0 |
| $FeCl_3$ | 133 | 0.5 | 0 | 57.6 | 0 |
| $CuCl_2$ | 110 | 24.3 | 7.5 | 18.5 | 64.3 |
| $PdCl_2$ | 86 | 35.2 | 84.5 | 13 | 0 |
| $TeCl_4$ | 82 | 50.2 | 85 | 11.4 | 3.2 |
| $PtCl_4$ | 55 | 9.1 | 51.7 | 38.4 | 5.3 |

[a]The quantity given here is the enthalpy of formation of the metal chloride divided by the number of Me—Cl bonds in the structure.

demonstrate a propensity to form a range of chlorinated hydrocarbons whose nature depended on the metal chloride in question. No reaction was observed if the Me—Cl bond strength exceeded ca 130 kJ mol$^{-1}$. As the Me—Cl bond strength diminished below this value, reaction started and chlorinated hydrocarbons were observed in the effluent. For the most part the chlorine was added to the allyl position of propene, consistent with the very weak C—H bonds observed at this position. Among the metal chlorides tested in this way $CuCl_2$ was exceptional because dichloropropane was observed as the main product.

Further evidence of the reactivity of copper chloride is presented in Figure 9.5. This figure compares the reactivity of $C_2H_4$–HCl–$O_2$ mixtures with $C_2H_4$–HCl and pure $C_2H_4$, over a $CuCl_2$–KCl/$SiO_2$ catalyst in a pulse microreactor at 563 K. Sustained reactivity was observed for successive pulses of a $C_2H_4$–HCl–$O_2$ mixture. A similar initial reactivity was observed with the $C_2H_4$–HCl mixture, but this reactivity declined rapidly after the first pulse. With pure $C_2H_4$ just a low and declining reactivity was observed. These data indicate that a reaction occurs between the gas phase ethylene and the chloride species present on this catalyst, but sustained reaction requires a continuous reoxidation and rechlorination of the catalysts with oxygen and HCl respectively (Bakshi *et al.*, 1984). The situation is shown schematically in Scheme 9.4, where ethylene reacts readily with a copper chloride surface yielding 1,2-dichloroethane but leaves behind a reduced and dechlorinated surface; HCl alone replenishes the $CuCl_2$ phase slowly, but this process is much more rapid when gas phase oxygen is present. As the pulse number increases in Figure 9.5, the pool of reactive chlorine in the catalyst diminishes

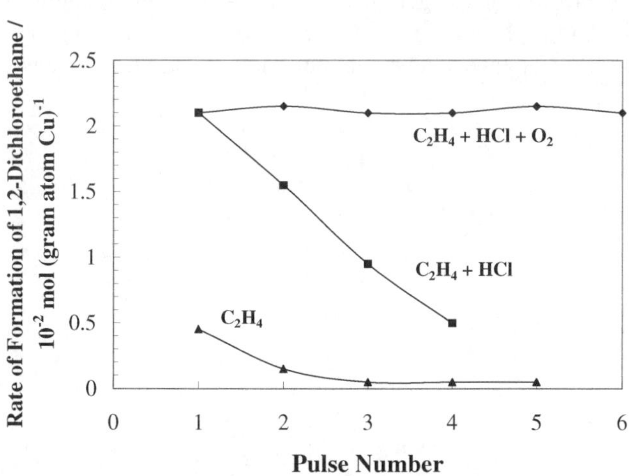


**Figure 9.5** Reaction of $C_2H_4$–HCl–$O_2$ and $C_2H_4$–HCl mixtures and pure $C_2H_4$ over a $CuCl_2$–KCl/$SiO_2$ catalyst in a pulse microreactor at 563 K (Bakshi *et al.*, 1984) (reproduced by permission of Kluwer Academic Publishers).

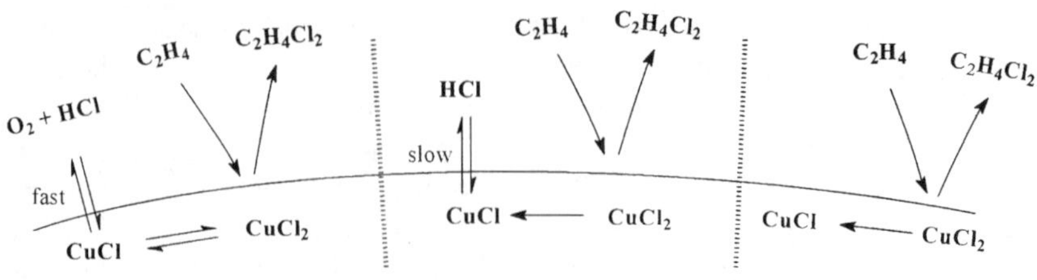


**Scheme 9.4** Schematic representation of the oxychlorination of ethylene over copper chloride catalysts.

when gas phase oxygen is absent. Oxygen reoxidizes the catalyst, probably via an oxychloride and this reoxidized form reacts with HCl to form $CuCl_2$. One proposed structure for the oxychloride intermediate is depicted in Scheme 9.5.

There is evidence from model studies of the Deacon reaction, carried out using molten salt mixtures of $CuCl_2$–KCl–$LaCl_3$ that oxygen uptake is the slow step in the overall oxychlorination reaction. The results of one such study are presented in Figure 9.6, and show that for the Deacon reaction the rates of HCl oxidation and of oxygen absorption by the melt are identical for a range of $LaCl_3$ compositions. These data are consistent with the concept that the slow step in the reaction is the absorption of oxygen into the melt phase. The role of $LaCl_3$ is also clear from this figure, because the reaction and oxygen absorption rates increase as the $LaCL_3$ content of the melt is increased (Ruthven and Kenney, 1968).

This effect was also shown in a separate study of oxygen uptake. Figure 9.7 shows the absorption of oxygen at 673 K by a number of melts of varying composition. Adding KCl to the melt enhanced the rate of oxygen uptake and the addition of $LaCl_3$ gave an even greater boost. In addition this work showed

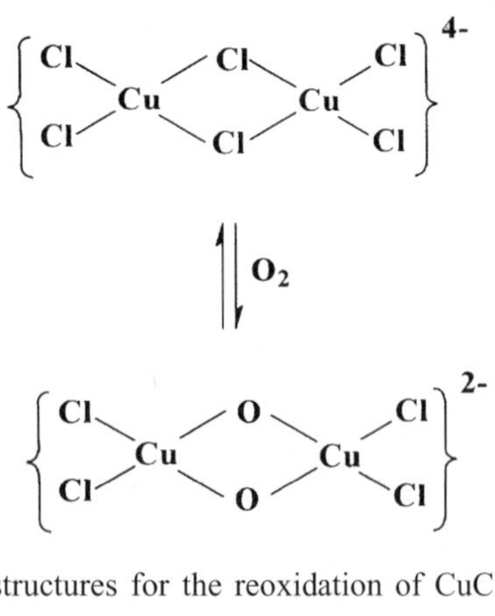


**Scheme 9.5** Proposed structures for the reoxidation of CuCl (Villadsen and Livbjerg, 1978).

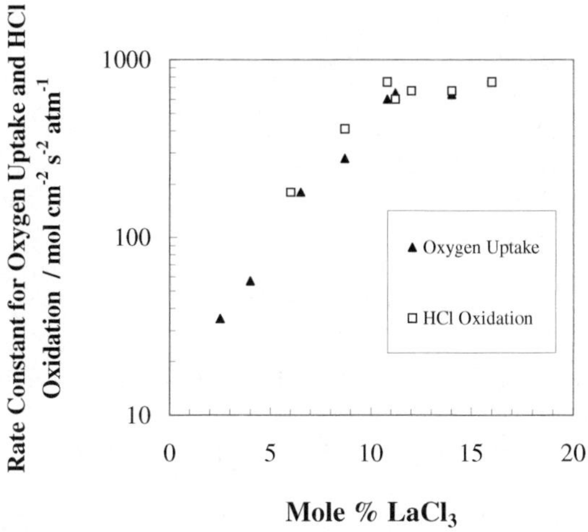


**Figure 9.6** Variation in the rate of HCl oxidation and oxygen absorption in $CuCl_2$–$KCl$–$LaCl_3$ melts. The $CuCl_2 : KCl$ ratio in the melts was 1 : 1 and the $LaCl_3$ content varied as shown (Ruthven and Kenney, 1968) (reproduced by permission of Elsevier Science).

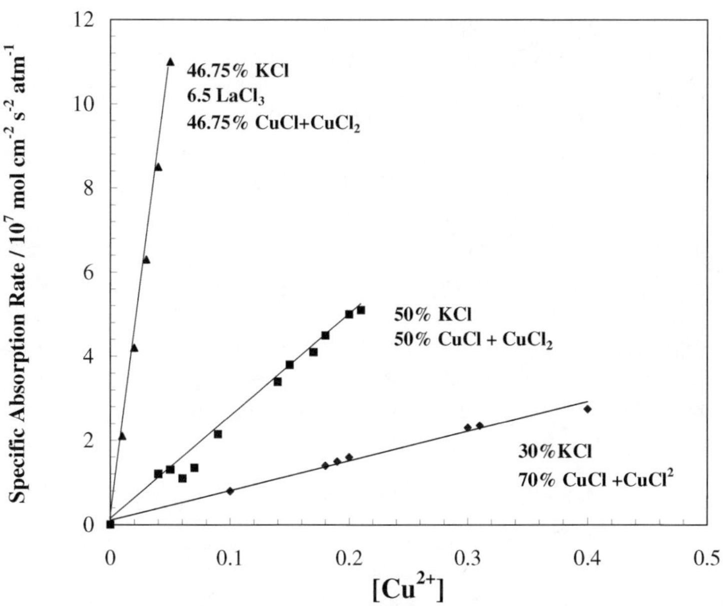


**Figure 9.7** Rate of oxygen absorption in molten salts of different compositions (Ruthven and Kenney, 1967) (reproduced by permission of Elsevier Science).

that the rate of oxygen absorption was second order in $Cu^+$ and first order in oxygen (Ruthven and Kenney, 1967).

The data in Figures 9.6 and 9.7 represent a rationale for the formulation of oxychlorination catalysts with KCl and $LaCl_3$. These materials are effective in promoting the uptake of oxygen by the $CuCl_x$ system.

The oxygen absorption step is often written in two stages. The first involves the incorporation of oxygen:

$$2CuCl + 0.5O_2 \rightarrow CuOCuCl_2 \tag{9.12}$$

and the second involves the reaction of this complex species with HCl:

$$CuOCuCl_2 + 2HCl \rightarrow 2CuCl_2 + H_2O \tag{9.13}$$

The chemical species involved in the reoxidation step have been represented as copper dimers with the structures shown in Scheme 9.5 (Villadsen and Livbjerg, 1978).

Thermodynamic calculations have also provided a valuable insight into the likely form and oxidation state of copper in oxychlorination conditions. Some indication of the stability of cuprous and cupric oxide and chloride species that are present in oxychlorination reactions can be got from a study of the thermodynamics of a few simple reactions (Allen, 1962). The reactions are presented in Table 9.7.

Figure 9.8 presents the Gibbs free energies as a function of temperature for the four reactions listed in Table 9.7. The data show a small negative value for reaction (2) and a small positive value for reaction (1) in the temperature range of interest for oxychlorination (473–573 K). This means that we can expect some amounts of CuO and $CuCl_2$ in oxychlorination conditions. For the $Cu^+$ reactions the formation of CuCl is favoured, but it is less likely that there will be appreciable quantities of $Cu_2O$ present in the reaction conditions. This approach is very simplistic because many of the species present in oxychlorination catalysts are probably complex oxychlorides, sometimes of variable oxidation state and complexed with potassium or lanthanum, but these simple calculations can be helpful. However, the approach is consistent with the results of a recent study of copper oxidation states in a $CuCl_2$–$KCl/Al_2O_3$ system during oxychlorination. The results of this study

**Table 9.7** Reactions of the Cu–O–Cl–HCl system

| | | |
|---|---|---|
| $CuCl_2 + 0.5O_2$ | $\rightarrow CuO + Cl_2$ | (1) |
| $CuO + 2HCl$ | $\rightarrow CuCl_2 + H_2O$ | (2) |
| $2CuCl + 0.5O_2$ | $\rightarrow Cu_2O + Cl_2$ | (3) |
| $Cu_2O + 2HCl$ | $\rightarrow 2CuCl + H_2O$ | (4) |

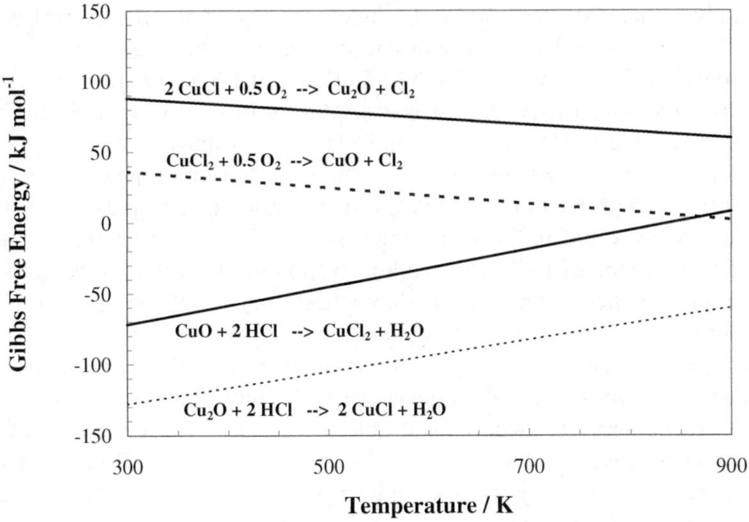


**Figure 9.8** Gibbs free energy for the reactions in Table 9.6 (Allen, 1962).

are shown in Figure 9.9. Below 533 K the $Cu^+ : Cu^{2+}$ ratio is never more than 0.04 : 1 and the catalyst comprises about 50 % CuO at this temperature. As the temperature increases so does the amount of $Cu^+$ present in the reactor (Sai Prasad *et al.*, 1997).

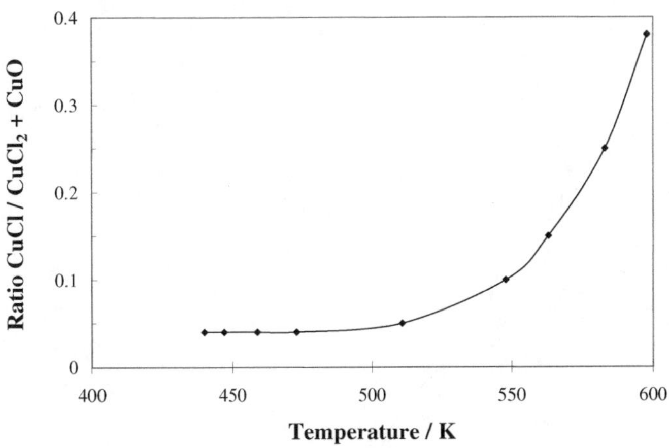


**Figure 9.9** Influence of temperature on the $Cu^+ : Cu^{2+}$ ratio in the $CuCl_x$–KCl–$Al_2O_3$ system (Sai Prasad *et al.*, 1997) (reproduced by permission of Kluwer Academic Publishers).

A study of heats of adsorption of ethylene, oxygen and HCl on copper (I) and (II) chlorides and on oxychlorination catalysts indicates that ethylene chemisorbs on $CuCl_2$ ($\Delta H = 65 \pm 5\ kJ\,mol^{-1}$) in the temperature range 320–420 K (Hall *et al*., 1984). Physical adsorption only was observed on CuCl ($\Delta H = 13\ kJ\,mol^{-1}$; $\Delta H_{liquefaction} = 14.5\ kJ\,mol^{-1}$). Values of isostearic heats of adsorption very close to 60 $kJ\,mol^{-1}$ were recorded for ethylene adsorption from mixtures with HCl or HCl and $O_2$. In addition, the amounts of ethylene chemisorbed were not influenced by the nature of the alumina support used nor by the presence of HCl. This work also showed that a double bond in the hydrocarbon is necessary for chemisorption, thus propene and ethylene chemisorbed on $CuCl_2$, but alkanes did not.

The overall mechanism of oxychlorination is summarized in Scheme 9.6. Ethylene is shown to react directly with $CuCl_2$, with the generation of 1,2-dichloroethene and a reduced copper chloride species. The remainder of the catalytic cycle involves the regeneration of the active site. Reoxidation of CuCl is by gas phase oxygen, to yield an oxychloride species (see Scheme 9.5). Thereafter, displacement of oxygen by chlorine completes the regeneration process.

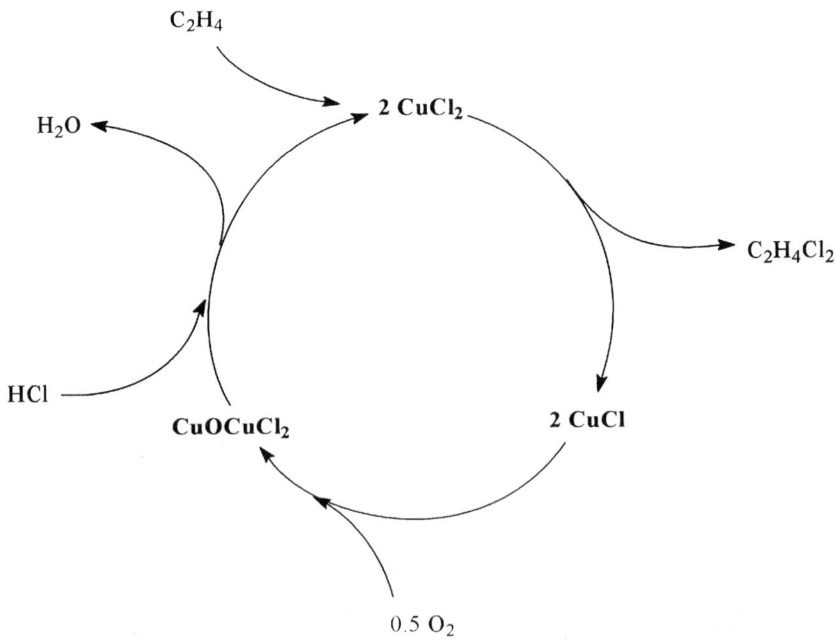


**Scheme 9.6** Catalytic cycle for oxychlorination over copper chloride.

## 9.6 DESTRUCTION OF CHLORINATED HYDROCARBONS BY CATALYTIC OXIDATION

The technology for the catalytic destruction of volatile organic compounds (VOCs), described in Chapter 6, can also be applied to the destruction of chlorinated hydrocarbons. Indeed there are encouraging signs that this strategy may already be at work in reducing levels of dioxin emissions. Field studies have shown that incinerators fitted with vanadia–titania catalysts, widely used in the reductive elimination of NOx by reaction with ammonia, also bring about a reduction in dioxin emissions (Krishnamoorthy *et al.*, 1998).

On a more general level there are many problems associated with the application of combustion technology to the destruction of chlorinated hydrocarbons. The reaction of choice for the destruction of chlorinated hydrocarbons by oxidation is:

$$C_2H_4Cl_2 + 3O_2 \rightarrow 2CO_2 + 2HCl + 2H_2O \tag{9.14}$$

Additional complications arise regarding the ultimate fate of the chlorine species. Many possible side reactions can occur including:

- the formation of $Cl_2$ via the Deacon reaction, with subsequent chlorination of hydrocarbon species;
- free radical insertion of chlorine into hydrocarbons;
- involvement of HCl in oxychlorination;
- formation of surface chlorides, such as $CuCl_2$, which react directly with hydrocarbons.

Figure 9.10 shows the performance of a Pt–Pd combustion catalyst that was operating at a steady state for the combustion of propene. When a small

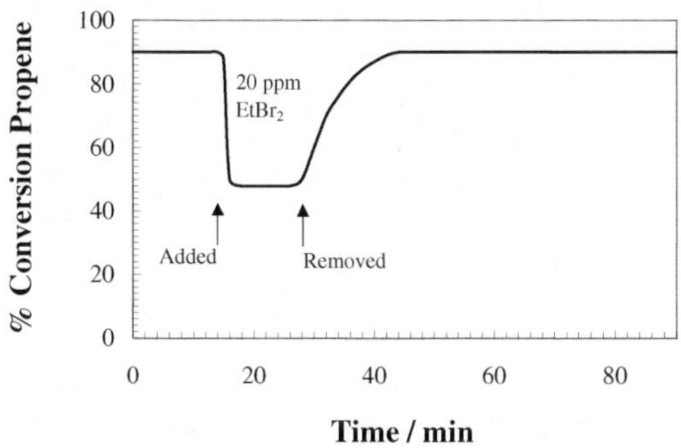


**Figure 9.10** Reversible poisoning of a precious metal catalysts by ethylene bromide (Freidel *et al.*, 1993) (reproduced by permission of Elsevier Science).

amount of $C_2H_4Br_2$ was added to the stream the conversion of propene immediately decreased, but also recovered quickly when the halogen source was cut off. This is an example of the well known reversible poisoning of precious metal catalysts by halogenated hydrocarbons and points to the ability of precious metal catalysts to form surface halogenated species (Freidel *et al.*, 1993).

Some insight into the mechanism of poisoning by halogens comes from the extensive literature on the interaction between platinum, oxygen and chlorine. Figure 9.11 shows the chemisorption of oxygen onto Pt(100) in the presence or absence of preadsorbed chloride (Schennach and Bechtold, 1996). The clean Pt(100) reconstructs into the hex-Pt(100) surface (a structure whereby the surface platinum layer buckles and rearranges) and there is no significant uptake of oxygen onto this structure below 573 K. (Note the very low sticking coefficient for oxygen on hex-Pt(100), Table 6.13 (Lauterbach *et al.*, 1993).) With a small amount of preadsorbed chloride there is a significantly improved oxygen uptake below 573 K, but the oxygen does not displace the chloride from the Pt surface. As the chloride preloading is increased so does the amount of oxygen which adsorbs below 473 K. Chloride preloadings in the range (1–6) $\times 10^{14}$ atoms per square centimetre of platinum surface promote oxygen chemisorption at low temperatures, but the maximum achievable

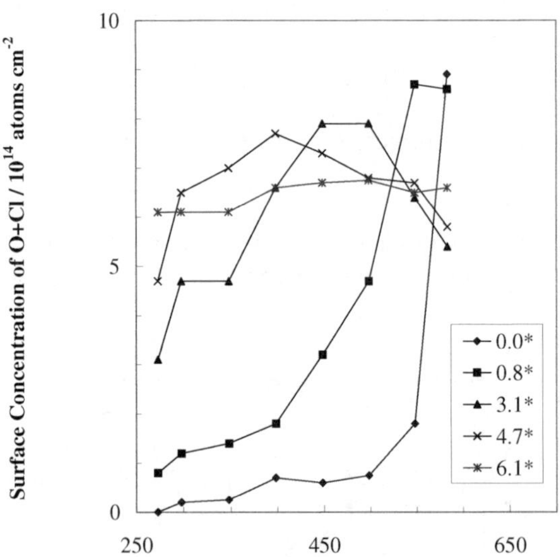


**Figure 9.11** Surface concentration of oxygen and chlorine on Pt(100) as a function of adsorption temperature. The levels of preadsorbed chlorine are indicated. Each surface was exposed to $1 \times 10^{-3}$ bar of $O_2$. The first point on each curve gives the level of preadsorbed chlorine (*preadsorbed chlorine $\times 10^{14}$ atoms per cm$^2$ of Pt surface) (Schennach and Bechtold, 1996) (reproduced by permission of Elsevier Science).

oxygen loading is reduced by the amount of chloride that is preadsorbed on the catalyst. Ultimately no oxygen chemisorption is observed when the platinum surface contains more than $6 \times 10^{14}$ chlorine atoms per square centimetre.

In TPD of the fully loaded structures, oxygen and chlorine desorbed separately (Schennach and Bechtold, 1996). Oxygen desorbs in the range 623–753 K and chlorine desorbs above 953 K. Hence the reversal of the poisoning effect shown in Figure 9.10 cannot be accounted for in terms of displacement of halogen by oxygen, but it seems more likely then that the Pt–halogen surface species is displaced by reaction with the propene in a manner analogous to the reactions reported in Tables 9.5 and 9.6.

A comprehensive testing of chromium oxide catalysts has been undertaken by Agarwal *et al.* (1992) for the destruction of chlorinated hydrocarbons. This work involved a chromium oxide catalyst (9 wt% Cr) supported on alumina and tested for the destruction of chlorinated hydrocarbons in fixed and fluid bed operations for up to 215 days. The test was designed to assess the suitability of this catalyst for the total combustion of differing mixtures of hydrocarbons and chlorinated hydrocarbons. The composition of the effluent gas was monitored as was catalyst deactivation.

Two gas mixtures were formulated each with 500 ppm of VOC: one mixture was composed exclusively of chlorinated $C_2$ hydrocarbons (the chlorinated stream) whereas the second mixture (the mixed stream) comprised mostly non-halogenated hydrocarbons (450 ppm) and trichloroethylene (ca 50 ppm). Water vapour, which is expected to constitute a part of any stream to which this catalyst would eventually be applied, was also present in each mixture. Full details of the mixture compositions are presented in Table 9.8. Two fluidized bed reactors and one fixed bed reactor were operated in the conditions indicated in Table 9.9.

The results of the test are shown in Figure 9.12 in terms of the temperature required to maintain greater than 99 % destruction of the feed gas mixture. Surprisingly, the catalyst deactivated more with the mixed stream than with the chlorinated stream. Just a 1 K increase in temperature was required to maintain the required level of conversion for the chlorinated feed over the 215 day test. The mixed feed required a higher starting temperature (658 K) and a steady increase in reactor temperature to 691 K to maintain the required 99 % conversion level.

The different chlorinated hydrocarbons showed differing reactivities in oxidative destruction. The order of oxidative destructability, wherein the easiest compound to destroy is listed first, was as follows:

$$CH_2{=}CCl_2 > CHCl + CCl_2 > CH_2Cl{-}CH_2Cl > CH_2Cl_2$$

This effect is shown in Figure 9.13. In general, dichloromethane conversion

**Table 9.8** Composition of feed streams for extended testing of chromium oxide based catalysts for VOC destruction
(Agarwal *et al.*, 1992)

| *Compound* | *Chlorinated stream/ppm* | *Mixed stream/ppm* |
|---|---|---|
| Pentane | — | 119 |
| 2,3-Dimethylbutane | — | 4.5 |
| Methyl cyclopentane | — | 21.5 |
| Cyclohexane | — | 185.5 |
| Benzene | — | 26 |
| Ethyl benzene | — | 23.5 |
| *m,p*-Xylene | — | 60 |
| Cumene | — | 9.5 |
| Trichloroethylene | 124 | 50.5 |
| Dichloromethane | 50.5 | — |
| 1,2-Dichloroethane | 105.5 | — |
| 1,1-Dichloroethylene | 220 | — |
| Total | 500 | 500 |
| Water vapour | 16 000 | 16 000 |



**Table 9.9** Reaction conditions for testing of $Cr_2O_3/Al_2O_3$ catalysts in the destruction of chlorinated hydrocarbons
(Agarwal *et al.*, 1992)

| *Catalyst* | *Stream used* | *Reactor type* | *Space velocity/$h^{-1}$* | *Initial temperature/K* |
|---|---|---|---|---|
| $Cr_2O_3/Al_2O_3$ | Chlorinated | Fluidized | 12 000 | 633 |
| $Cr_2O_3/Al_2O_3$ | Mixed | Fluidized | 12 000 | 658 |
| $Cr_2O_3/Al_2O_3$ | Chlorinated | Fixed | 12 000 | 623 |



was about 97 % during the 215 day test. Conversion of all other components in the stream was greater than 99 %.

Table 9.10 presents some characterization points for the fresh and used catalysts. It appears that the greater loss in surface area was experienced with the catalyst tested in the mixed stream, consistent with the higher reactor temperature. Chromium loss was recorded from the chlorinated stream test, but this did not lead to deactivation. The explanation given was that a volatile oxychloride of chromium ($CrO_2Cl_2$) was formed in the chlorinated stream.

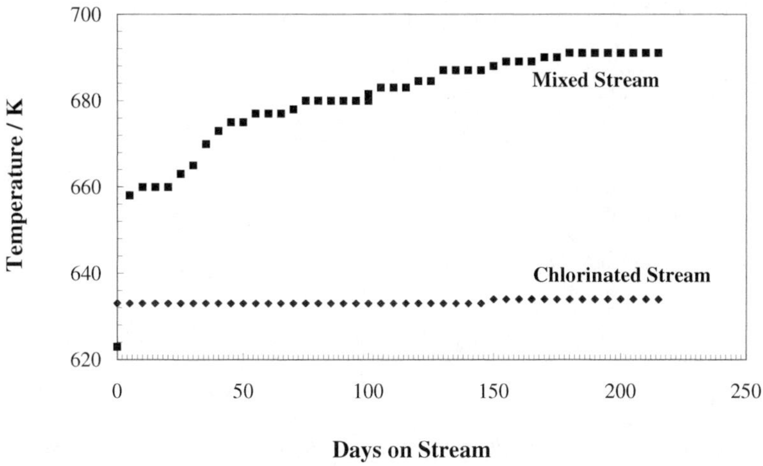


**Figure 9.12** Temperature increase required to maintain a constant conversion above 99 % in fluid bed reactors over chromia/alumina catalysts for the chlorinated and mixed streams detailed in Table 9.8 (Agarwal *et al.*, 1992) (reproduced by permission of Elsevier Science).

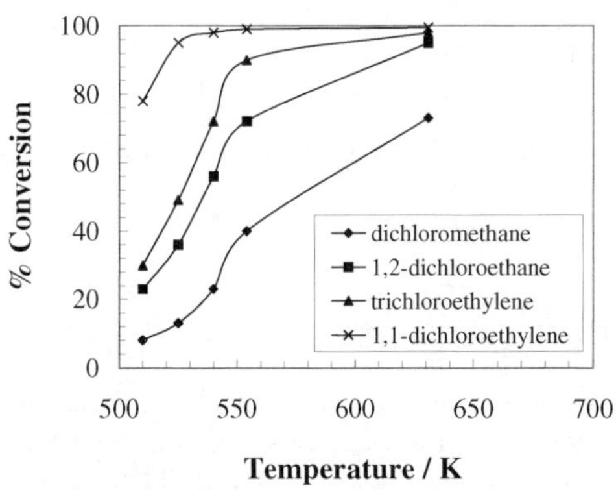


**Figure 9.13** Conversion of individual components in the chlorinated stream over chromia/alumina as a function of temperature. Total chlorohydrocarbon content was 200 ppm, space velocity = 23 970 $h^{-1}$ (Agarwal *et al.*, 1992) (reproduced by permission of Elsevier Science).

Loss of this material continuously renewed the surface layer of chromium oxide, and deactivation was not noticed on the time scale of the test.

The chlorine balance was carefully monitored during the 215 day test. The results are summarized in Table 9.11. The time required to arrive at a satisfactory chlorine balance was close to the duration of the test. The authors offer no explanation for this finding except to point to some uncertainties in the chlorine analysis technique. By the end of the test period all the chlorine

**Table 9.10** BET surface area and chromium content for fresh and used fluid bed catalysts
(Agarwal *et al.*, 1992)

| *Catalyst* | *BET surface area/$m^2\ g^{-1}$* | *Bulk Cr content/%* |
|---|---|---|
| Fresh | 188 | 9.2 |
| Used in chlorinated stream | 187 | 8.3 |
| Used in mixed stream | 151 | 9.3 |

Reproduced by permission of Elsevier Science.

**Table 9.11** Chlorine balance for the chlorinated stream treated by a $Cr_2O_3/Al_2O_3$ catalyst
(Agarwal *et al.*, 1992)

| *Time on stream/days* | *Effluent concentration/ppm* | | *Chlorine balance/%* |
|---|---|---|---|
| | *HCl* | *$Cl_2$* | |
| 117 | 613 | 13 | 59 |
| 153 | 735 | 19 | 71 |
| 165 | 879 | 22 | 85 |
| 214 | 1042 | 23 | 99.9 |

Reproduced by permission of Elsevier Science.

fed to the reactors was accounted for in terms of the HCl and $Cl_2$ observed in the effluent stream.

## 9.7 BY-PRODUCT FORMATION DURING THE CATALYTIC OXIDATION OF CHLORINATED HYDROCARBONS

Attempts have been made to use the excellent chlorine handling capacity of copper chlorides as a total oxidation catalyst for chlorinated hydrocarbons. In one such attempt a catalyst designated $CuCl/KCl/SiO_2$ was used for the catalytic oxidation of 1,2-dichloroethane. The results for substrate conversion and selectivity to other chlorinated hydrocarbons are presented in Figure 9.14. Although the catalyst appeared to operate effectively above 723 K, large amounts of other chlorinated organics were observed below this temperature. When 1,2-dichlorobenzene was used as feed in the temperature range 473–673 K a dark coloured product collected at the exit of the reactor. A polychlorinated biphenyl of formula $C_{12}HCl_9$ was detected in this material by mass spectral analysis (Lago *et al.*, 1996).

Polychlorinated benzenes were also reported at intermediate temperatures when chlorobenzene was oxidized over a $Pt/Al_2O_3$ catalyst. At 673 K, upwards of 20 % of the feed chlorine was converted into polychlorinated hydrocarbons. The distribution of polychlorinated products is indicated in

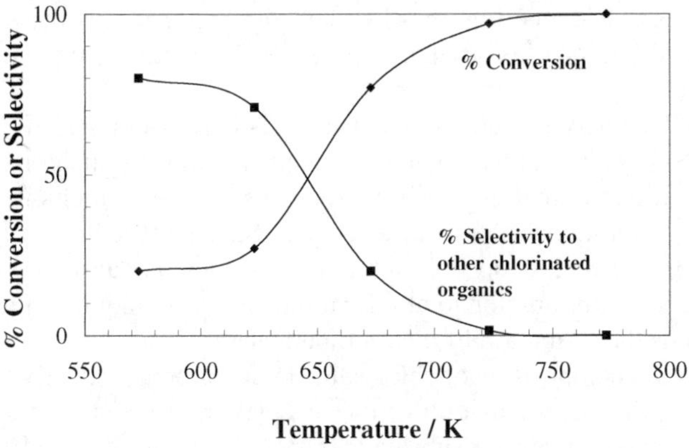


**Figure 9.14** Conversion of 1,2-dichloroethane over CuCl/KCl/$SiO_2$ and selectivity to other chlorinated organics (Lago *et al.*, 1996) (reproduced by permission of Elsevier Science).

Figure 9.15. Water vapour in the feed reduced the amounts of polychlorinated products detected without significantly influencing the activity of the catalyst. The species believed to be responsible for the production of polychlorinated benzenes was a platinum oxychloride of indeterminate structure ($PtO_xCl_y$) (van den Brink *et al.*, 1998).

In general a difficulty arises when a product is formed which is itself more difficult to destroy by oxidation over the catalyst being investigated than the original substrate in the feed. The point can be illustrated by reference to the reactivity patterns observed for the chlorinated stream used to assess the

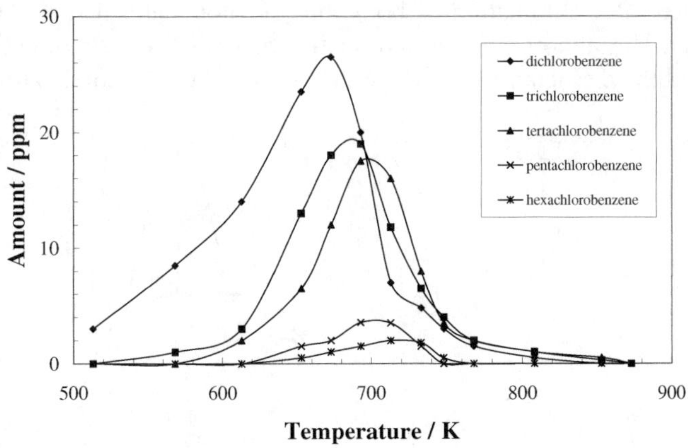


**Figure 9.15** Distribution of polychlorinated benzene from the reaction of chlorobenzene over Pt/$Al_2O_3$ at the temperatures indicated (van den Brink *et al.*, 1998) (reproduced by permission of Elsevier Science).

$Cr_2O_3/Al_2O_3$ catalyst (see Figure 9.13). Dichloromethane was the least reactive chlorinated hydrocarbon in the mixture and 1,1-dichloroethylene was the most reactive.

Scheme 9.7 depicts the situation where 1,1-dichloroethylene is oxidized over a catalyst (catalyst 1) which shows a tendency to form dichloromethene. More severe reaction conditions will be required to oxidize dichloromethane, corresponding to a higher activation energy in Scheme 9.7. In effect dichloromethane represents a kinetic sink in the route to total oxidation products. By contrast milder conditions for total oxidation should be required for catalyst 2, which avoids the formation of dichloromethane.

The mode of decomposition of chlorinated hydrocarbons is clearly important in determining the decomposition of the original chlorinated hydrocarbon in the feed and for any material subsequently formed. Yang *et al.* (1997) have carried out a fundamental study of the mode of activation of the series of chlorinated hydrocarbons listed in Table 9.12, over a Cu(100) surface. The study concentrated on the mode of adsorption of the chlorinated hydrocarbon, i.e., whether H or Cl elimination occurred from the possible reaction intermediate. In Table 9.12 dissociation is expressed as the fraction of a monolayer that decomposes at the temperature indicated in column 3. The first point to note in the data in this table is that the dissociation products appearing predominantly in the gas phase do not contain any chlorine or, for $CH_2ClCHCl_2$ just one remaining chlorine atom. This fact points to dissociation by rupture of C—Cl bonds rather than C—H or C—C bonds. A second point is that even from dichlorinated reactants no chloride was present in the reaction product, indicating that Cl elimination was favoured over H elimination from the primary dissociation product. The $C_4$ and $C_6$ products formed from the 1,1-dichloroethane and 2,2-dichloropropane respectively arise from coupling products formed after elimination of both chlorine atoms attached to the same carbon atom. The authors point to a correlation between C—Cl homolytic bond energies and the dissociation yield. Some C—Cl and C—H bond dissociation

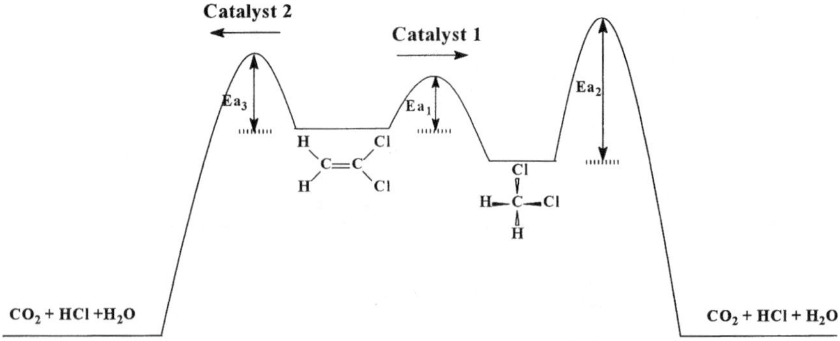


**Scheme 9.7** Schematic diagram of activation energies for the decomposition of 1,1-dichloroethane by different paths.

**Table 9.12** Degradation of chlorinated hydrocarbons on Cu(100) (Yang *et al.*, 1997)

| *Compound* | *Dissociation* | | | *Product formation* | |
|---|---|---|---|---|---|
| | *Extent of degradation*[a] | *Temp/K* | *Ea/ kJ mol$^{-1}$* | *Product* | *Temp/K*[b] |
| Chlorinated ethanes | | | | | |
| $CH_3CHCl_2$ | 0.15 | 190 | 11.5 | $CH_2CH{=}CHCH_3$ | 190 |
| $CH_2ClCHCl_2$ | 0.25 | 170 | 10.2 | $CH_2{=}CHCl$ | 170 |
| Chlorinated propanes | | | | | |
| $CH_2CHClCH_2Cl$ | 0.10 | 200 | 12.1 | $CH_3CH{=}CH_2$ | 200 |
| $CH_3CCl_2CH_3$ | 0.62 | 140 | 8.4 | $(CH_3)_2C{=}C(CH_3)_2$ | 140 |

[a]Extent of dissociation upon heating to the temperatures indicated in column 3, following monolayer saturation by adsorption at 110 K, expressed as molecules dissociated per surface copper atom.
[b]Peak maximum in TPD profiles.


**Table 9.13** Gas phase homolytic C—Cl and C—H bond energies (Yang *et al.*, 1997; Cassidy, 1999)

| *Compound* | *ΔH°/kJ mol$^{-1}$* | *Compound* | *ΔH°/kJ mol$^{-1}$* |
|---|---|---|---|
| Cl—$CH_3$ | 350 | H—$CH_3$ | 438 |
| Cl—$CH_2Cl$ | 329 | | |
| Cl—$CHCl_2$ | 317 | | |
| Cl—$CCl_3$ | 282 | | |
| Cl—$CH_2CH_3$ | 352 | H—$CH_2CH_3$ | 420 |
| Cl—$CH_2CH_2Cl$ | 346 | | |
| Cl—$CH(CH_3)_2$ | 350 | H—$CH(CH_3)_2$ | 401 |
| Cl—$C(CH_3)_3$ | 345 | H—$C(CH_3)_3$ | 380 |



energies are listed in Table 9.13 and show that in general the C—Cl bonds are weaker than the corresponding C—H bonds. Some exceptions to the general correlation between bond energies and dissociation reactivity were noted; 1,2-dichloroethane did not dissociate at all over the Cu(100) surface. This finding was associated with a greater tendency for this material to desorb rather than to dissociate in the reaction conditions. Scheme 9.8 presents a schematic diagram for the mode of interaction of 2,2-dichloropropane with the Cu(100) surface, showing the retention by the Cu(100) surface of chlorine and the appearance of a dimer product in the gas phase (Yang *et al.*, 1997).

An important point reported throughout this work is that residual Cl remained on the surface and was eventually desorbed as CuCl at 750 K during

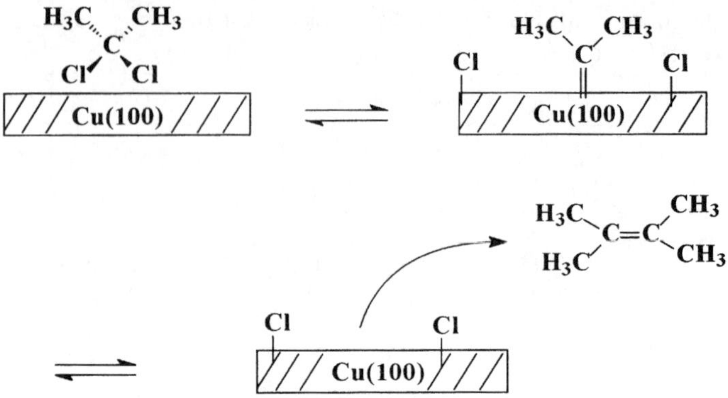


**Scheme 9.8** Hydrodechlorination of 2,2-dichloropropane on a clean Cu(100) surface (Yang *et al*., 1997) (reproduced by permission of the American Chemical Society).

TPD. These surface CuCl species must be considered to possess a reactivity similar to that of $CuCl_2$ in a standard oxychlorination or Deacon catalyst, hence being capable of inserting chlorine into non-chlorinated substrates (see Tables 9.5 and 9.6). This type of reaction would be particularly favoured in the packed bed of a catalytic reactor where multiple encounters are possible allowing, for example, an olefin formed by dechlorination on one catalyst particle to react with the $CuCl_2$ on another particle, and end up as a different chlorohydrocarbon. This type of encounter would be less favoured in the low pressure condition ($10^{-6}$ Torr) used by Yang *et al*.

In spite of the difficulties listed above, there are claims that a number of proprietary precious metal and oxide based catalyst formulations can successfully oxidize chlorinated hydrocarbons without by-product formation. The

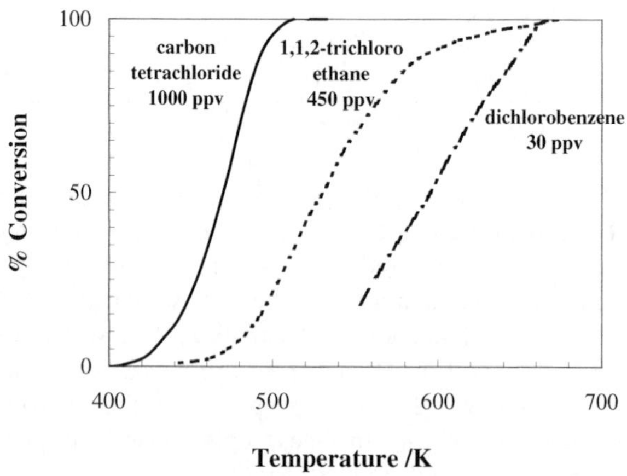


**Figure 9.16** Destruction of $CCl_4$, $C_3H_3Cl_3$ and $C_6H_4Cl_2$ over a precious metal catalyst (Freidel *et al*., 1993) (reproduced by permission of Elsevier Science).

operation of one such catalyst is shown in Figure 9.16 for the destruction of carbon tetrachloride, 1,1,2-trichloroethane and 1,2-dichlorobenzene (Freidel *et al.*, 1993).

## 9.8 POLYCHLORINATED BIPHENYLS, DIOXINS AND THE ENVIRONMENT

Many chlorinated compounds, including polychlorinated biphenyls (PCB), dibenzo-*p*-dioxins (PCDDs), dibenzofurans (PCDFs), dibenzoethers (PCDEs), naphthalenes (PCNs) and benzenes (PCBzs) are no longer in commercial production because of their proven toxicity. Scheme 9.9 presents some typical structures for a number of these compounds, which were used in the past as heat transfer agents, ink lubricants, plasticizers, fire retardants, paint additives, sealing fluids, adhesives, waxes and dielectric fluids. Past global production of these materials is estimated at 1 200 000 tonne with about 400 000 tonne persisting in the environment (*Safe*, 1993).

The toxicity of these materials is not a simple function of their structure and amount. For example 3,3′,4,4′,5,-pentaCB is regarded as the most toxic of the polychlorinated biphenyls. 2,3,7,8-TCDD is between 8 and 125 times more potent again depending on the toxicity test which is applied. The potency ratio for 2,3,7,8-TCDD : 3,3′4,4′5,5′-pentaCB is 66 : 1 for body weight loss in rats and 125 : 1 for aryl hydrocarbon hydroxylase (AHH) induction in rats. Notwithstanding these difficulties, a potency scheme has been devised whereby the toxicity of 2,3,7,8-TCDD is given a value of unity; other compounds are then given values accordingly. Some are listed in Table 9.14 (Safe, 1993).

Retrospective studies of mortality rates in workers from capacitor manufacturing plants, where PCBs were used as dielectric fluids, do not indicate any significant incidence of liver or rectal cancers compared with a control group. However, there is evidence of a correlation between *in utero* exposure to low levels of PCBs and developmental defects in children, including reduced birth weight, neonatal behaviour anomalies and recognition memory (Safe, 1993). Scientists are agreed that exposure of laboratory animals to dioxins causes a variety of toxicological effects, which are mediated via the Ah receptor.

PCBs, PCDDs and PCDFs have been distributed in a widespread manner throughout the environment. One study of these chemicals in sediment cores in the Baltic has allowed a profile of their deposition to be established, going back to 1882. The trend in deposition is presented in Figure 9.17 and shows increasing levels of deposition, with a spike in 1978, and a decrease since 1981 (Kjeller and Rappe, 1995).

Another study of PCDD/F levels was carried out on vegetation samples taken from the control plot of the Grass Park experiment started in 1856 at Rothamsted Experimental Station, a semi-rural location in southeast England. The results, summarized in Figure 9.18, point again to reduced levels of

Polychlorinated Biphenyl (PCB)

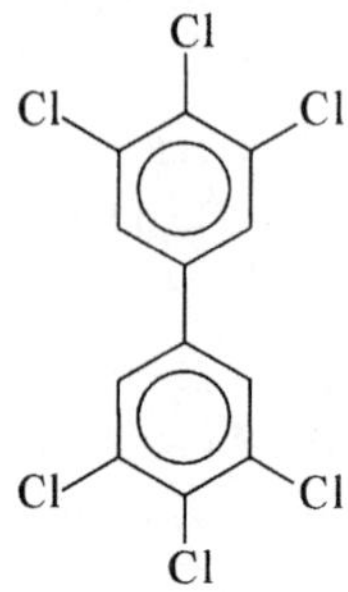

3,3',4,4',5,5'-hexaCB

Polychlorinated Dibenzo-p-dioxin(PCDD)

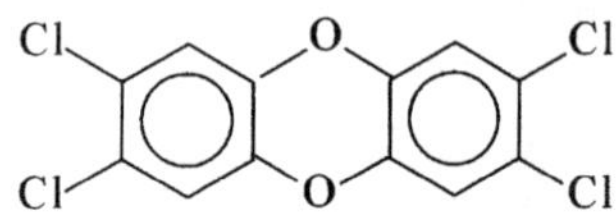

2,3,7,8-tetrachlorodibenzo-p-dioxin

Polychlorinated Dibenzofuran(PCDF)

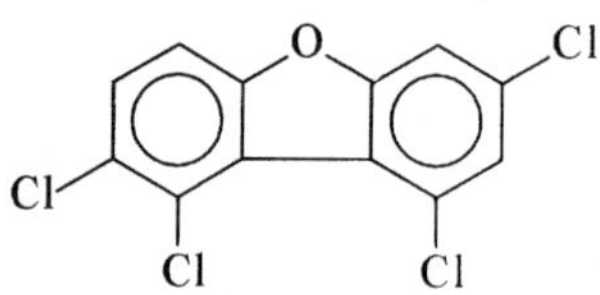

2,4,5,6-trtrachlorodibenzofuran

**Scheme 9.9** Structure of some polychlorinated aromatic compounds.

**Table 9.14** Some toxic equivalent factors (Safe, 1993)

| *Compound* | *Toxic equivalent factor (TEF)* |
|---|---|
| 2,3,7,8-tetrachlorodibenzo-*p*-dioxin | 1 |
| 3,3′4,4′5,-pentaCB | 0.1 |
| 3,3′4,4′5,5′-hexaCB | 0.05 |
| 3,3′4,4′-tertraCB | 0.01 |
| 2,3,3′4,4′,5,5′-septaCB | 0.001 |

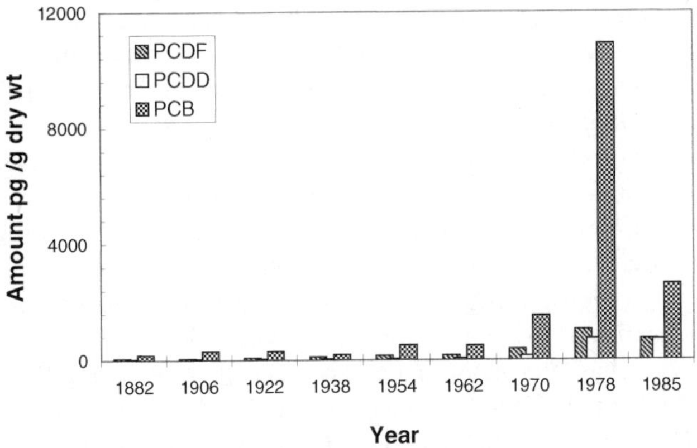


**Figure 9.17** Levels of PCDD, PCDF and PCB in a sediment core taken from the Baltic Sea (Kjeller and Rappe, 1995).

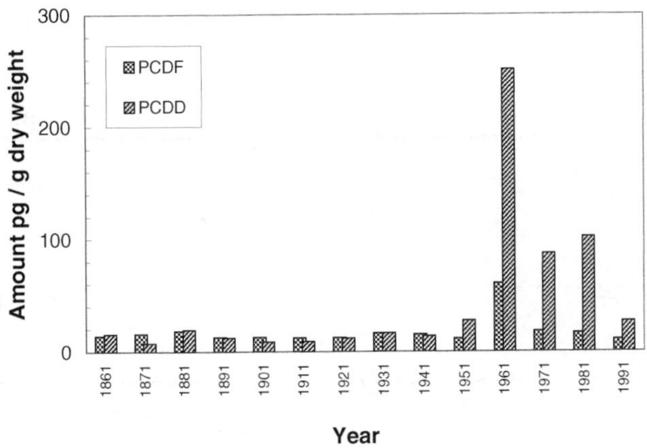


**Figure 9.18** Levels of PCDD and PCDF in archived vegetation samples taken from the Rothamsted Experimental Station, UK (Kjeller *et al.*, 1996).

PCDD/F deposition in recent years, following a sharp increase in emissions in the 1960s (Kjeller *et al.*, 1996).

Since the suspension of industrial production, combustion and incineration represent by far the biggest causes of continuing dioxin emissions, with municipal and medical waste incineration being identified as the principal sources. In the USA emission estimates for 1992 have been put in the range 3–23 kg per year for all anthropogenic sources, with a working mean given as 9 kg. In the same year German emissions were estimated at 0.5 kg. These figures are in toxic equivalents (TEQ) for all dibenzodioxins and dibenzofurans based on the toxicity of 2,3,7,8-tetrachlorodibenzo-*p*-dioxin. The amounts emitted are miniscule by comparison with past production figures and a continuing difficulty in terms of researching this area is the analytical costs involved in dioxin determination. US and European environmental regulations now specify dioxin emissions below 0.1 ng m$^{-3}$ in gaseous effluent (Rose, 1994; Johnson, 1995a and b).

Dioxin formation during combustion processes has been attributed to two routes, namely the *de novo* route from particulate carbon, and dioxin formation from aromatic precursors, such as chlorinated phenols. In addition, there is a body of evidence to support the hypothesis that fly-ash (particulate material in the combustor composed of a mixture of inorganic and organic compounds) catalyses the formation of dioxins. Scheme 9.10 outlines a mechanism whereby chlorinated hydrocarbons can form during the combustion of a mixtures of aliphatic hydrocarbons, HCl and air. Essentially, products of incomplete combustion such as $C_2H_2$, $C_4H_6$ and $C_2H_5\bullet$ can recombine, by a variety of mechanisms, to form cyclic structures which in turn can be chlorinated by Cl• radicals present in the combustion environment. The presence of copper in the fly-ash is also thought to catalyse strongly the formation of chlorinated hydrocarbons. In fact the copper species appear to be

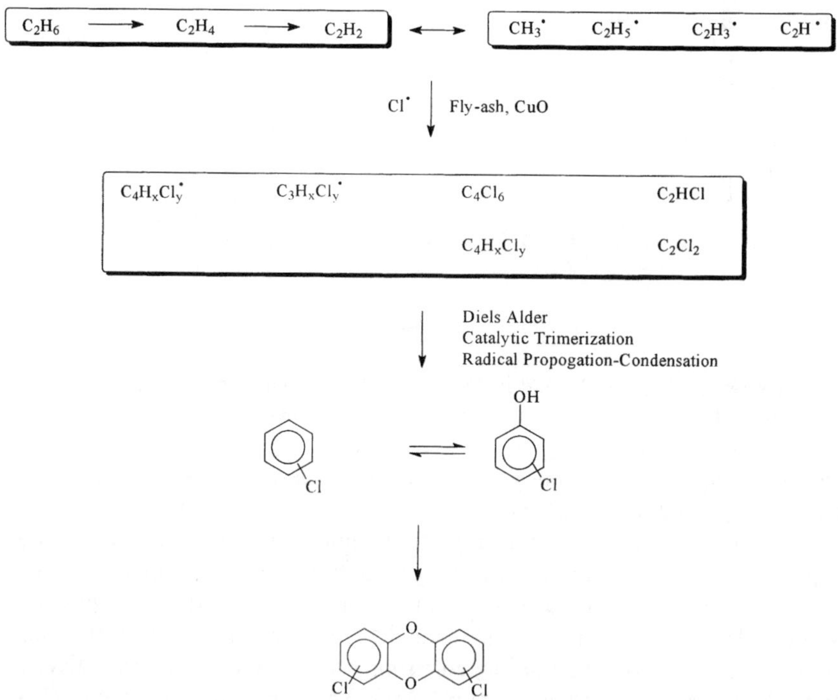


**Scheme 9.10** Mechanism of formation of polychlorinated aromatics during combustion (Froese and Hutzinger, 1997) (reproduced by permission of the American Chemical Society).

capable of chlorinating the aliphatic species which are present in the combustor and ring formation follows (Froese and Hutzinger, 1997).

A number of proprietary catalyst systems have been developed to tackle the dioxin problem. There is evidence that vanadium oxide–titanium oxide catalysts in widespread use for the reduction of nitrogen oxides by ammonia are also effective for the decomposition of dioxins (Krishnamoorthy *et al.*, 1998). A number of proprietary catalysts, based on vanadia/titania or on precious metals, have been developed specifically for dioxin destruction from incinerator sources, although some doubt exists whether the dioxins are completely degraded to $CO_2$ and HCl (Freidel *et al.*, 1993).

## REFERENCES TO CHAPTER 9

Agarwal SK, Spivey JJ and Butt JB (1992) Catalyst deactivation during deep oxidation of chlorohydrocarbons. *Appl. Catal. A General* **82**: 259–275.

Allen JA (1962) Energetic criteria for oxychlorination catalysts. *J. Appl. Chem.* **12**: 406–412.

Arganbright RP and Yates WF (1962) Chlorination with cupric chloride. *J. Organic Chem.* **27**: 1205–1208.

Bakshi Yu M, Dmitrieva MP and Gel'bshtein AI (1984) Study of the mechanism and kinetics of the additive oxychlorination of olefins: I. The conversion of ethylene on $CuCl_2$–KCl–$SiO_2$ catalysts. *Kinet. Catal. (Engl. Trans.)* **25**: 111–115.

Cassidy F (1999) Thesis, University of Limerick.

Cowfer JA and Gorensek MB (1997) Vinyl chloride. In *Kirk–Othmer Encyclopedia of Chemical Technology*, Vol. 24. Wiley, New York, pp 851–882.

Cox JD and Pitcher G (1970) *Thermochemistry of Organic and Organometallic Compounds*. Academic Press, New York.

Fontana CM, Gorin E, Kidder GA and Meredith CS (1952) Chlorination of methane with copper chloride melts. *Ind. Eng. Chem.* **44**: 363–368.

Freidel IM, Frost AC, Herbert KJ, Meyer FJ and Summers JC (1993) New catalyst technologies for the destruction of halogenated hydrocarbons and volatile organics. *Catal. Today* **17**: 367–382.

Froese KL and Hutzinger O (1997) Mechanism of the formation of polychlorinated benzenes and phenols by heterogeneous reactions of $C_2$ aliphatics. *Environ. Sci. Technol.* **31**: 542–547.

Hall PG, Heaton P and Rosseinsky DR (1984) Adsorption and conductivity studies of oxychlorination catalysis. Part 2: Gas-adsorption chromatography with catalytic copper (I, II) chlorides. *J. Chem. Soc., Faraday Trans. 1* **80**: 2785–2802.

Johnson J (1995a) Incineration targeted by EPA. New limits proposed for biggest dioxin generators. *Environ. Sci. Technol.* **29**: 33A–35A.

Johnson J (1995b) Dioxin risk: Are we sure yet. *Environ. Sci. Technol.* **29**: 24A–25A.

Kjeller L-O, Jones KC, Johnston AE and Rappe C (1996) Evidence for a decline in atmospheric emissions of PCDD/F in the UK. *Environ. Sci. Technol.* **30**: 1398–1403.

Kjeller L-O and Rappe C (1995) Time trends in levels, patterns and profiles for polychlorinated dibenzo-*p*-dioxins, dibenzofurans and biphenyls in a sediment core from the Baltic proper. *Environ. Sci. Technol.* **29**: 346–355.

Krishnamoorthy S, Parker JP and Amiridis MD (1998) Catalytic oxidation of 1,2-dichlorobenzene over $V_2O_5/TiO_2$-based catalysts. *Catal. Today* **40**: 39–46.

Lago RM, Green MLH, Tsang SC and Odlyha M (1996) Catalytic decomposition of chlorinated organics in air by copper chloride based catalysts. *Appl. Catal. B: Environ.* **8**: 107–121.

Lauterbach J, Haas G, Rotermund HH and Ertl G (1993) Spatio-temporal pattern formation on polycrystalline platinum surfaces during catalytic CO oxidation. *Surf. Sci.* **294**: 116–130.

Markeloff RG (1984) Convert oxychlorination from air to oxygen. *Hydrocarbon Process.* November 91–94.

Miyake T and Hanaya M (1995) Screening of metal chloride catalysts for oxychlorination of propene. *Appl. Catal. A General* **121**: L13–L17.

Naworski JS and Velez ES (1983) Oxychlorination of ethylene. In *Applied Industrial Catalysis*, Vol. 1, Leach BE (Ed). Academic Press, New York, pp 239–273.

Reed DJ (1993) Chlorocarbons and chlorohydrocarbons. In *Kirk–Othmer Encyclopedia of Chemical Technology*, Vol. 5. Wiley, New York, pp 1017–1028.

Rose J (1994) Incineration regulations set to tighten. *Environ. Sci. Technol.* **28**: 521A.

Ruthven DM and Kenney CN (1967) The kinetics of the oxygen absorption in molten salts containing cuprous chloride. *Chem. Eng. Sci.* **22**: 1561–1570.

Ruthven DM and Kenney CN (1968) The kinetics of the oxidation of hydrogen chloride over molten salt catalysts. *Chem. Eng. Sci.* **23**: 981–990.

Safe SH (1993) Toxic aromatics. In *Kirk–Othmer Encyclopedia of Chemical Technology*, Vol. 6. Wiley, New York, 4th edition, pp 127–139.

Sai Prasad PS, Prasad KBS, Kanta Rao P and Kaushik VK (1997) Redistribution of copper in KCl-promoted $CuCl_2/\gamma\text{-}Al_2O_3$ catalyst during oxychlorination of ethylene. *J. Mater. Sci.* **32**: 1479–1482.

Schennach R and Bechtold E (1996) Adsorption of oxygen on chlorine modified Pt(100) surfaces. *Surf. Sci.* **369**: 277–288.

Van den Brink RW, Louw R and Mulder P (1998) Formation of polychlorinated benzenes during the catalytic combustion of chlorobenzene using a $Pt/\gamma\text{-}Al_2O_3$ catalyts. *Appl. Catal. B: Environ.* **16**: 219–226.

Villadsen J and Livbjerg H (1978) Supported liquid-phase catalysts. *Catal. Rev.-Sci. Eng.* **17**: 203–272.

Yang MX, Sarkar S, Bent BE, Bare SR and Holbrook MT (1997) Degradation of multiply-chlorinated hydrocarbons on Cu(100). *Langmuir* **13**: 229–242.

Zabetakis MC (1965) *Flammability Characteristics of Combustible Gases and Vapors*. US Dept of the Interior, Bureau of Mines, Washington, DC.

# 10

# Oxidation Reaction Catalysed by Zeolites and Structured Mesoporous Materials


10.1 Zeolites 302
  10.1.1 Structural Features of Zeolites 302
  10.1.2 Preparation of Zeolites 308
10.2 Redox Molecular Sieves 313
  10.2.1 The Preparation of Titanium Silicalite 318
  10.2.2 The Structure of Titanium Silicalite 320
  10.2.3 Other Zeolites Containing Framework Titanium 322
  10.2.4 Mechanism of Operation of TS-1 in Oxidation Reactions 323
  10.2.5 Other Redox Molecular Sieves 326
10.3 Structured Mesoporous Silicates in Oxidation Reactions 328
  10.3.1 Synthesis and Characterization of Structured Mesoporous Silicates 328
  10.3.2 Oxidation Reactions Catalysed by Ti-MCM-41 333
  10.3.3 MCM as a Catalyst Support 335
References to Chapter 10 336



Summary

Zeolites are in widespread use for cracking of hydrocarbons in the petrochemical industry, but the invention of the titanium silicalite family of redox molecular sieves has led to the use of these material as selective oxidation catalysts, mainly with hydrogen peroxide as oxidizing agent. In addition, the invention of structured mesoporous silicates with regular pore dimensions considerably larger than those of zeolites, has stimulated many investigations in which large molecules are selectivity oxidized within these materials.

This chapter describes

Structural features and methods of preparation of zeolites,
The properties of redox molecular sieves, including the mechanism of oxidation reactions in titanium silicalite,
Structured mesoporous silicates in oxidation reactions.

## 10.1 ZEOLITES

### 10.1.1 Structural features of zeolites

During the 1960s a number of oil refinery operators introduced zeolites as cracking catalysts, to replace amorphous silica–alumina materials. The zeolite materials give better yields of gasoline from cracking units and their application has become widespread throughout the petrochemical industry (Weisz, 1982; Avidar and Shinnar, 1990). Zeolites are microcrystalline, act as solid acids and are structurally microporous, meaning that they have pores smaller than 2 nm in diameter, which are a direct result of their crystal structure. By contrast, microporosity in amorphous materials arise from cracks and voids within these materials. Zeolites were also found to have very high surface areas, and most of this arises from the microporosity. In fact, model building confirmed by electron microscopy and other structural studies reveals that zeolites possess a crystalline array of channels, and cages which make up a three-dimensional network in the interior of these materials. High resolution microscopy evidence has been available for a number of years to support this model of zeolite structures and Figure 10.1 shows one such micrograph of zeolite-$\beta$ epitaxially grown onto nanofibres of the one-dimensional zeolite SSZ-31 (Nair *et al.*, 1999).

A key feature of zeolites is that small molecules can diffuse through the interior and react at specific sites located on the internal surface. Typically, pore sizes in silica–alumina zeolites range from 0.4 to 1.4 nm in diameter, whereas molecular diameters are typically: propane $\approx$ 0.42 nm; isobutane $\approx$ 0.5 nm; $CCl_4 \approx$ 0.59 nm; $(CuH_9)_3N \approx$ 0.81 nm and $(C_4F_9)_3N \approx$ 1.02 nm (Newsam, 1992). These materials have assumed enormous importance to our way of life and it is probably no exaggeration to say that nearly every organic molecule we use in everyday life (in fuels, polymers etc.) has, at some stage in its preparation, diffused through and been transformed inside a zeolite.

Initially all known zeolites were based on aluminosilicates possessing the structures of well known minerals (e.g. faujasite, mordenite, erionite). The first of these minerals was discovered in 1756 by AF Cronstedt. Later, synthetic aluminosilicate zeolites (e.g. ZSM-5, ZSM-11, Zeolite theta-1) were developed which have no known naturally occurring analogues.

From a structural point of view all zeolites possess a number of common features. The simplest zeolite formula is $SiO_2$. A zeolite with this formula

**Figure 10.1** Electron micrograph of a zeolite-$\beta$ epitaxially grown onto SSZ-31 nanofibres (Nair *et al.*, 1999) (reproduced by permission of the Royal Society of Chemistry).

does exist and is called silicalite. The primary building block of all zeolites is the $SiO_4$ tetrahedron. Other metal atoms can also be introduced, most commonly Al. Substitution of a trivalent aluminium atom for a tetravalent silicon atom generates a net negative charge on the zeolite structure. This is compensated by the introduction of a cation, most often sodium, in the preparation step, to maintain charge neutrality. A generalized empirical formula for zeolites can be written as

$$Na_xAl_xSi_yO_{2(x+y)} \cdot yH_2O$$

or

$$Na_x[(AlO_2)_x(SiO_2)_y] \cdot mH_2O$$

Exchange of these cations for ammonium ions, and thermal treatment leads to the well known acidic properties of zeolites, as shown in Scheme 10.1 (Whan, 1981). Suitable treatment with water vapour can interconvert Lewis and Brønsted acid sites in zeolites. In fact, an important feature, to be referred to

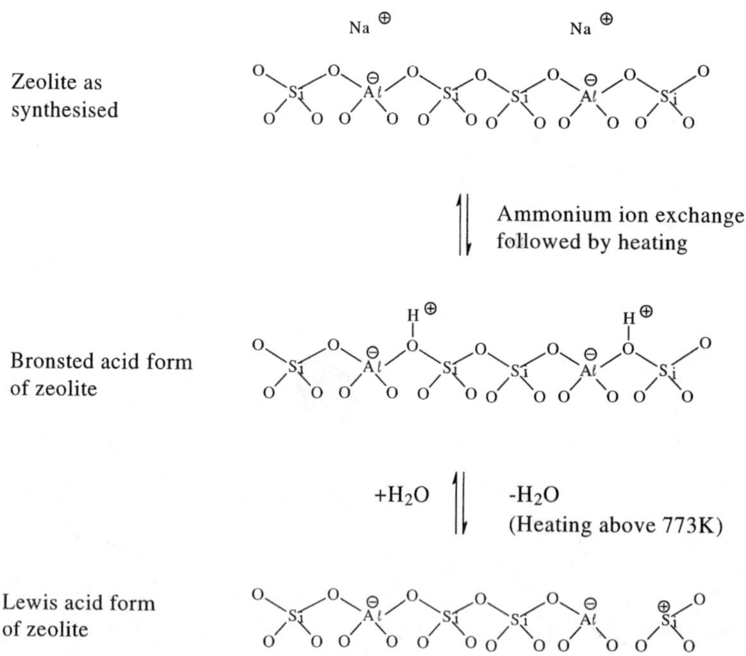


**Scheme 10.1** The generation of Brønsted and Lewis acid sites in zeolites (Whan, 1981) (reproduced by permission of the Royal Society of Chemistry).

below, is that without aluminium in the structure these material are non-acidic.

More than 200 synthetic and 50 naturally occurring zeolites are known and the number is continuously growing. A system of nomenclature for zeolite structure types has been developed by the International Union of Pure and Applied Chemistry commission on zeolite nomenclature. A small number of these structures are presented in Table 10.1 (Meier *et al.*, 1996). The material in this table was extracted from the *Atlas of Zeolite Structure Types* published by the International Zeolite Association. The material in Table 10.1 is identified in three ways. The first is by its commonly used name, then by its chemical formula. Finally each zeolite is associated with a unique framework topology. This topology is identified by three capital letters assigned by the International Union of Pure and Applied Chemistry. The three letters of the codes are intended to be mnemonic (easy to remember) and are chosen to be mineral names following the nomenclature rules of the International Mineralogical Society if they conform to commonly accepted synthetic material types. Otherwise workers who first determined the structure have priority in assigning the name (Meier *et al.*, 1996).

The origin of common codes for individual zeolites can be confusing to the newcomer to this field, but generally they relate to the companies, universities and research institutions where the zeolites were first discovered. Accordingly ZSM stands for **Z**eolite **S**ocony **M**obil; SAPO-40 is derived from **s**ilico-**a**lumino **pho**sphate-forty; and VPI-5 refers to **V**irgina **P**olytechnic **I**nstitute-5.

Synthetic zeolites can now be prepared where a variety of other trivalent atoms are incorporated into the structure. There are examples where Fe, Ga or B are incorporated into ZSM-5 by replacement of tetravalent silicon atoms; some examples are listed in Table 10.1. Not all literature claims for replacement of silicon atoms in the structure of zeolites are credible and it is important to make the distinction between framework and extraframework substitution. In the former silicon or aluminium atoms in the zeolite structure are replaced, whereas in the latter only the charge compensating ions are replaced.

Alternatively silicon can be replaced resulting in a new family of aluminophosphates, based on $AlPO_4$. Here acidity can be generated by replacement of some trivalent aluminium atoms by divalent cations such as cobalt, zinc or manganese. These materials are called MeALPO (**me**tal-**al**uminium-**pho**sphate). Alternatively where tetravalent silicon replaces some of the pentavalent phosphorus the material is called SAPO (**s**ilicon-**a**luminium-**pho**sphate). Some examples are listed in Table 10.1.

Zeolites are made up of primary and secondary structures. $TO_4$ is the common building unit for these materials, and the variety of structures is possible by virtue of the way a small number of secondary units are put together in the structure. The basic $TO_4$ unit is shown in Scheme 10.2 and the standard method of representation of 4, 6 and 8 membered rings is also

**Table 10.1** Structural properties of zeolites (Meier *et al.*, 1996)

| *Material* | | | *Largest channel* | | | *Unit cell* | | | |
|---|---|---|---|---|---|---|---|---|---|
| *Common* | *Formula* | *IUPAC* | *No of $TO_4$ units in ring* | *Dimension/ nm* | *Direction* | *Type* | *Dimensions/ nm* | *Space group* | *Other materials with the same topology* |
| ALPO-11 | $Al_{20}P_{20}O_{80}$ | AEL | 10 | 0.39 × 0.63 | [100] | Orthorhombic | $a = 0.84$; $b = 1.85$; $c = 1.35$ | *Imma* | MnAPO-11<br>SAPO-11 |
| SAPO-40 | $Si_7Al_{29}P_{28}O_{128} \cdot 4(C_3H_7)_4NOH$ | AFR | 8<br>12 | 0.37 × 0.37<br>0.67 × 0.69 | [010]<br>[001] | Orthorhombic | $a = 2.19$; $b = 1.37$; $c = 1.42$ | *Pccn* | |
| Zeolite beta[a] | $Na_n[Al_nSi_{64-n}O_{128}]$ ($n < 7$) | BEA | 12<br>12 | 0.76 × 0.64<br>0.55 × 0.55 | [100]<br>[001] | Tetragonal | $a = 1.27$; $c = 2.64$ | $P4_122$ | Tschernichite |
| **Clo**verite | $[\{Ga_{96}P_{96}O_{372}(OH)_{24}\} \cdot 24RF]_8$[b] | CLO | 20<br>8 | 1.32 × 0.40<br>0.38 × 0.38 | [100]<br>[100] | Cubic | $a = 5.17$ | *Fm-3c* | |
| **Fau**jasite | $(Na_2,Ca,Mg)_{29}[Al_{58}Si_{134}O_{384}] \cdot 240H_2O$ | FAU | 12 | 0.74 × 0.74 | [111] | Cubic | $a = 2.47$ | *Fd-3m* | Linde X<br>Linde Y<br>SAPO-37 |

| | | | | | | | | | |
|---|---|---|---|---|---|---|---|---|---|
| Ferrierite | $Na_2Mg_2[Al_6Si_{30}O_{72}] \cdot 18H_2O$ | FER | 10<br>8 | 0.42 × 0.54<br>0.35 × 0.48 | [001]<br>[010] | Orthorhombic | $a = 1.92$;<br>$b = 1.41$;<br>$c = 0.75$ | *Immm* | ISI-6<br>ZSM-35 |
| Linde Type A | $Na_{12}\{Al_{12}Si_{12}O_{48}\} \cdot 27H_2O$ | LTA | 8 | 0.41 × 0.41 | [100] | Cubic | $a = 2.46$ | *Fm-3c* | N-A<br>SAPO-42 |
| ZSM-11 | $Na_n[Al_nSi_{96-n}O_{192}] \cdot 16H_2O$ ($n < 16$) | MEL | 10 | 0.53 × 0.54 | [100] | Tetragonal | $a = 2.01$;<br>$c = 1.34$ | *I-4m2* | Silicalite-2<br>TS-2 |
| ZSM-5 | $Na_n[Al_nSi_{96-n}O_{192}] \cdot 16H_2O$ ($n < 27$) | MFI | 10 straight<br>10 sinusoidal | 0.53 × 0.56<br>0.51 × 0.55 | [010]<br>[100] | Orthorhombic | $a = 2.01$;<br>$b = 1.99$;<br>$c = 1.34$ | *Pnma* | Silicalite-1<br>TS-1 |
| Mordenite | $Na_8[Al_8Si_{40}O_{96}] \cdot 24H_2O$ | MOR | 12<br>8 | 0.65 × 0.70<br>0.26 × 0.57 | [001]<br>[010] | Orthorhombic | $a = 1.81$;<br>$b = 2.05$;<br>$c = 0.75$ | *Cmcm* | |
| Zeolite theta-1 | $Na_n[Al_nSi_{24-n}O_{48}] \cdot 4H_2O$ | TON | 10 | 0.44 × 0.55 | [001] | Orthorhombic | $a = 1.38$;<br>$b = 1.74$;<br>$c = 0.50$ | *Cmc2₁* | ZSM-22<br>KZ-2 |

[a]BEA is a mixture of two closely related polymorphs.
[b]Where R is quinuclidinium.

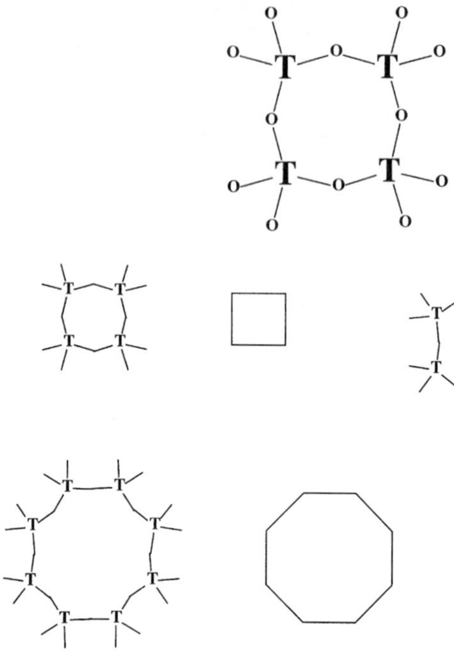


**Scheme 10.2** $TO_4$ basic building unit in zeolites and reduction of this unit to simple line drawings.

indicated in this Scheme. Some secondary structures are presented in Scheme 10.3 (Thomas *et al.*, 1997). Table 10.1 also indicates the size of the largest channels in the zeolites in terms of the number of $TO_4$ units in the ring, the channel dimensions and crystallographic direction. Basic crystallographic data and other zeolites in the same topology class complete Table 10.1. A number of complete zeolites structures are presented in Scheme 10.4

### 10.1.2 Preparation of zeolites

Most synthetic zeolites are prepared as microcrystalline materials by prolonged reaction of a silicon source and an aluminium source in the presence of alkali metals held in an alkaline solution. The reaction is usually carried out in an autoclave and at autogeneous pressure. Organic compounds are added to the reacting mixture as templating agents for the preparation of high silica zeolites. Some typical synthesis conditions are reported in Table 10.2 (Kuhl and Kresge, 1995).

The first step in the zeolite synthesis process is gel formation, which occurs rapidly when the reactants are mixed. The gel results from the destabilization

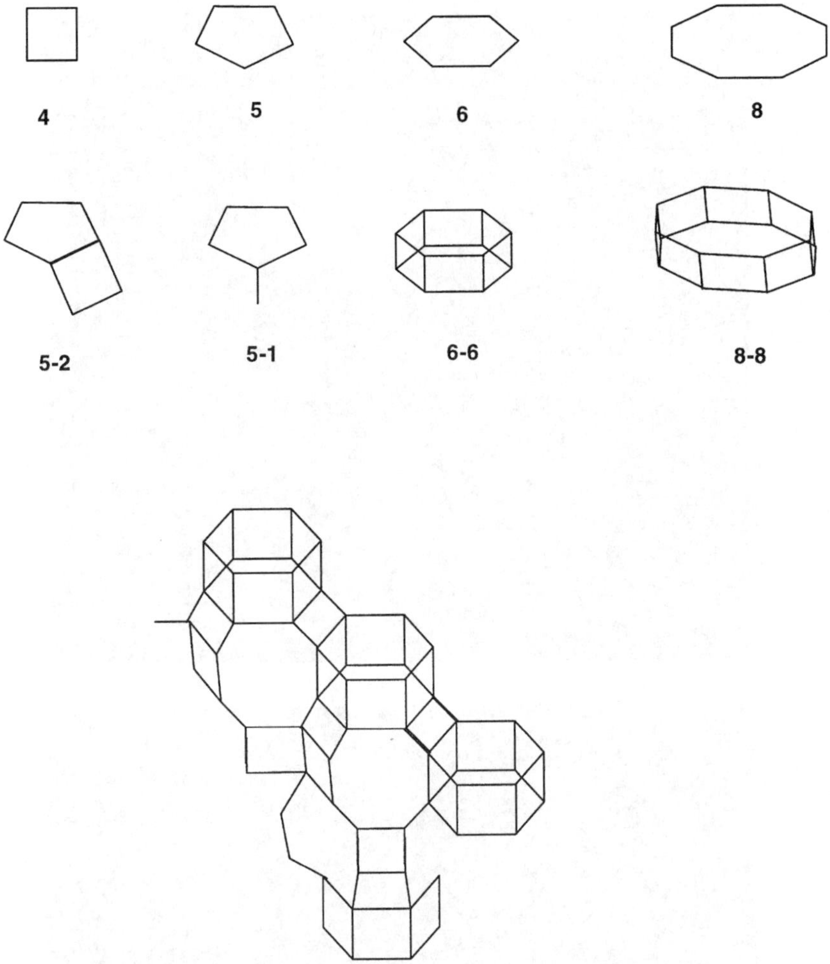


**Scheme 10.3** Secondary building units commonly encountered in zeolites and their coalescence into a zeolite structure.

of a solution and the formation of a three-dimensional network of colloidal particles (Feijen *et al.*, 1997). The process may be represented as follows:

$$NaOH_{(aq)} + NaAl(OH)_{4\,(aq)} + Na_2SiO_{3\,(aq)} \rightarrow [Na_a(AlO_2)_b(SiO_2)_c \cdot NaOH \cdot H_2O]_{gel} \qquad (10.1)$$

The gel which forms then serves as a source of silicate and aluminate monomers and oligimers which eventually form the secondary building units shown in Scheme 10.3. This step in the process may be represented as:

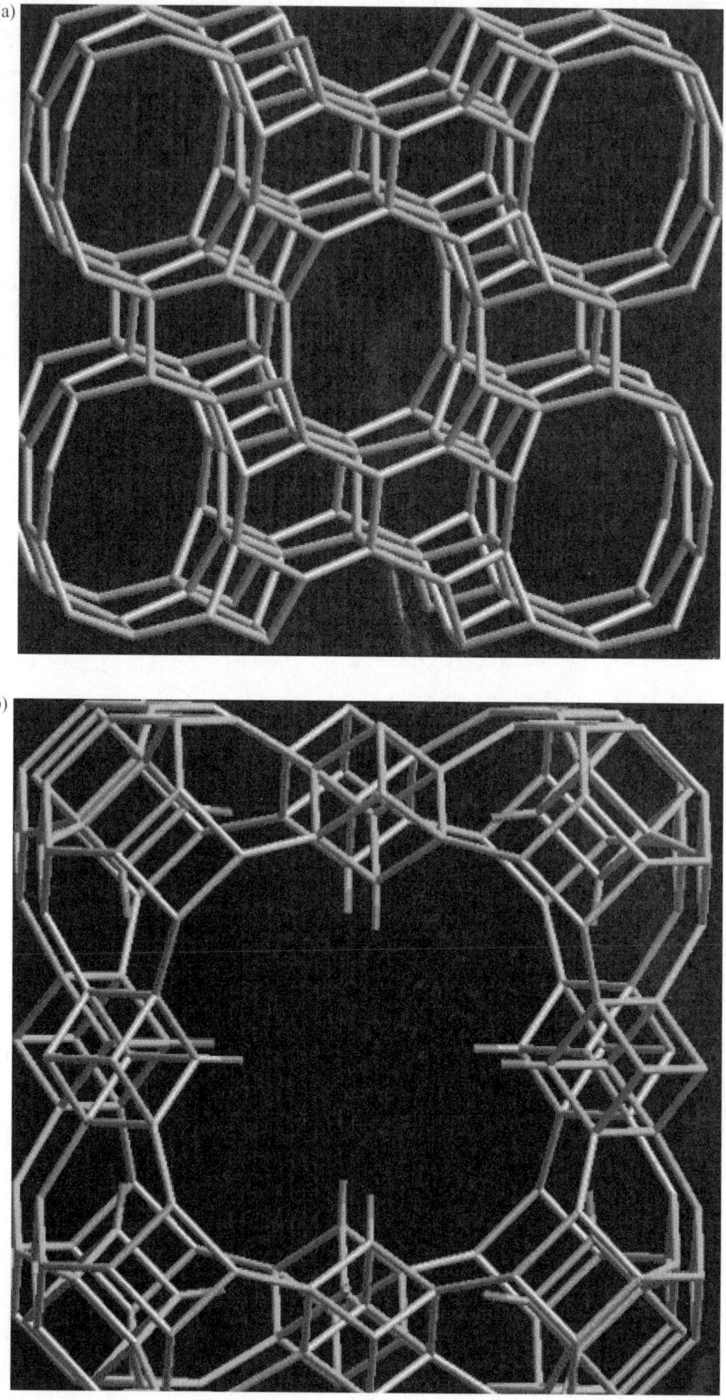
(a)
(b)

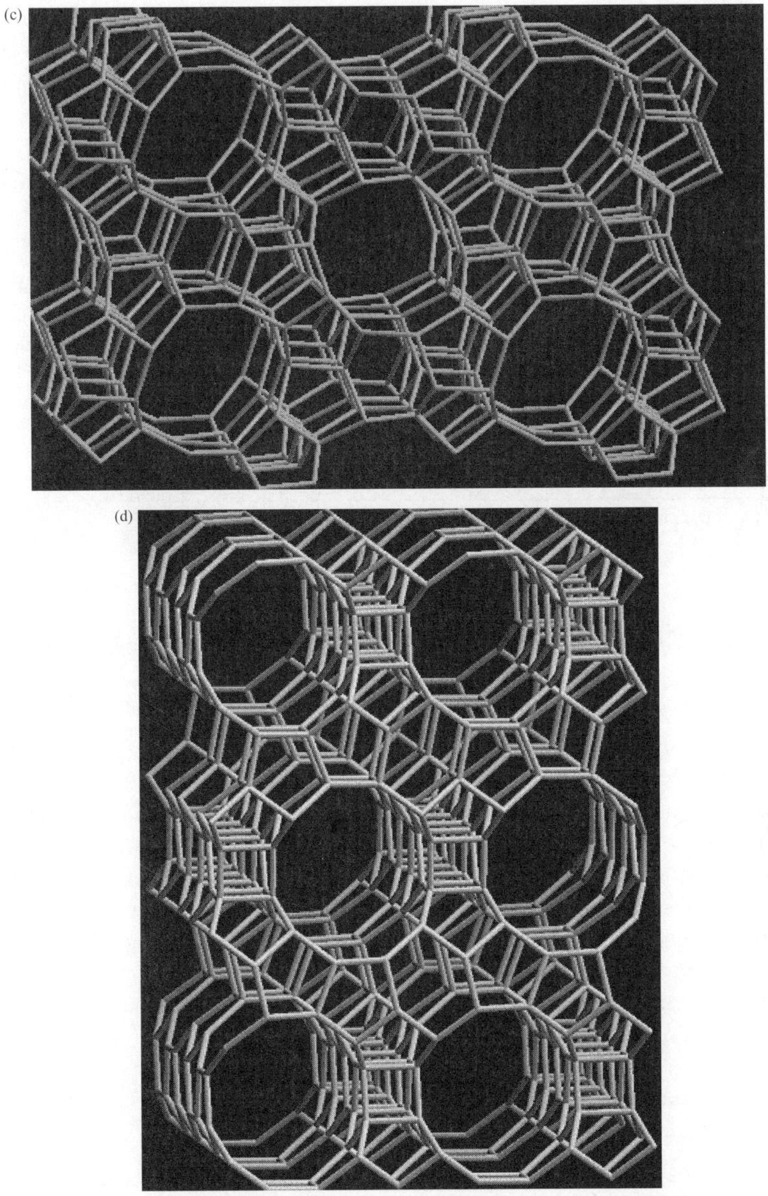


**Scheme 10.4** Some zeolite structures: (a) AEL viewed along the [001] direction, (b) CLO viewed along the [100] direction, (c ) MFI viewed along the [010] direction and (d) BEA viewed along the [100] direction. The corresponding dimensions are given in Table 10.1. (Ch. Baerlocher and L.B. McCusker, *Atlas of Zeolite Structures Types on the Web*, http://www.iza-sc.ethz.ch/IZA-SC/Atlas/AtlasHome.html). Reproduced by permission.

**Table 10.2** Conditions for the hydrothermal synthesis of sodium aluminosilicates (Kuhl and Kresge, 1995)

| *Zeolite* | *Reactant composition/ mol/mol $Al_2O_3$* | | | *Reactants* | *Synthesis temp/K* | *Final zeolite composition* | | |
|---|---|---|---|---|---|---|---|---|
| | *$Na_2O$* | *$SiO_2$* | *$H_2O$* | | | *$Na_2O$* | *$SiO_2$* | *$H_2O$* |
| A | 2 | 2 | 35 | $NaAlO_2$<br>NaOH<br>$Na_2SiO_3$ | 293–448 | 1 | 2 | 4.5 |
| Mordenite | 6.3 | 27 | 61 | $NaAlO_2$<br>Diatomite<br>$Na_2SiO_3$ | 373 | 1 | 9–10 | 6.7 |
| ZSM-5 | 10 | 7.7 | 453 | $NaAlO_2$<br>$SiO_2$<br>TPAOH[a] | 423 | 0.89 | 31.1 | 2.0 |

[a]Tetrapropylammonium hydroxide.
Reproduced by permission of John Wiley and Sons.

$$[Na_a(AlO_2)_b(SiO_2)_c \cdot NaOH \cdot H_2O]_{gel} \longrightarrow Na_x[(AlO_2)_x(SiO_2)_y] \cdot mH_2O \quad (10.2)$$

These secondary structures in turn grow into the specific zeolite structures which are favoured by the synthesis conditions.

For zeolites with low Si : Al ratios, for example zeolite A in Table 10.2, the secondary building units are complexed to alkali metals. With high silica : alumina ratios, for example ZSM-5 in Table 10.2, the mechanism involves templating where the alkylammonium cation complexes with silica by hydrogen bonding. The structure replicates by hydrogen bonding of the organic cation with framework oxygen atoms.

There are three basic steps involved in zeolite formation (Feijen *et al.*, 1997). These are:

- supersaturation,
- nucleation,
- crystal growth.

Supersaturation is the condition whereby the concentration of a solute exceeds the normal equilibrium concentration. Without supersaturation the solute species are stable in solution and crystallization cannot be achieved. Crystallization is most easily achieved in normal laboratory practice by cooling a solution, which often results in precipitation of the solute. In zeolite synthesis, superaturation is achieved by providing a highly alkaline environment in which silicate and aluminate monomers and secondary building units are generated at a faster rate than their incorporation into evolving crystals.

Nucleation is the process whereby the first stable entity of a new phase is established. This usually means that a minimum size of nucleus must be achieved. Prior to this condition being achieved the germs of the new phase which form are unstable and will redissolve in the solution rather than proceed to the crystal growth phase. The rate of nucleation (expressed as the number of viable nuclei formed per unit volume and unit time) depends on the degree of supersaturation and the reaction temperature (Feijen *et al.*, 1997).

Nuclei grow into crystals by multiple addition of precursor species (de Moor *et al.*, 1999). A typical curve for the evolution of the crystalline zeolite material is presented in Figure 10.2 and the overall process is presented schematically in Scheme 10.5. The crystal growth curve is usually characterized by an induction period, representing the time required for the formation of the first viable nuclei. Nucleation is followed by a period of rapid growth and a further period when the reaction approaches completion at a slower rate. Increasing the reaction temperature or the pH reduces the induction period and accelerates the rate of crystal formation.

The form of the zeolite which crystallizes from a given reaction system depends to a large extent on the composition of the reaction mixture, but temperature and time can also play a role. Phase diagrams have been established for many systems. However, there are circumstances in which two zeolites may form from the same reaction mixture. In such circumstances, the Ostwald rule applied whereby the least thermodynamically stable is formed first, but if allowed to react for a sufficiently long time it will be transformed into the more thermodynamically favoured phase (Dwyer and Chu, 1979). This feature of zeolite synthesis is illustrated in Figure 10.2 for ULM-5, a microporous oxy-fluorinated gallophosphate containing 16 T-atom pores (Francis and O'Hare, 1998). In this synthesis gallium oxide, phosphorus pentoxide, hydrofluoric acid and 1,6-diaminohexane as templating agent were employed. The crystallization was monitored by *in situ* energy dispersive X-ray diffraction. The course of the reaction at 453 K is shown in the three-dimensional plot in Figure 10.2. Two phases were observed; the first to appear had a Bragg reflection at 13.1°. Its intensity reached a maximum at ca 20 min at 453 K and at ca 60 min at 423 K, then declined. Peaks characteristic of ULM-5 started to appear after ca 20 min at 453 K. The (002) reflection of ULM-5 was used to construct the plot in Figure 10.2.

## 10.2 REDOX MOLECULAR SIEVES

One of the most significant developments in oxidation by heterogeneous catalysts was the announcement in 1983 of titanium-containing crystalline silicas (Taramasso *et al.*, 1983). These materials, commonly known as

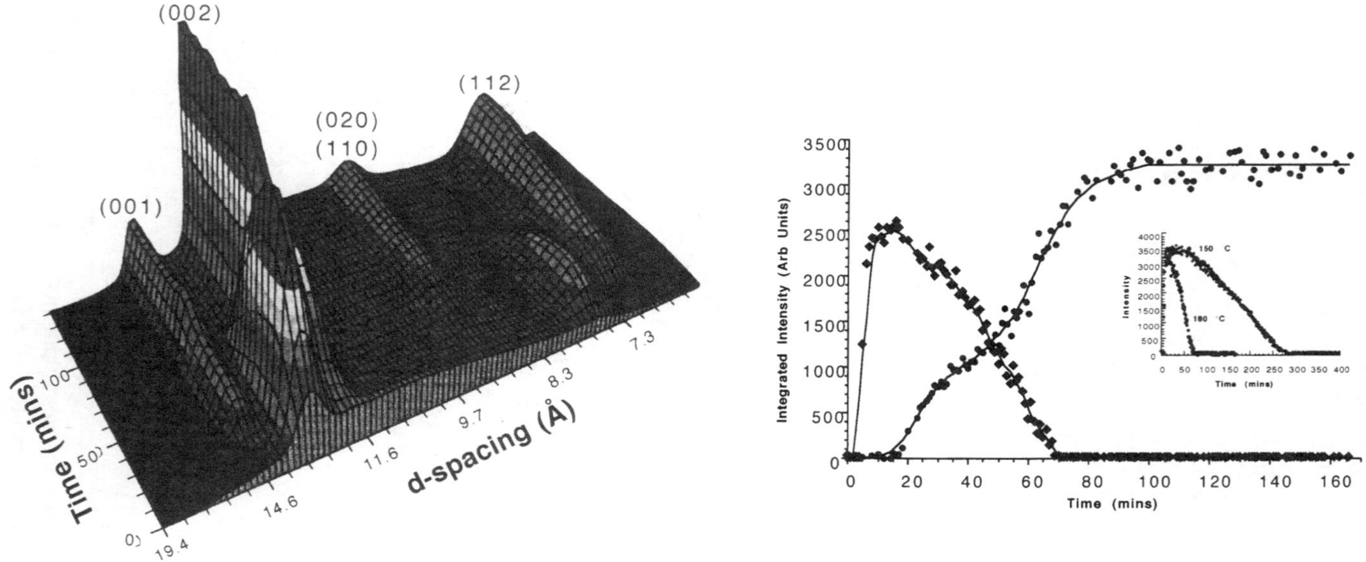


**Figure 10.2** Evolution of phases during the synthesis of ULM-5. The three-dimensional plot shows the evolution of the *in situ* X-ray diffraction patterns at 453 K. The two-dimensional plot shows initial formation of an unstable intermediate at $2\theta = 13.1°$ followed by the emergence of peaks associated with ULM-5. The insert in this plot shows the development of the intermediate structure at 423 and 453 K (Francis and O'Hare, 1998) (reproduced by permission of the Royal Society of Chemistry).

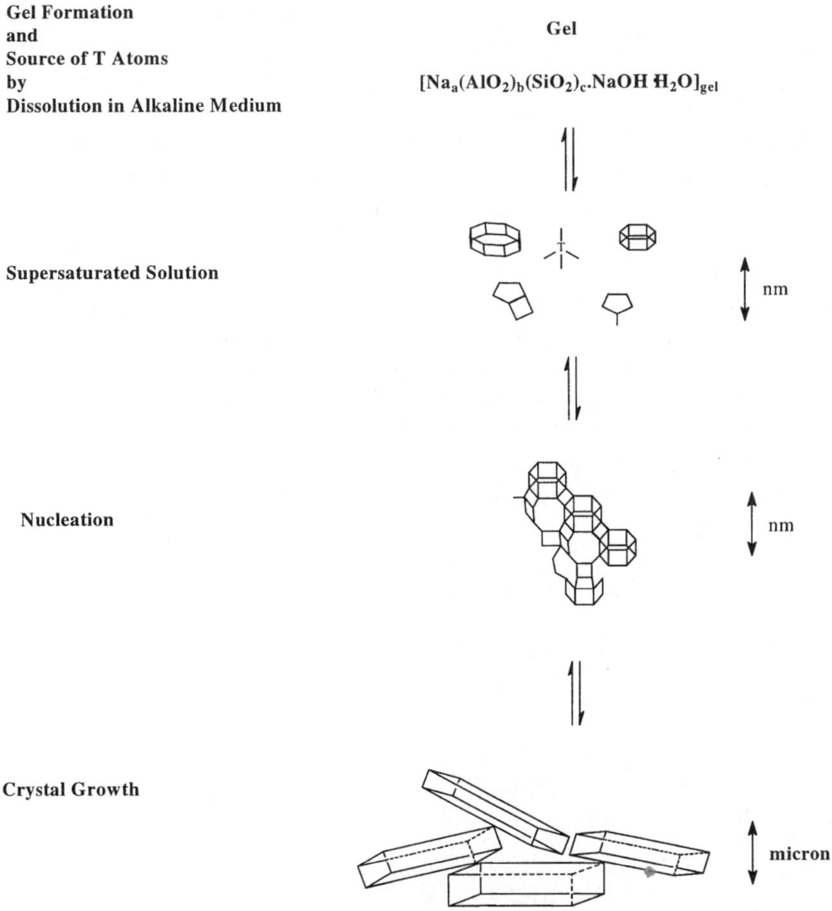


**Scheme 10.5** Schematic representation of supersaturated solution formation, nucleation and crystal growth applied to zeolite synthesis.

titanium silicalites, are effective for numerous oxidation reactions with hydrogen peroxide as the oxidant.

Titanium-containing oxidation catalysts were known prior to the discovery of the titanium silicalites, principally in the form of Shell's amorphous $TiO_2/SiO_2$ catalyst used for the epoxidation of propene with hydroperoxides, such as ethylbenzene hydroperoxide or *tert*-butyl hydroperoxide. However, just as the crystalline aluminosilicates have replaced the amorphous analogues in the cracking operations so the crystalline titanium silicates offer many more possibilities than the amorphous analogues in oxidation reactions. In addition, titanium compounds in homogeneous solution were known as reasonably effective epoxidation catalysts. The announcement of the TS-1 material has inspired a large number of complementary studies wherein researchers have sought to modify existing zeolites and zeotypes by preparation methods directed at introducing other metals into the framework. In this regard we will concentrate here on those studies which seek to give zeolites some redox character. These materials are sometimes referred to as redox molecular sieves, in deference to their combined redox properties and zeolitic structures.

Some of the reactions catalysed by titanium silicalite are shown in Table 10.3. A feature is that they operate in very mild reaction conditions, usually below 373 K. In addition, they are not limited in terms of selectivity in the way that epoxidation reactions are, where molecular oxygen is the oxidant. This point is illustrated in the first line of Table 10.3, which shows a very high selectivity to propylene oxide in the epoxidation of propene with hydrogen peroxide as oxidant. By contrast, the corresponding selectivity with oxygen as oxidant is always very poor, as shown in Figure 6.12. The reason is the tendency for oxidizing species derived from molecular oxygen to preferentially attack the very weak allylic C—H bond where present in the molecules as shown in Table 6.6. Figure 10.3 presents a plot of selectivity in epoxidation of a number of substrates by hydrogen peroxide over titanium silicalite catalysts, some with and some without allylic C—H bonds. The

**Table 10.3** Reactions catalysed by TS-1 (Notari, 1991)

| *Reaction* | $H_2O_2$ %*conv* | %*sel* |
|---|---|---|
| 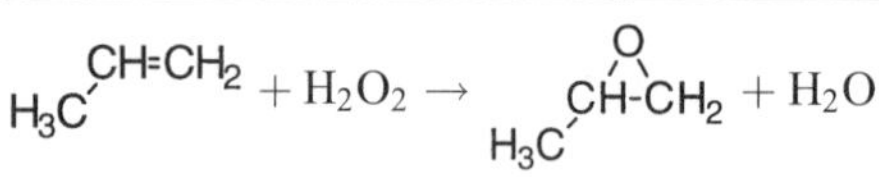 | 99 | 98 |
| 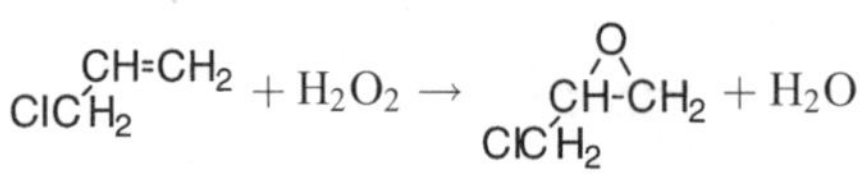 | 98 | 93 |
| 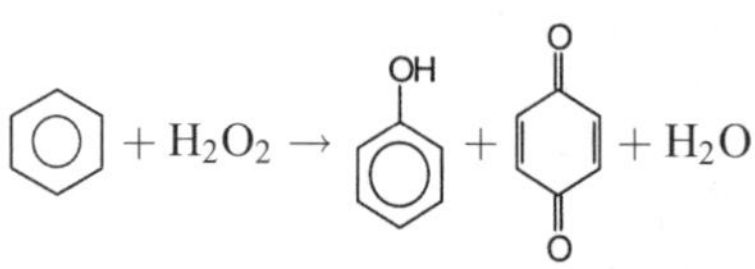 | 100 | 76 + 24 |
| 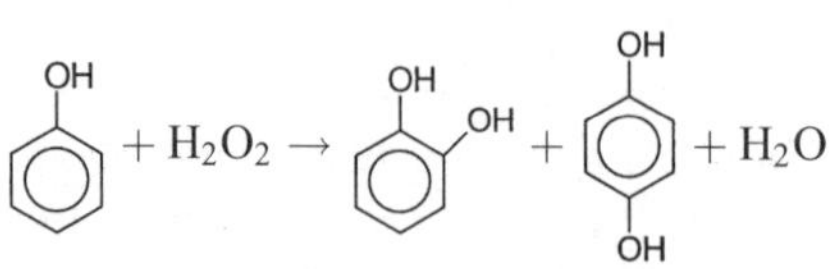 | 98 | 90 |
| 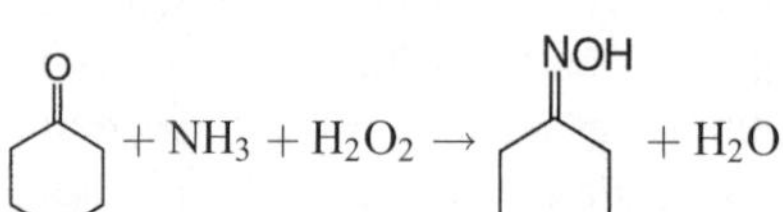 | 99.9 | 98 |

Reproduced by permission of Elsevier Science.

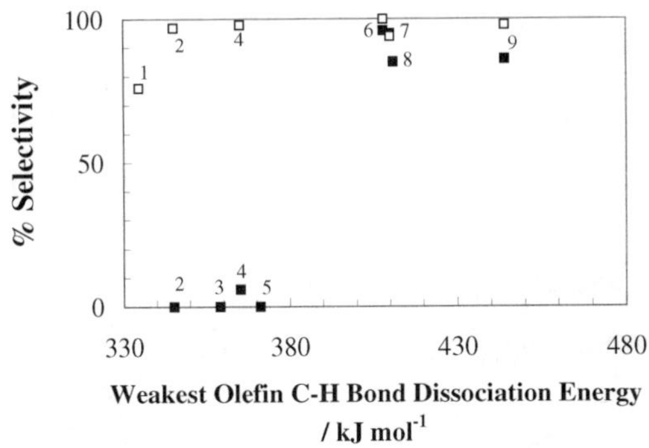


**Figure 10.3** Selectivity in epoxidation for a range of substrates plotted against the weakest C—H bond in the substrate structure. (◆) Epoxidations over TS-1 with $H_2O_2$; (○) epoxidations with molecular oxygen. 1, 1-octene; 2, 1-butene; 3, 2-butene; 4, propene; 5, 4-vinylpyridine; 6, 1,3-butadiene; 7, styrene; 8, 4-vinylpyridine; 9, ethylene. (Cassidy, 1999).

corresponding plot for epoxidations with molecular oxygen is overlaid on this figure. The contrast is striking. Epoxidation involving hydrogen peroxide and titanium silicalite extends the applicability of epoxidation reactions to a much wider variety of substrates, including those with allylic C—H bonds. This is consistent with the epoxidizing species present in the peroxide–TS-1 system being much more electrophilic than the corresponding species in the molecular oxygen–silver system.

The reactivity of the hydrogen peroxide–TS-1 system towards a range of unsaturated alcohols has been determined by Tatsumi *et al.* (1993) and their results are presented in Table 10.4. Two reactions are possible here, namely epoxidation at the double bond or oxidation at the alcohol group to yield the corresponding ketone or aldehyde. The results are presented as the yield of each product per titanium atom in the structure. The fraction of epoxide in the total product slate is also presented in this table. The influence of substituents on the double bond was investigated in this study. Column 1 of Table 10.4 indicates the level of substitution around the double bonds so that 1–2 indicates one substituent on one side and two on the other side. The nature of the 0–1 substituent on the C=C double bond is not a crucial factor in determining reactivity and for all of these structures epoxidation is favoured over oxidation of the OH group. With a substituent on each side of the double bond the overall reaction rates appear to increase and the ratio of epoxidation to carbonyl formation is a little less than that observed with just one substituent. Two substituents on one side of the double bond dramatically reduce the epoxide fraction, but overall reaction rates are considerable higher than those observed with the least substituted double bonds. A general conclusion is that in the absence of steric effects the oxidizing species on titanium silicalite derived from hydrogen peroxide preferentially attack the double bond.

**Table 10.4** Oxidation of unsaturated alcohols on TS-1 (Tatsumi *et al.*, 1993)

| *Substituents* | *Substrate* | *Product yield/mol (mol Ti)$^{-1}$* | | |
|---|---|---|---|---|
| | | *Ketone (K) or Aldehyde (A)* | *Epoxide (E)* | *E/E + (K or A)* |
| 0–1 |  | 0 | 19 | 1 |
| 0–1 |  | 0 | 16 | 1 |
| 0–1 | 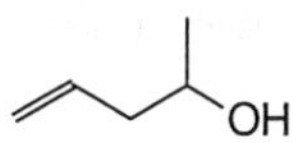 | 7 | 27 | 0.8 |
| 0–1 | 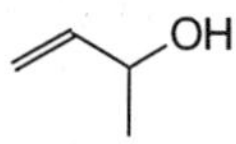 | 0 | 30 | 1 |
| 0–1 | 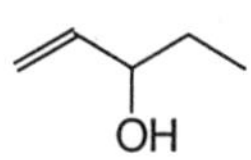 | 43 | 65 | 0.6 |
| 1–1 | 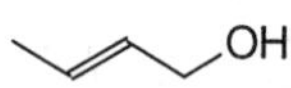 | 31 | 95 | 0.7 |
| 1–1 | 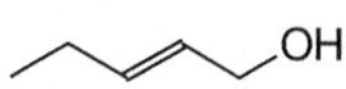 | 44 | 141 | 0.75 |
| 1–1 | 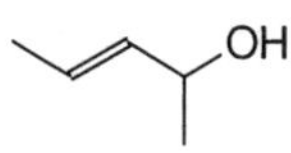 | 98 | 94 | 0.5 |
| 0–2 | 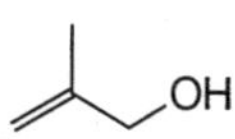 | 37 | 4 | 0.1 |
| 0–2 | 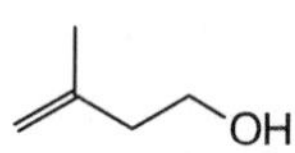 | 18 | 10 | 0.35 |
| 1–2 | 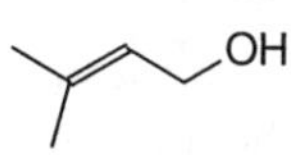 | 75 | 17 | 0.18 |



A further example of the selectivity associated with the TS-1 system can be seen in the epoxidation of *cis*- and *trans*-1-hydroxy-3,7-dimethylocta-2,6-diene with hydrogen peroxide. This reaction is chemiselective at the 2 position, as shown in Scheme 10.6. This result has been described as a hydroxyl assisted epoxidation (Kumar *et al.*, 1995).

### 10.2.1 The preparation of titanium silicalite

The preparation of titanium silicalite (TS-1) is very demanding and a brief outline of the preparation procedure is as follows:

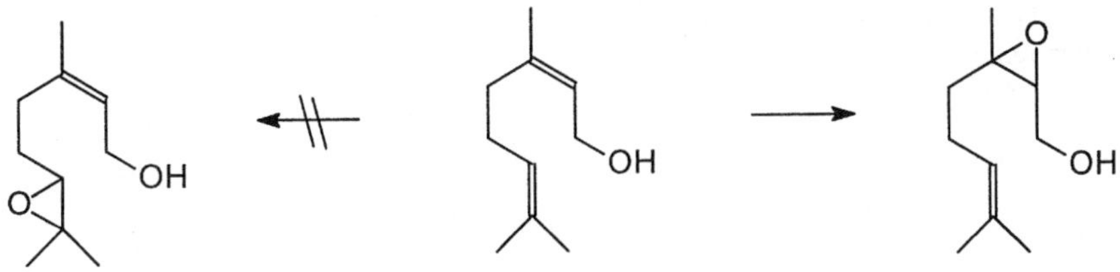


**Scheme 10.6** Epoxidation of *trans*-1-hydroxy-3,7-dimethylocta-2,6-diene with hydrogen peroxide over TS-1 (Kumar *et al.*, 1995).

> 455 g of tetraethylorthosilicate are placed in a Pyrex glass vessel fitted with a stirrer, kept under a $CO_2$-free atmosphere, and 15 g of tetraethyltitanate are added followed gradually by 800 g of 25 % weight solution of tetrapropylammonium hydroxide (free from inorganic alkali). The mixture is stirred for about 1 h. Then heating is commenced carefully in order to accelerate hydrolysis and evaporate the ethanol which is generated. After about 5 h at 353–363 K the alcohol is completely eliminated. The volume is increased to 1.5 l with distilled water and the opalescent homogeneous solution is transferred to a titanium autoclave fitted with a stirrer. The mixture is heated to 548 K and is stirred at this temperature under its own pressure for 10 days. When the treatment is finished, the autoclave is cooled, the contents are discharged and the mass of fine crystals obtained is recovered. This is carefully washed on a filter with hot distilled water many times (Taramasso *et al.*, 1983).

A few comments are pertinent here to indicate a number of pitfalls to be avoided in this preparation. When the silica source tetraethylorthosilicate (TES) and the titanium source tetraethyltitinate (TET) are mixed with tetrapropylammonium hydroxide (TPAOH) a precipitate can form immediately, and when this happens silicalite, rather than TS-1, is formed. The reason is that the TET can be hydrolysed by the action of TPAOH and the product formed does not redissolve during the hydrothermal treatment step. In effect the titanium source becomes unavailable to participate in the reaction. A procedure has been developed to prevent the rapid hydrolysis. This involves cooling the mixture of TES and TET below 273 K prior to the addition of the TPAOH. Vigorous stirring is also reported to be effective. These means ensure that a clear solution of TES, TET and TPAOH is in the reactor at the start of the hydrothermal step and this is essential for a successful synthesis.

The presence of trace amounts of alkali metals is known to interfere with the TS-1 synthesis. In fact extreme care must be exercised to ensure that all reagents are free of trace amounts of alkalis. Table 10.5 shows the effect of small additions of sodium or potassium to the synthesis. The orthorhombic structure was observed for all preparations except that with the highest $K^+$content, where a monoclinic structure was produced. The alkali metal has a drastic effect on the performance of the material produced. One possible explanation is the formation of an insoluble sodium or potassium titanate

**Table 10.5** Influence of alkali metals addition during the preparation of TS-1 on yield in the hydroxylation of phenol
(Notari, 1991)

| $Na^+$/ppm | $K^+$/ppm | Yield based on $H_2O_2$ |
|---|---|---|
| 0 | 0 | 79 |
| 0 | 1060 | 55 |
| 0 | 3530 | 23 |
| 0 | 7060 | 0 |
| 1765 | 0 | 42 |
| 3529 | 0 | 22 |

Reproduced by permission of Elsevier Science.

during the synthesis. In normal preparations, care must be taken to remove all non-framework titanium oxides and this is normally achieved by extensive washing.

The best performance has been reported for TS-1 crystals in the size range 0.2–0.3 μm; larger crystals give slower reaction rates and poorer selectivities. Crystal size is determined by suitable choice of the hydrothermal time and temperature. Table 10.6 shows the effect of varying the particle size of TS-1 on the hydroxylation of phenol.

A further point is that for industrial applications the crystals must be made into an agglomerate of size 20–30 μm. This is carried out by suspending the crystals in a solution containing TPA-silicate and spray drying. This produces a binder made up of very finely divided silica, which does not influence the catalytic properties.

### 10.2.2 The structure of titanium silicalite

The structure of TS-1 greatly resembles that of silicalite-1 (MFI) and can be described as a silicalite in which isomorphous substitution of Si for Ti has occurred. It is in fact a solid solution of titanium in the silicalite-1 host.

**Table 10.6** Influence of TS-1 particle size on catalytic performance for the hydroxylation of phenol
(Van der Pol *et al.*, 1992)

| Average particle size/μm | % $H_2O_2$ converted after 1 h | Selectivity to dihydroxybenzene |
|---|---|---|
| 0.2–0.3 | 50 | 95 |
| 0.3–0.5 | 44 | 93 |
| 5.0–6.0 | 6 | 15 |
| 10.0–11.7 | 2.5 | 8 |

Reproduced by permission of Elsevier Science.

Silicalite-1 has the same structure type as ZSM-5. The extent to which the titanium atoms can be incorporated into the silicalite-1 structure is very limited. Detailed analysis of the X-ray diffraction patterns of TS-1 is possible by a technique called Rietveld analysis. This technique makes use of the whole diffraction pattern, to calculate unit cell dimensions. In TS-1 the unit cell is orthorhombic, with a very slight distortion for some compositions in titanium. With increasing incorporation of titanium there is a small linear increase in unit cell length along the *a*, *b* and *c* axes. The results are shown in Figure 10.4 as the unit cell volume ($a \times b \times c$) as a function of the incorporation of titanium. The expansion is only 1 % for incorporation of 2.5 % titanium but the data are clearly characteristic of the formation of a solid solution (Millini *et al.*, 1992). About 2.5 % also appears to be the limiting solubility of titanium in silicalite. Preparation methods that use more titanium do not generate any further increase in unit cell volume and $TiO_2$ phases begin to form. There is evidence also that the titanium is distributed very evenly throughout the silicalite host, with no evidence for the formation of titanium dimers. This extreme dilution level is one of the factors believed to be responsible for the catalytic properties of TS-1.

Another characteristic feature of TS-1 is the appearance of an absorption band in its infrared spectrum at 960 $cm^{-1}$. This band is absent in pure silicalite, but it has been detected in a number of other silicalites including vanadium-silicalite and chromium-silicalite. It is now felt that this peak is not associated directly with the titanium component of the catalyst, but is assigned to the stretching vibration of $SiO^-$ groups, generated by the perturbing action of the solute cations (Ti, V or Cr) on the silicalite structure (Notari, 1996).

Further characterization is possible through diffuse reflectance UV–vis spectroscopy. A peak appears at 48 000 $cm^{-1}$ in the spectrum of TS-1, but not

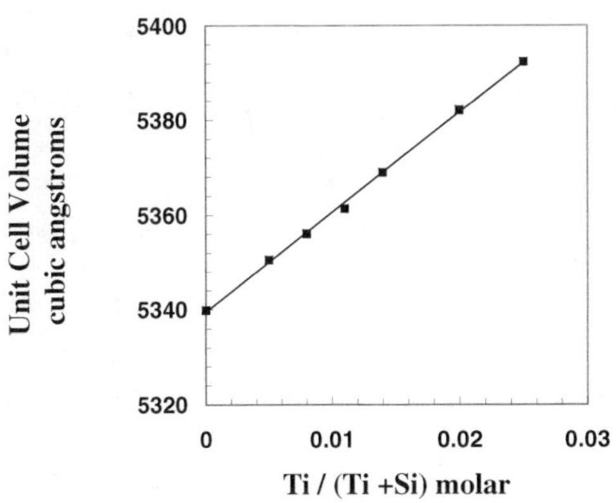


**Figure 10.4** Variation in unit cell volume with added titanium for titanium-silicalite (Millini *et al.*, 1992) (reproduced by permission of Academic Press).

in silicalite. As exposure to water vapour is increased this peak broadens and shifts to 42 000 $cm^{-1}$. These values are very close to those calculated for the $Ti^{4+}$ tetrahedrally and octahedrally coordinated to oxygen respectively (Notari, 1996).

### 10.2.4 Other zeolites containing framework titanium

There are now a large number of literature claims for the synthesis of other microporous materials with incorporation of titanium into the framework. These include TS-2, Ti-beta and Ti-Mordenite. A large number of other successful incorporations have also been claimed, but the experimental evidence to support these claims is not always convincing.

TS-2 is reported to form the MEL structure (isostructural with ZSM-11, see Table 10.1), and its unit cell volume expands with increasing incorporation of titanium into the framework in a similar way to that reported for TS-1. The synthesis of TS-2 involves the use of alkoxides of silicon and titanium and the organic base, tetrabutylammonium hydroxide. The maximum amount of incorporation of titanium into TS-2 is very similar to that found for TS-1 and their catalytic properties are very similar (de Lucas *et al.*, 1999).

Ti-beta is an interesting material, principally because it contains framework aluminium, hence has some associated acidity. Slightly more framework titanium can be incorporated into this structure than into TS-1 or TS-2. The acidity means that epoxides formed in this material tend to react further by acid-catalysed solvolysis. This effect is shown in Table 10.7 which compares the activity and selectivities of TS-1 with Ti-beta for a number of epoxidation reactions. In terms of activity, as measured by the $H_2O_2$ conversion, the TS-1

**Table 10.7** Comparison of TS-1 and Ti-beta in epoxidation reactions[a] (Corma *et al.*, 1994)

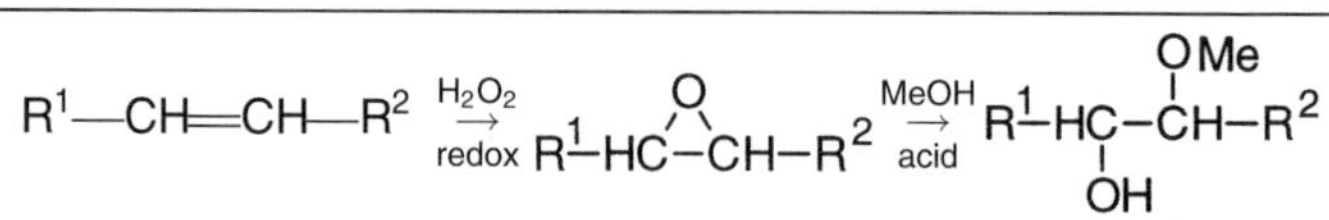


I II

| Olefin | Catalyst | $H_2O_2$ conv/% | Product selectivity/% I | Product selectivity/% II |
|---|---|---|---|---|
| 1-Hexene | TS-1 | 98 | 96 | 4 |
| | Ti-beta | 80 | 12 | 80 |
| Cyclohexene | TS-1 | <5 | 100 | 0 |
| | Ti-beta | 80 | 0 | 100 |

[a] 35 % $H_2O_2$ in methanol at 298 K.
Reproduced by permission of Academic Press.

and Ti-beta are similar, but cyclohexene oxidation proceeds much more rapidly over Ti-beta, presumably because of its somewhat larger pore size. However, the selectivities to epoxide and methyl ether products are completely different (Corma *et al.*, 1994). The acid sites in Ti-beta catalyse solvolysis of the epoxide and when these sites are neutralized by treating the zeolite with an alkali metal (note this procedure involves treating the fully formed Ti-beta or TS-1 with alkali metal and does not involve the use of alkali metals in the hydrothermal treatment stage), the selectivity to the epoxide improved. These results are summarized in Table 10.8. There is even an improvement for the TS-1, which in principle does not contain any alumina, hence should not contain any acidic sites (Sheldon *et al.*, 1998).

### 10.2.4 Mechanism of operation of TS-1 in oxidation reactions

Any discussion of the mechanism of action of TS-1 must start from the known structure of this material. The most pertinent features to be recalled are the incorporation of titanium into the framework of the silicalite, and the very homogeneous distribution of the titanium throughout the structure. In addition, this material is stable in the reaction conditions, so that hydrolysis of titanium—oxygen bonds, which would generate non-framework titanium, must not be extensive. A reasonable representation of the interaction between framework titanium species and water, which causes a change in coordination of the titanium atoms from tetrahedral to octahedral can be represented in Scheme 10.7. At all stages of the hydrolysis some Ti—O—Si linkages remain intact, hence the titanium remains within the framework. In addition Ti═O is

**Table 10.8** Influence of alkali metals added to preformed TS-1 and Ti-beta on the selectivity of epoxide formation
(Sheldon *et al.*, 1998)

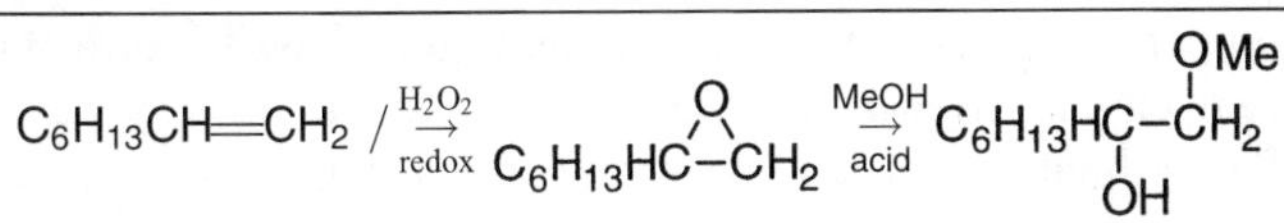


I II

| *Catalyst* | *$H_2O_2$ conv/%* | *Selectivity* | |
|---|---|---|---|
| | | I | II |
| TS-1 | 95 | 76 | 24 |
| Li-TS-1 | 85 | 98 | 0 |
| Ti-beta | 48 | 0 | 97 |
| Li-Ti-beta | 31 | 87 | 5 |
| Na-Ti-beta | 22 | 84 | 6 |
| K-Ti-beta | 25 | 63 | 0 |

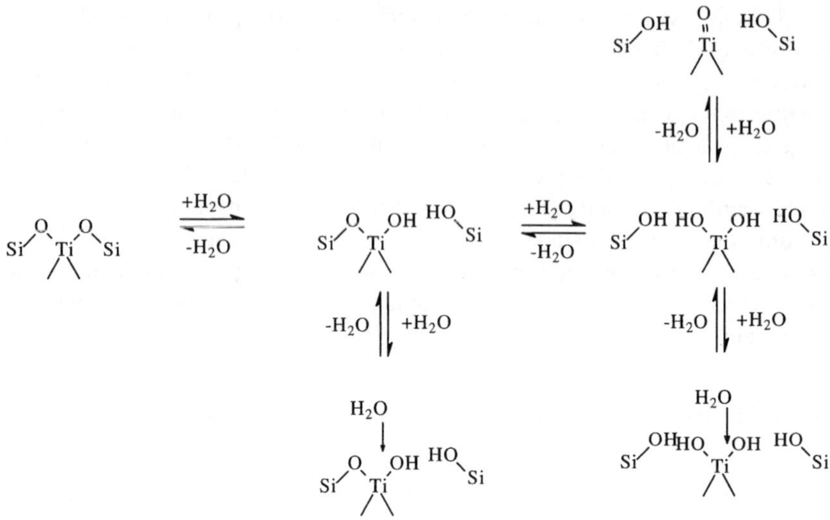

**Scheme 10.7** Interaction between framework titanium species and water (Notari, 1996).

chemically very similar to $Ti(OH)_2$, and in the scheme these species can interchange readily (Notari, 1996).

*Mode of interaction of hydrogen peroxide with TS-1*

The first evidence for the mode of interaction of hydrogen peroxide with TS-1 comes from a comparison of the UV–vis spectra of solution phase species and species generated when hydrogen peroxide is exposed to TS-1. A peak in the UV–vis spectrum of TS-1 at 26 000 $cm^{-1}$, when exposed to hydrogen peroxide, has been compared with a similar band observed in the peroxo complex, $[TiF_5(O_2)]^{3-}$.

Hydrogen peroxide reacts with $Ti^{4+}$ in the liquid phase to form stable side-on peroxo species as shown in Scheme 10.8. On the other hand hydroperoxides can coordinate by an end-on arrangement, but in some cases a more stable side-on arrangement is possibly by displacement of a solvent molecule.

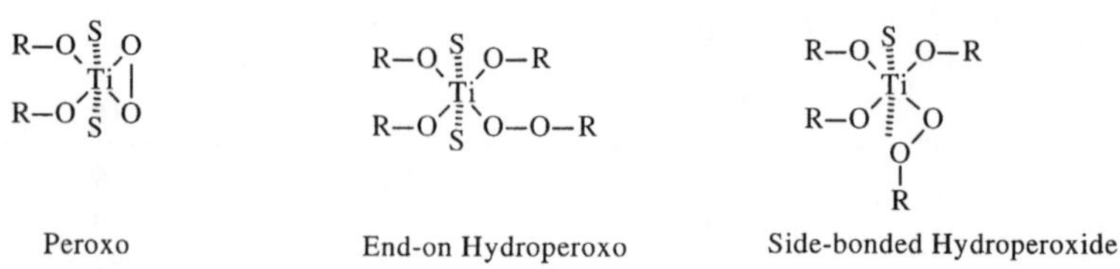

**Scheme 10.8** Stable titanium-containing peroxo and hydroperoxide species derived from reaction with hydrogen peroxide in the liquid phase (Conte *et al.*, 1992), where S is a solvent molecule.

The peroxo arrangement is more stable than the hydroperoxo, as indicated by their respective dissociation constants (Notari, 1996).

In TS-1 the peroxo species decomposes at 373 K, but its stability is improved in the presence of bases. Careful calibration of the UV–vis spectra indicates that the ratio of peroxo to titanium atoms is very close to 1 : 1, so that every titanium atom in the structure is accessible for reaction with hydrogen peroxide (Notari, 1996).

Detailed mechanisms for the epoxidation and other oxidation reactions over TS-1 have been proposed. However, these rely heavily on mechanistic information derived from reactions carried out in homogeneous solution. In general two types of mechanism have been proposed, namely those that propose a concerted mechanism and those that favour a radical type mechanism.

Notari has proposed the so called butterfly mechanism, wherein there is an electrophilic–nuclepohilic interaction between the electrophilic titanium peroxy species and the electron rich double bond as shown in Scheme 10.9 (Notari, 1996).

A modified version of this mechanism has also been proposed wherein the reactive species is the end-on hydroperoxide structure. In his case the two oxygen atoms of the hydrogen peroxide are not equivalent, so that either can act as the electrophile. However, the other steps in the mechanisms are not significantly different from that proposed by Notari.

A further concerted mechanism is the so called metallacycle mechanism shown in Scheme 10.10 (Huybrechts *et al.*, 1992). Here a cyclic intermediate is formed consisting of the titanium atom, the peroxo species and the reactant molecule.

Mechanisms based on radical mechanisms were originally proposed to explain the oxidation of less reactive alkanes. In these mechanisms it is

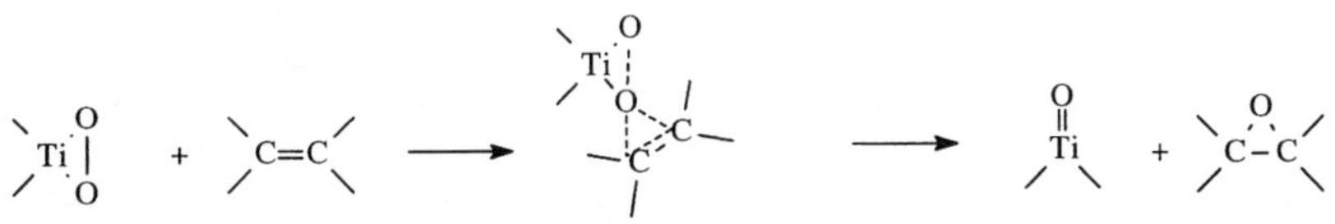

**Scheme 10.9** Butterfly mechanism with interaction between the electrophilic titanium peroxy species in TS-1 and the electron rich double bond (Notari, 1996).

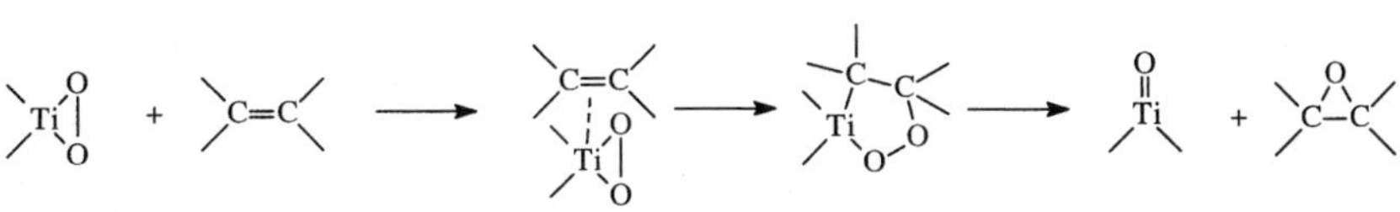

**Scheme 10.10** Metallacycle mechanism involving a cyclic intermediate consisting of the titanium atom, the peroxo species and the reactant molecule (Huybrechts *et al.*, 1992).

proposed that a radical species is generated by decomposition of the titanium peroxo species. Thus formed, the radical reacts with the saturated hydrocarbon as shown in Scheme 10.11 (Mimoun, 1987).

### 10.2.4 Other redox molecular sieves

In spite of the large number of studies devoted to the development of redox molecular sieves there still remains considerable doubt as to whether any truly heterogeneous redox molecular sieve exists apart from the titanium-containing materials. The primary difficulty relates to the stability of these materials in the reaction environment, in particular their stability against leaching in the presence of oxidizing agents such as hydrogen peroxide (Sheldon *et al.*, 1998).

Table 10.9 lists some zeolites and aluminophosphate ALPO materials to which have been added a number of metal compounds with known redox properties, including vanadium, cobalt and chromium. A variety of substitutions were claimed, including chromium for silicon in silicalite, chromium for aluminium in AlPO-5, vanadium for aluminium in ALPO-5 and ALPO-11 and cobalt and manganese for aluminium in ALPO-5. These materials were

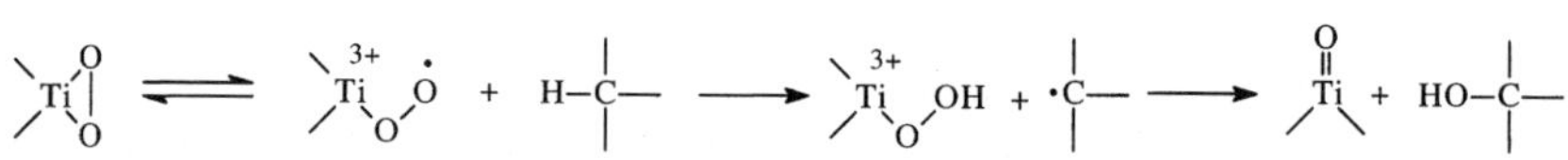

**Scheme 10.11** Mechanism of radical species generation by the decomposition of titanium peroxo species (Mimoun, 1987).

**Table 10.9** Decomposition of cyclohexyl hydroperoxide (CHHP) over redox molecular sieves at 343 K
(Chen *et al.*, 1994)

| *Catalyst* | *CHHP conv/%* | *Selectivity/%* | |
|---|---|---|---|
| | | *Cyclohexanone* | *Cyclohexanol* |
| TS-1 | 0 | — | — |
| VS-1 | 0 | — | — |
| CrS-1 | 98 | 64 | 36 |
| CrAPO-5 | 87 | 86 | 13 |
| VAPO-5 | 17 | 51 | 43 |
| VAPO-11 | 76 | 50 | 50 |
| CoAPO-5 | 2 | 50 | 50 |
| VS-1 | 0 | – | – |
| MnAPO-5 | 2 | 50 | 50 |
| Co-ZSM-5 | 24 | 43 | 50 |

used as catalysts for the decomposition of cyclohexyl hydroperoxide. Two products were possible, namely cyclohexanone and cyclohexanol. TS-1 was also included for comparison purposes and was inactive in the reaction conditions. Generally the CrAPO-5 material was selected for further study, whereas the cobalt and manganese analogues were reported as inactive for the test reaction.

The CrAPO-5 material was not active when hydrogen peroxide was the oxidizing agent, but could be used with a variety of organic hydroperoxides, such a *tert*-butyl hydroperoxide for the oxidation of alkylaromatics. Good yields of the corresponding ketones were reported at 353 K. However, subsequent work has cast some doubts on the stability to leaching in the reaction conditions of the catalysts used in these reactions. Analysis of the liquid phase after the reaction indicated some small level of chromium leaching from the catalyst during operation (Sheldon *et al.*, 1998). In a study of alpha-pinene oxidation (see Scheme 10.12), Sheldon *et al.* studied the activity of a number of chromium-containing redox molecular sieves. The catalysts studied included CrAPO-5, CrAPO-11 and CrS-1. In the reactions conditions specified in Scheme 10.12 each of these catalysts gave a conversion of close to 100 % and a selectivity to the ketone product of 70–80 %. To test for leaching the reaction medium containing CrAPO-5 was filtered at the reaction temperature (353 K) and the filtrate was allowed to react further. The hot filtrate reacted at a similar rate to that observed when the charge of CrAPO-5 was present. When the same experiment was repeated with cold filtering, the filtrate did not exhibit any reactivity. The conclusions are clear. CrAPO-5 is subject to leaching, and the leached chromium species in solution are capable of catalysing the reaction in Scheme 10.12. When the reacting solution is allowed to cool prior to filtration, the leached chromium species readsorb on the molecular sieve and do not appear in the filtrate.

A further test for leaching involved a comparison of the rate of reaction of a bulky hydroperoxide (PMPHP) with alpha-pinene on the one hand and a bulky olefin (valencene) with TBHP on the other over CrAPO-5 (Scheme 10.13). In each case one of the reactants is too big to enter into the microporosity of CrAPO-5. The implication is that if the reaction is occurring exclusively in the microporosity then no reaction should be observed since valencene and PMPHP are both too big to access the pores. In fact, the

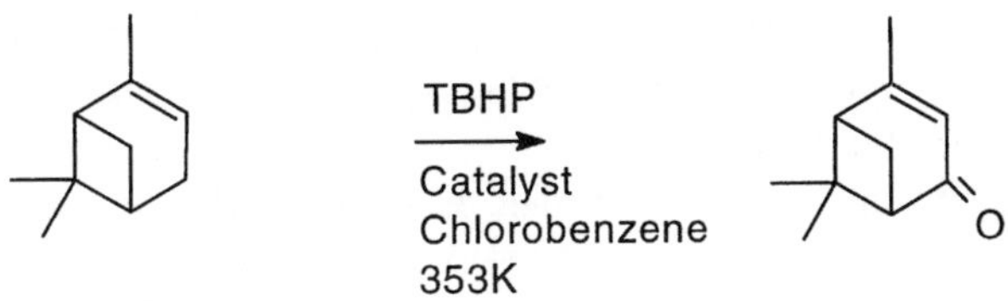


**Scheme 10.12** Reaction of $\alpha$-pinene with TBHP over a chromium redox molecular sieve (Sheldon *et al.*, 1998).

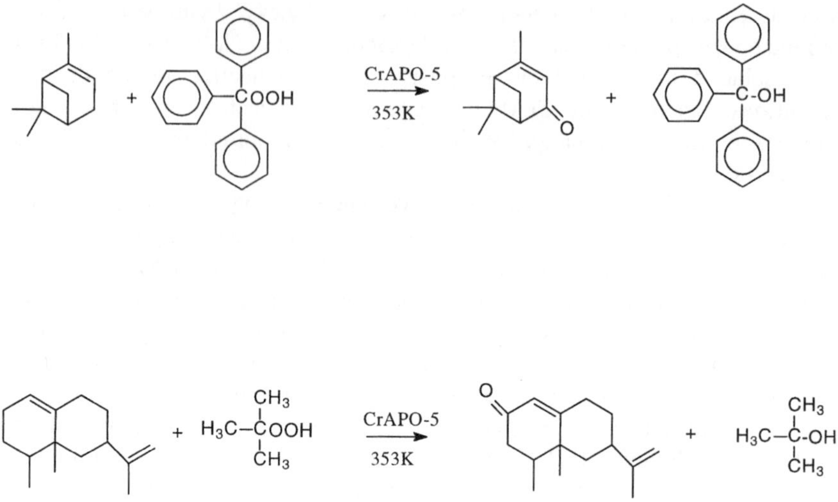


**Scheme 10.13** Reaction of a bulky hydroperoxide (PMPHP) with $\alpha$-pinene and a bulky olefin (valencene) with TBHP (Sheldon *et al.*, 1998).

valencene reacted readily with the TBHP in the presence of CrAPO-5, whereas no reaction was observed with the alpha-pinene and TMPHP. These observations are consistent with the hypothesis that the soluble chromium is responsible for catalysing these reactions and that it is leached by the action of the hydroperoxide. When a bulky hydroperoxide was used no reaction occurred, because no leaching had occurred (Sheldon *et al.*, 1998).

## 10.3 STRUCTURED MESOPOROUS SILICATES IN OXIDATION REACTIONS

### 10.3.1 Synthesis and characterization of structured mesoporous silicates

The synthesis of mesoporous silicate and aluminosilicate materials with regular structures and a mesoporosity which is a consequence of the structure has been described in recent years (Beck *et al.*, 1992; Kresge *et al.*, 1992). These materials typically have surface areas above 700 $m^2\,g^{-1}$ and feature large channels of dimensions 1.5–10 nm (Beck *et al.*, 1992). Hexagonal (MCM-41), cubic (MCM-48) and laminar (MCM-50) arrays of channels have been prepared, but most interest to date has been shown in the hexagonal structure.

Preparation typically involves addition of a surfactant, such as $C_{16}H_{33}(CH_3)_3NOH/Cl$, to sodium silicate dissolved in acid solution. To this is added silica, tetramethylammonium silicate and tetramethyl ammonium hydroxide. The mixture is heated for 24 h in an autoclave at 373 K. The final structure of the silicate that emerges is determined by micelle structures that form within the solution. The structure of a simple surfactant is shown in Scheme 10.14. It features hydrophobic (water insoluble) and hydrophilic (water soluble) ends. In practice cationic, anionic and neutral surfactants have been used in the preparation of mesoporous silicates. In appropriate conditions surfactants gather into micelle structures. In aqueous solutions the micelle can be spherical, with the hydrophilic ends of the surfactant pointing out into the aqueous solution, and the hydrophobic ends gathered at the centre of the sphere. This is an idealized representation, and in reality the micelles

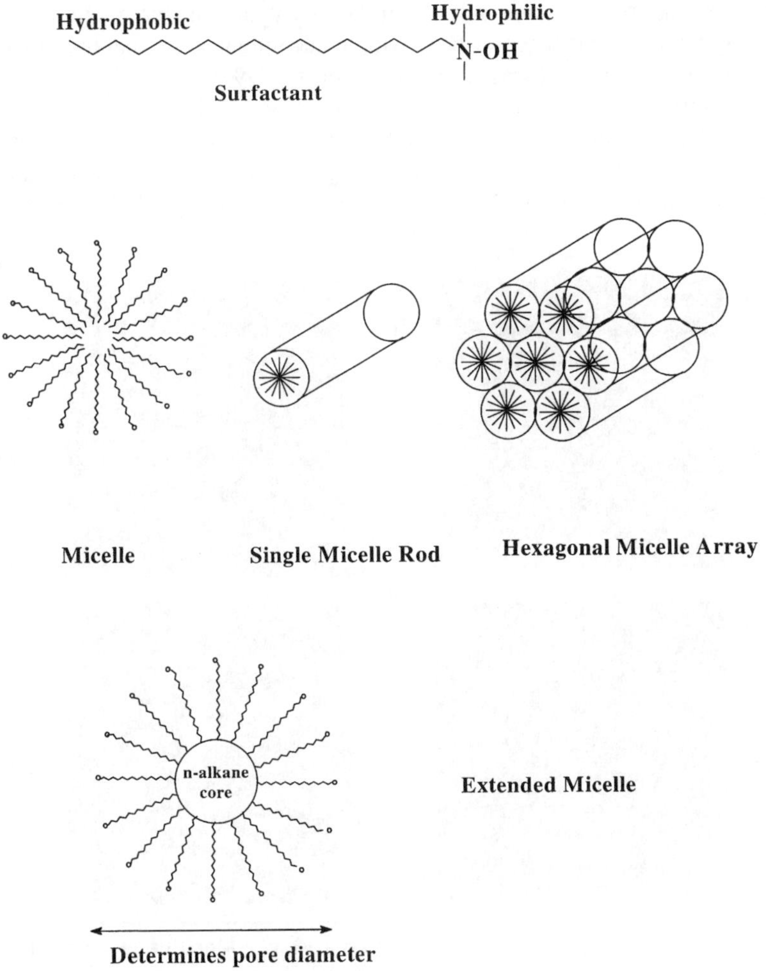


**Scheme 10.14** Surfactant and micelle structures important for the preparation of structured mesoporous silicates (Beck *et al.*, 1992).

are less ordered than the way they are represented in Scheme 10.14. In the conditions which prevail for the synthesis of MCM-41 the micelles congregate into a cylindrical shape, with the hydrophobic ends still pointing out into the aqueous solution. The result is called a micelle rod, and these rods can gather into hexagonal arrays as shown in Scheme 10.14. Again the arrays will be less ordered than the representation shown. The siliceous material (water soluble) then reacts at the outside of each micelle rod and the basic MCM structure starts to build up. Clearly the dimensions of the MCM material are determined by the micelle dimensions, in particular the diameter of the micelle rods. The pore diameters of MCM materials can be tailored to some extent by changing the chain length of the surfactant in use and in others by adding an appropriate amount of an auxiliary organic compound (1,3,5-trimethylbenzene has been used), which accumulates in the hydrophobic centre of the micelle, causing it to swell (Corma, 1997).

Transmission electron micrographs of some MCM-41 materials, prepared using different chain length surfactants and auxiliary organics, are presented in Figure 10.5 (Beck *et al.*, 1992). Hexagonal arrays of uniform channels ranging in diameter from 2 to 10 nm can be observed. The most regular

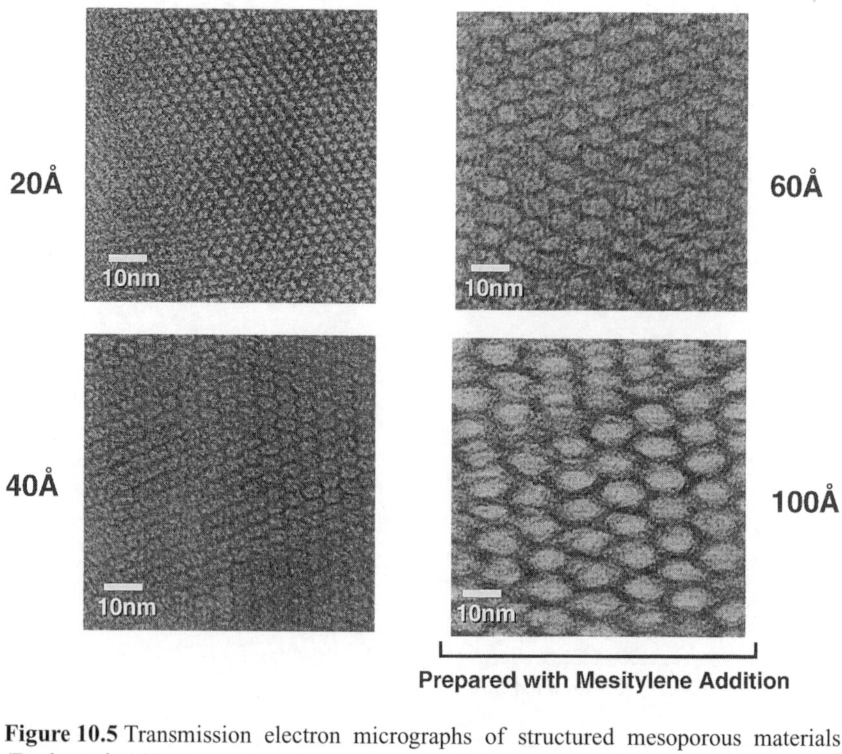


**Figure 10.5** Transmission electron micrographs of structured mesoporous materials (Beck *et al.*, 1992) (reproduced by permission of the American Chemical Society).

hexagonal arrangement of uniform pores is observed for the smaller pore size materials (<4 nm).

A view of the regular MCM-41 structure perpendicular to the pore axis is presented in Figure 10.6 (Mulukutla *et al.*, 1998). The electron micrograph is of a sample of MCM-41 with rhodium oxide particles, of average size 6 nm, dispersed along the mesoporous channels.

These materials can also be characterized by a variety of other methods. For example, the hexagonal array (MCM-41) gives a rather broad and poorly defined X-ray diffraction pattern, and low angles ($\theta < 1°$) are necessary to observe the (100) reflection, usually the strongest peak. This in itself is an indication that large features have been created because the Bragg equation indicates that the $d_{100}$ interlattice spacing is more than 4.4 nm when $\theta$ is less than 1°. The position of this peak varies as the porosity is altered, for example by changing the surfactant used. By comparison the other peaks in the XRD pattern tend to be very low in intensity. A typical example is shown in Figure 10.7 for two MCM-41 samples prepared to different degrees of crystallinity. The sample labelled 'primary' MCM-41 in this figure is the less well crystallized of the two. For each sample the (100), (110), (200) and (210) peaks could be resolved but in addition the (300) peak and a diffuse (220), (310) peak were observed with the 'secondary' MCM-41 sample. The presence of higher order peaks is an indication of improved long range order associated with the 'secondary' MCM-41 (Mokaya *et al.*, 1999).

An important consideration for these types of materials is a traditional measure of porosity. A typical adsorption isotherm for nitrogen adsorption onto MCM-41 at 77 K is presented in Figure 10.8 and for comparison purposes the corresponding isotherms for two zeolites and two amorphous silicons are also presented in this figure (Storck *et al.*, 1998). Note the very

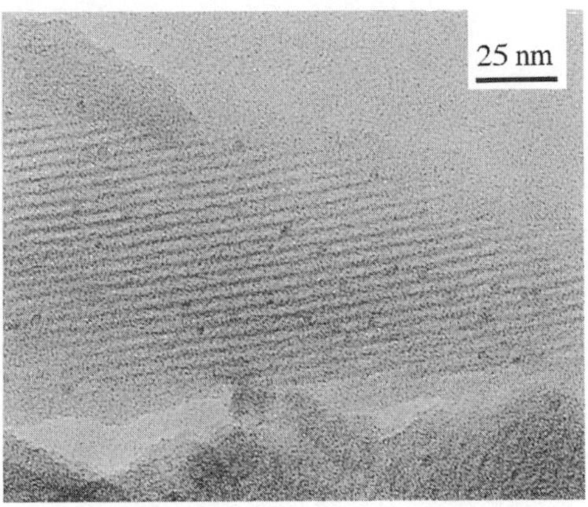


**Figure 10.6** Transmission electron micrograph of Rh-MCM-41 viewed perpendicular to the pore axis (Mulukutla *et al.*, 1998) (reproduced by permission of the Royal Society of Chemistry).

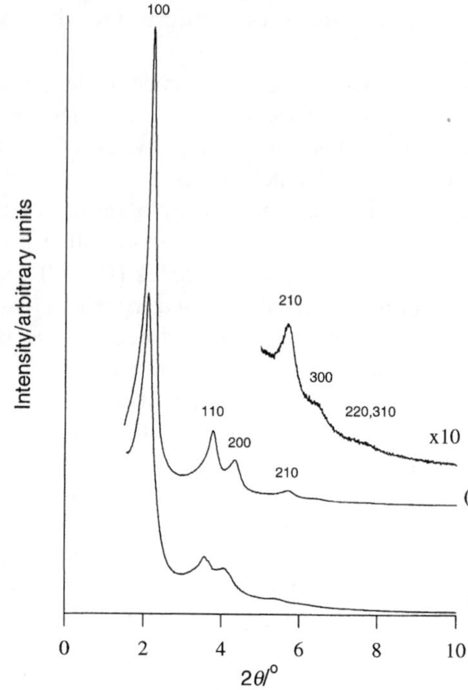


**Figure 10.7** Powder X-ray diffraction patterns of calcined (a) primary MCM-41 and (b) secondary MCM-41 (Mokaya *et al.*, 1999) (reproduced by permission of the Royal Society of Chemistry).

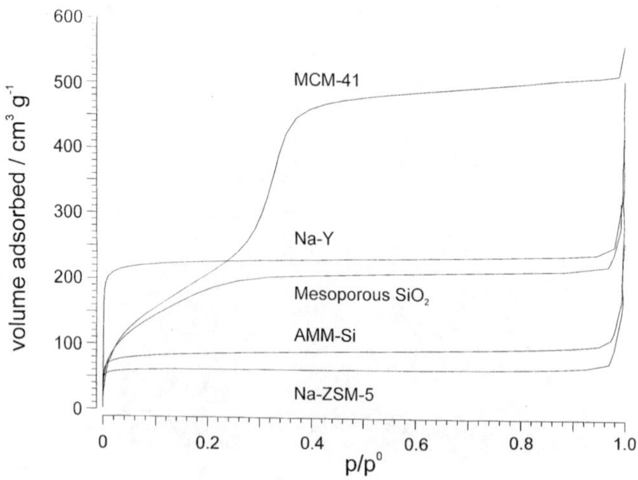


**Figure 10.8** Adsorption isotherms for Ar onto the materials indicated at 87.5 K (Storck *et al.*, 1998) (reproduced by permission of Elsevier Science).

sharp increase in adsorption at $p/p_o = 0.3$–$0.5$. This corresponds to the pressure at which nitrogen condenses into the available pores in the MCM-41 material by the process of capillary condensation. It is possible to estimate from the Kelvin equation (see Section 2.7.2) that the corresponding pore diameter is about 3 nm, consistent with the X-ray diffraction and TEM data.

A very uniform distribution of pore sizes and shapes is expected from a material in which porosity is a consequence of its ordered structure. This is demonstrated in Figure 10.9 where the adsorption isotherms of MCM-41 and the amorphous mesoporous silica of Figure 10.8 are analysed by the BJH method (see Section 2.7.2) to determine pore size distribution in the mesopore range (Barrett *et al.*, 1951). Comparison of the samples in Figure 10.9 points to a very narrow pore size distribution for the MCM-41 material (Storck *et al.*, 1998).

### 10.3.2 Oxidation reactions catalysed by Ti-MCM-41

A key feature of structured mesoporous materials is their ability to accommodate metals other than silicon in the structure. To generate acidity it is necessary to add aluminium, and useful oxidation properties can be generated by the addition of metals such as titanium, chromium or vanadium. Hence a number of MCM type materials have been prepared with incorporation of aluminium or titanium in the preparation procedure. With titanium alkali metals must be rigorously excluded from the preparation because of the tendency to form alkali titanates (Notari, 1996).

Attempts at the preparation of chromium and manganese substituted

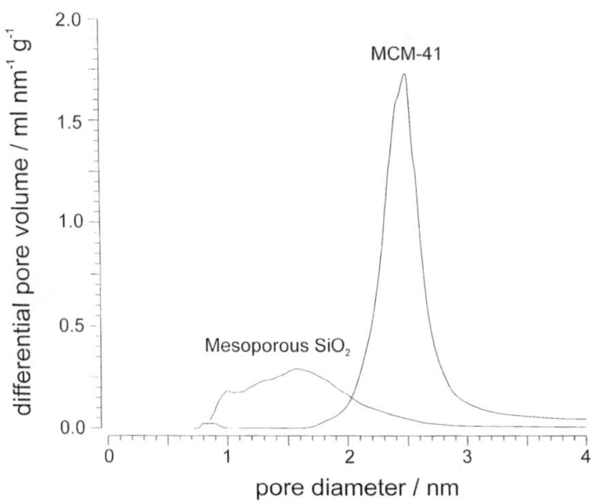


**Figure 10.9** Mesopore size distribution of MCM-41, and an amorphous mesoporous silicate according to the BJH method (Storck *et al.*, 1998) (reproduced by permission of Elsevier Science).

mesoporous materials have not been successful and a vanadium structure can be prepared but in general it is not stable in reaction conditions where hydrogen peroxide is present.

The intrinsic reactivity of Ti-MCM-41 materials for the epoxidation of small olefins is less than that of TS-1, but the main advantage of the mesoporous material is that much larger molecules can be accommodated within the pore volume. Hence molecules such as norbornene can be oxidized using *tert*-butylhydroperoxide on the MCM materials, according to the reactions shown in Scheme 10.15, at a faster rate than can be achieved on the microporous materials. A potential disadvantage, however, is that the larger size of the pores precludes the expectation of any shape selectivity effects.

Similar results have been reported for a comparison of the oxidation of sulfides on Ti-MCM-41 and Ti-beta zeolite. The bulkier of the two sulfides shown in Scheme 10.16 is oxidized more rapidly on the mesoporous material, presumably because the larger molecule can enter into the mesopore volume

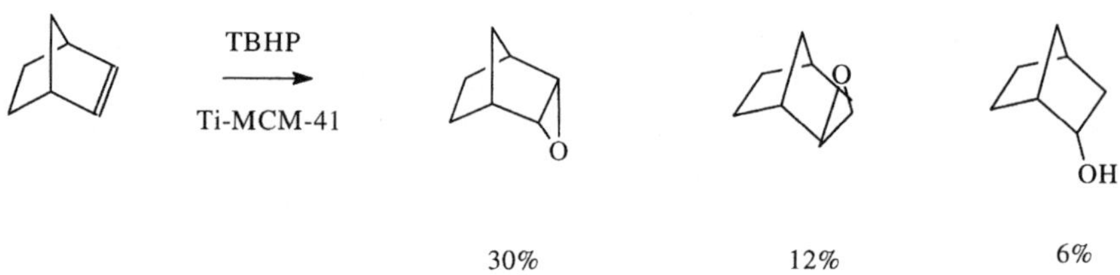


**Scheme 10.15** Oxidation reactions with bulky substrates catalysed by Ti-MCM-41 (Corma, 1997).

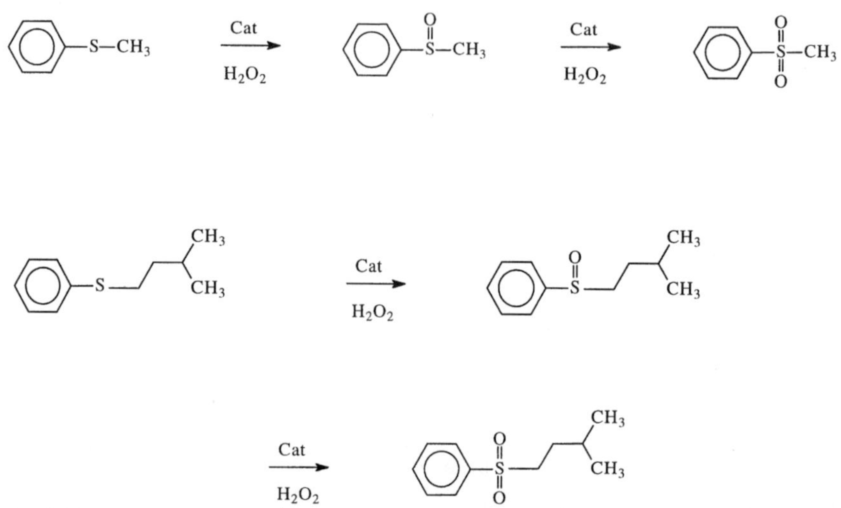


**Scheme 10.16** Oxidation of bulky sulfides catalysed by Ti-MCM-41 (Corma, 1997) (reproduced by permission of the American Chemical Society).

to react on the titanium centres, whereas the microporous nature of the beta zeolite precludes this possibility and the reaction is restricted to those titanium sites on the external surface of the zeolites (Corma, 1997).

A further example of a complex organic reaction carried out within the mesoporosity of an MCM-41 material is presented in Scheme 10.17. In this case the MCM-41 material contains framework titanium and aluminium; the former was added for its redox properties and the latter to generate acidity. The test reaction is the one pot synthesis of cyclic furans and pyran hydroxy ethers by the oxidation and acid-catalysed rearrangement of linalool (Corma, 1997).

### 10.3.2 MCM as a catalyst support

A number of attempts have been made to use mesopororous materials as supports for oxidation catalysts. On such effort was by Maschmeyer *et al.* (1995), who grafted a titanocene complex onto the walls of MCM-41. The result is different from the situation when titanium is incorporated into the pore walls during the preparation. The grafting technique results in a titanium species which protrudes into the mesopore volume. A high dispersion was achieved by allowing a solution of chloroform titanocene dichloride to diffuse into the pores of MCM-41. When triethylamine was added the surface silanols became activated and grafting occurred. Analysis by X-ray absorption spectroscopy indicated that the titanium was tetrahedrally coordinated.

Transition metal complexes have also been immobilized on MCM-41 by anchoring the ligands to the silanol groups of the walls. One such example is shown in Scheme 10.18. Here the silanol groups of the MCM material are functionalized by reaction with (3-chloropropyl)triethoxysilane. This material was then treated with an excess of 3-[*N*,*N*′-bis-3,3-(3,5-di-*tert*-butylsalicylidenamino)propyl]amine (*t*-slapr), which attached to the silane group. The attached ligand then reacts with $Mn^{2+}(acac)_2$ and oxidation results in the formation of the $Mn^{3+}$ complex (Sutra and Brunel, 1996). This material has

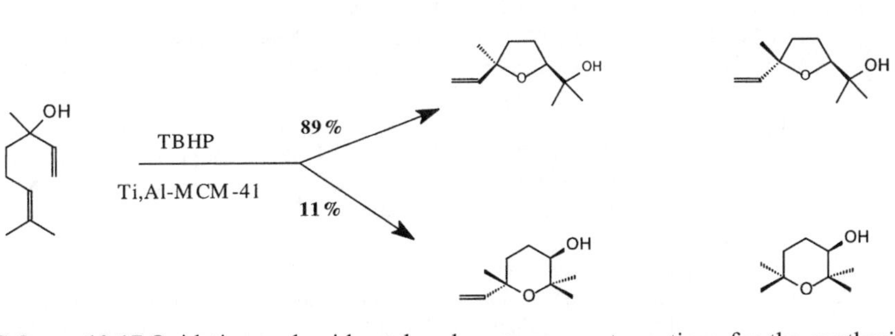


**Scheme 10.17** Oxidation and acid-catalysed rearrangement reactions for the synthesis of cyclic furans and pyran hydroxy ethers on Ti,Al,MCM-41 (Corma, 1997) (reproduced by permission of the American Chemical Society).

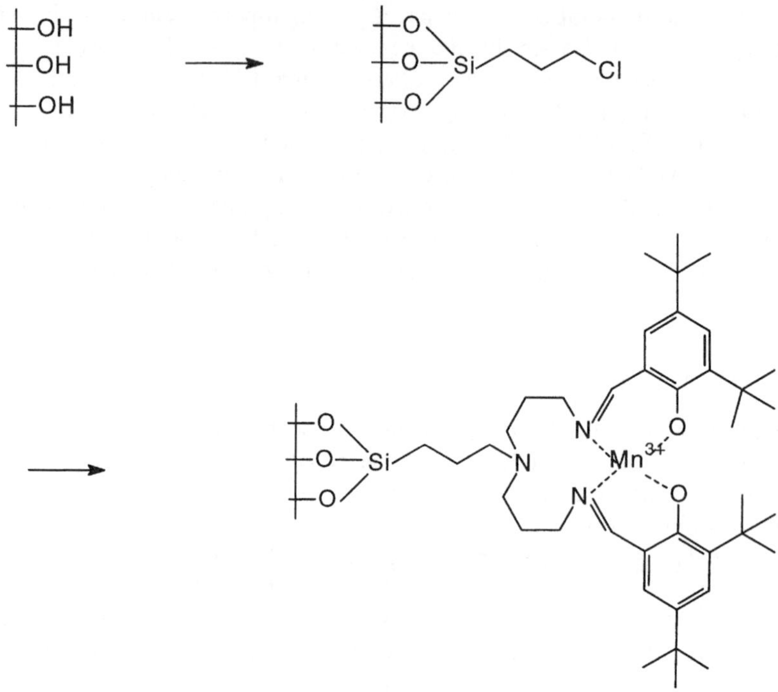


**Scheme 10.18** Grafting of a manganese complex onto MCM-41 (Sutra and Brunel, 1996) (reproduced by permission of the Royal Society of Chemistry).

been used in homogeneous conditions for the enantioselective epoxidation of olefins.

## REFERENCES TO CHAPTER 10


Avidar AA and Shinnar R (1990) Development of catalytic cracking technology. A lesson in chemical reactor design. *Ind. Eng. Chem. Res.* **29**: 931–942.

Barrett EP, Joyner LG and Halenda (1951) The determination of pore volume and area distributions in porous substances: 1. Computations from nitrogen isotherms. *J. Amer. Chem. Soc.* **73**: 373–380.

Beck JS, Vartuli JC, Roth WJ, Leonowicz ME, Kresge CT, Schmitt KD, Chu CT-W, Olson DH, Sheppard EW, McCullen SB, Higgins JB and Schlenker JL (1992) A new family of mesoporous molecular sieves prepared with liquid crystal templates. *J. Am. Chem. Soc.* **114**: 10 834–10 843.

Baerlocher Ch. and McCusker LB *Atlas of Zeolite Structures Types on the Web* (http://www.iza-sc.ethz.ch/IZA-SC/Atlas/AtlasHome.html).

Cassidy F (1999) Thesis, University of Limerick.

Chen JD, Dakka J and Sheldon RA (1994) Selective decomposition of cyclohexyl hydroperoxide to cyclohexanone catalysed by chromium aluminophosphate-5. *Appl. Catal. A: General* **108**: L1–L6.

Conte V, Di Furia F and Modena G (1992). In *Organic Peroxides*, Ando W (Ed.). Wiley, New York, p 559.

Corma A (1997) From microporous to mesoporous molecular sieve materials and their use in catalysis. *Chem. Rev.* **97**: 2373–2419.

Corma A, Camblor MA, Esteve P, Martinez A and Perez-Pariente J (1994) Activity of Ti-beta catalysts for the selective oxidation of alkenes and alkanes. *J. Catal.* **145**: 151–158.

De Lucas A, Rodriguez L and Sanchez P (1999) Synthesis of TS-2 in the system $SiO_2$–$TiO_2$–$H_2O_2$–TBAOH. Influence of synthesis variables. *Appl. Catal. A: General* **180**: 375–383.

De Moor PPEA, Beelen TPM and van Santen RA (1999) In situ observation of nucleation and crystal growth in zeolite synthesis. A small-angle X-ray scattering investigation on Si–TPA–MFI. *J. Phys. Chem. B* **103**: 1639–1650.

Dwyer FG and Chu P (1979) ZSM-4 crystallization via faujasite metamorphosis. *J. Catal.* **59**: 263–271.

Feijen EJP, Martens JA and Jacobs PA (1997) Hydrothermal zeolite synthesis. In *Handbook of Heterogeneous Catalysis*, Vol. 2. VCH, Weinheim, pp 311–324.

Francis RJ and O'Hare D (1998) The kinetics and mechanism of the crystallization of microporous materials. *J. Chem. Soc., Dalton. Trans.*: 3133–3148.

Kresge CT, Leonowicz ME, Roth WJ, Vartulli WJ and Beck JS (1992) Ordered mesoporous molecular sieves synthesised by a liquid-crystal template mechanism. *Nature* **359**: 710–712.

Kumar R, Pais GCG, Pandey B and Kumar P (1995) Hydroxyl-assisted chemo- and stereoselective epoxidation catalysed by a titanium silicalite molecular sieve (TS-1)/$H_2O_2$ system. *J. Chem. Soc., Chem. Commun.* 1315–1316.

Kuhl GH and Kresge CT (1995) Molecular sieves. In *Kirk–Othmer Encyclopedia of Chemical Technology*, Vol. 16. J. Wiley and Sons, New York, 4th edition, pp 888–925.

Huybrechts DRC, Buskens PL and Jacobs PA (1992) Physicochemical and catalytic properties of titanium silicalites. *J. Mol. Catal.* **71**: 129–147.

Maschmeyer T, Rey F, Sankar G and Thomas JM (1995) Heterogeneous catalysts obtained by grafting metallocene complexes onto mesoporous silicates. *Nature* **378**: 159–162.

Meier WM, Olson DH and Baerlocher Ch (1996) *Atlas of Zeolite Structure Types*. International Zeolite Association.

Millini R, Previde Massara E, Perego G and Bellussi G (1992) Framework composition of titanium silicalite-1. *J. Catal.* **137**: 497–503.

Mimoun H (1987) Do metal peroxides as homolytic and heterolytic reagents: Mechanism of the Halcon epoxidation process. *Catal. Today* **1**: 281–295.

Mokaya R, Zhou W and Jones W (1999) A method for the synthesis of high quality large crystal MCM-41. *Chem. Commun.*: 51–52.

Mulukutla RS, Asakura K, Nambna S and Iwasawa Y (1998) Nanosized rhodium oxide particles in the MCM-41 mesoporous molecular sieve. *Chem. Commun.*: 1425–1426.

Newsam JM (1992) Zeolites. In *Solid State Chemistry, Compounds*, Cheetham AK and Day P (Eds). Clarendon Press, Oxford.

Nair S, Villaescusa LA, Camblor MA and Tsapatsis M (1999) Zeolite-$\beta$ grown epitaxially on SSZ-31 nanofibers. *Chem. Commun.*: 921–922.

Notari B (1991) Titanium silicalite: A new selective oxidation catalyst. In *Structure–Activity and Selectivity Relationships in Heterogeneous Catalysis. Stud. Surf. Sci. Catal.* **68**: 243–256.

Notari B (1996) Microporous crystalline microporous silicates. *Adv Catal.* **41**: 253–334.

Sheldon RA, Arends IWCE and Lempers HEB (1998) Heterogeneous catalysts for liquid phase oxidations. In *Supported Reagents and Catalysts in Chemistry*, Hodnett BK *et al.* (Eds). Royal Society of Chemistry, Cambridge, pp 37–46.

Storck S, Bretinger H and Maier WF (1998) Characterization of micro- and mesoporous solids by physiorption methods and pore size analysis. *Appl. Catal. A: General* **174**: 137–146.

Sutra P and Brunel D (1996) Preparation of MCM-41 type silica-bound manganese(III) schiff-base complexes. *J. Chem. Soc., Chem. Commun.*: 2485–2486.

Taramasso M, Perego G and Notari B (1983) Preparation of porous crystalline synthetic material comprised of silicon and titanium oxides. US Patent 4 410 501, assigned to Snamprogetti S.p.A.

Tatsumi T, Yako M, Nakamura M, Yuhara Y and Tominaga H (1993) Effect of alkene structure on selectivity in the oxidation of unsaturated alcohols with titanium silicalite-1 catalyst. *J. Mol. Catal.* **78**: L41–45.

Thomas JM, Bell RG and Catlow CRA (1997) Zeolites and related molecular sieves. In *Handbook of Heterogeneous Catalysis*, Vol. 2. VCH, Weinheim, pp 286–310.

Van der Pol AJHP, Verduyn AJ and van Hooff JHC (1992) Why are some titanium silicalite-1 samples active and others not. *Appl. Catal. A General* **92**: 113–130.

Weisz PB (1982) The science of the possible. *Chemtech* July: 424–425.

Whan DA (1981) Structure and catalytic properties of zeolites. *Chem. Br.* **17**(11): 532–535.

# Index


Acrolein 1, 69, 74, 78, 80, 85, 88, 103
  crude 103
  deuterium content 118
  oxygen concentration 125
  propene oxidation to. *See* Propene oxidation to acrolein
  refined 103
Acrylonitrile 1, 72, 75, 85, 241
  annual production 241
  principal uses 241
  *see also* Propane ammoxidation to acrylonitrile; Propene ammoxidation to acrylonitrile
Activation energies 69
  oxidation of organic compounds over Pt and Pd 220
  reoxidation of bismuth molybdate catalysts 123
Active oxygen species involved in oxidation catalysis 81–90
Active sites 72
  discriminating mechanisms 78–81
  heterogeneous and enzymatic catalysis 97
  oxide catalysts 98
  selective oxidation catalysts 96–9
Additives 110
Adsorbed oxygen species, nature and reactivity 85–6
Adsorbed species 42–4
Adsorption
  *n*-butane on $(VO)_2P_2O_7$ 156
  gas–solid interface 29–44
  stoichiometry 59
Adsorption isotherms 31–6
  catagories 30
  Langmuir 33, 34, 54
  $N_2$, $C_3H_8$ and $O_2$ 32
$Ag/\alpha$-$Al_2O_3$ catalyst 91
Air as oxidant 273–4
$Al_2O_3$ 11, 250
$\alpha$-$Al_2O_3$ 165, 166, 167, 205
$\gamma$-$Al_2O_3$ 210
Aldehydes, ammonia addition to 259
Alkali metals
  addition during preparation of TS-1 320
  in selectivity of epoxide formation 323
  silver surfaces 177–9
Alkanes 23, 85, 86, 218–19, 228, 325
  oxidation, kinetic parameters 225
  oxidative activation 132–59
Alkenes 1, 85
  epoxidation of 160–88
Allyl iodide 126
Allylic oxidation 102–31
ALMA process 136, 137
Alumina 205
  sintering mechanism 205–11
  stabilization 211
Aluminium hydroxides and oxides 209
Aluminophosphate ALPO materials 326
Aluminophosphates 305
Ammonia 6
  addition to aldehydes 259
Ammoxidation 2, 5–7
  catalysts used in 249–50
  nitrogen insertion mechanism 254–60

Ammoxidation (*cont.*)
  principal uses 6
  production capacity 6
  reaction network 250–3
  reactions 1, 76, 98
Ammoximation
  cyclohexanone 260
  ketones 261
  nitrogen insertion in 260–2
Anhydride functional group 133
Anion vacancies 209
Antimony oxide 99
Arhenius expression 69
Autoignition temperature (AIT) 22, 24

Barium carbonate 169
$BaTiO_3$ 108
Beckmann rearrangement 260
Benzaldehyde 251, 253
Benzenes (PCBs) 295–9
Benzonitrile 252, 256–8
Benzylamine 251–3
BET surface area 122
Bi:Mo ratios 116
$Bi_2(Mo^{16}O_4)_3$ 124
$Bi_2(MoO_4)_3$ 112–14
$Bi_2MoO_6$ 20, 122
$Bi_2O_3$ 126, 128
$\alpha$-$Bi_2Mo_3O_{12}$ 117
$\beta$-$Bi_2Mo_2O_9$ 119, 120
$\tau$-$Bi_2MoO_6$ 117, 123
Bidentate formate 223
Bismuth molybdate catalysts
  compositional range 107
  oxygen transport within 124–6
  preparation method 106
  propene oxidation to acrolein 105–16, 249
BJH method 56, 57, 333
Bond dissociation enthalpy 81, 82
Bragg angle 52
Brønsted acid sites 304
Brunauer, Emmet and Teller (BET) equation 53–4
Bulk organic chemicals 6
1,3-Butadiene 135, 153
Butane conversion 18
*n*-Butane 74, 79, 80, 108
  activation 149–53
  adsorption and reaction on $(VO)_2P_2O_7$ 156
  initial distribution of oxygen isotopes in reaction products 151
*n*-Butane oxidation to maleic anhydride 132–59
  generalized scheme 154
  kinetic isotope effect 150
  location of active sites 153–6
  multiple selectivity-conversion plot 80, 134, 149
  oxygen transport and incorporation 151
  reaction intermediates 135
  reactive network 153
  reactor inlet and outlet compositions 136
  role of catalyst microstructure 153–6
  summary information 134
Butene 135, 153
1-Butene 115
Butterfly mechanism 325

Caprolactam 260
Carbon dioxide 223
Carbon monoxide 197
Carbon tetrachloride 295–9
Catalyst particles 11
Catalyst pellets, composition 11
Catalyst supports, mesopororous materials 335–6
Catalytic combustion 189–239
  environmental applications 191–9
  nitrogen oxides (NOx) 192–6
  principles 199–201
  reaction kinetics 223–5
  substrate reactivity 218–22
  *see also* Combustion catalysts
Catalytic cracking technology 138
Cation vacancy 108, 125, 209, 249
$CaWO_4$ 108
C—C bond 149, 292
C═C double bond 161, 186, 317
$CCl_4$ 294
C—Cl bond 292, 293
C—D bond 73, 222
Ceramic honeycomb structure 201
$C_2H_4{}^{16}O$ 184
$C_2H_4{}^{18}O$ 184
$C_2H_4$-HCl 279

$C_2H_4$-HCl-$O_2$ 279
$C_3H_3Cl_3$ 294
$C_3H_8$-$CO_2$-air mixture 24
$C_3H_8$-$N_2$-air mixture 24
$C_6H_4Cl_2$ 294
C—H bond 1, 72, 73, 75, 77, 78, 80, 81, 84, 85, 149, 179, 180, 186, 219, 221, 243, 254, 255, 275, 292, 293, 316, 317
C—H stretching vibration 74
Chemical industry
  major components 5
  overview 4
Chemisorption 30, 36, 39
  characteristics 31
  *see also* Selective chemisorption
Chlorinated hydrocarbons 169
  catalytic oxidation by-product formation 290–5
  destruction by catalytic oxidation 285–90
Chlorine 6
  catalytic oxidation reactions involving 264–300
  on silver surface 185–6
Chlorobenzene 291
Chromium oxide catalysts 287, 288
Chromium redox molecular sieve 327
Claperyon-Clausius equation 38
$Co_{11/12}Fe_{1/12}MoO_4$ 112, 126
$\beta$-$Co_{11/12}Fe_{1/12}Mo^{18}O_4$ 124
Combustibles, homogeneous mixtures in air 22
Combustion catalysis
  oxygen species in 231–8
  reactivity of surface species 227–30
Combustion catalysts
  chemical reactivity of active phase 200
  choice of materials 202–16
  choice of support acid-base properties 214–16
  poisoning 216–18
  power law kinetic expressions 234–5
  stability of active phase 202–4
  stability of support materials 204
  *see also* Catalytic combustion
$CoMoO_4$ 112, 113
Complex oxides 108–9
Copper chloride 265, 275
Cordierite 200
CrAPO-5 material 327–8
$Cr_2O_3/Al_2O_3$ 288, 289, 292
Crystal growth, zeolites 313
Crystallographic model of active sites (CMAS) 155
$CuCl_2$—CuCl—KCl 275–7
CuCl—KCl 275
$CuCl_x$—KCl—$Al_2O_3$ 283
$CuCl_2$—KCl—$LaCl_3$ 280
$CuCl_2$—KCl/$SiO_2$ 279
CuCl/KCl/$SiO_2$ 290
Cu—O—Cl—HCl system 282
Cyclohexanone
  ammoximation 260
  hydroxylamine reaction with 262
Cyclohexanone oxime 260
Cyclohexyl hydroperoxide (CHHP) 326–9
Cyclopropane hydrogenolysis 96
Cyclopropane type intermediate 116

Deacon reaction 265, 275, 276, 280, 285
deNOx problem 2–3
Desorption 39
Deuterated propenes 116
Deuterium content of acrolein 118
Dibenzo-*p*-dioxins (PCDDs) 266, 295–9
Dibenzoethers (PCDEs) 295–9
Dibenzofurans (PCDFs) 266, 295–9
1,2-Dichlorobenzene 295–9
Dichlorobutenes 277
Dichloroethane 274, 292
Dichloromethane 287–8
Differential heats of adsorption 37, 38
Diffusion coefficients 14
Diffusion processes 25–9
Dioxins 295–9
Direct chlorination of ethylene 268
Dissociative chemisorption 34
Dopants 109–10
Dual site adsorption 39
DuPont circulating solids riser reactor 138

Effectiveness factor-Thiele modulus curve 28
Effectiveness factors 27–9

Electron acceptor double bond 133
Electron spin resonance (ESR) 43, 75, 84–6
Electrophilic-nuclepohilic interaction 325
Eley-Rideal mechanism 45–8
End-on hydroperoxide structure 325
Energy dispersive X-ray analysis (EDXA) 110, 112
Environmental problems 2
Epoxidation of alkenes 160–88
Epoxide formation, alkali metals in selectivity of 323
Ethane 75, 79
Ethylene 85, 91
  direct chlorination 268
  oxychlorination 265, 270–5, 280
Ethylene bromide 285
Ethylene dichloride 273–4
Ethylene epoxidation to ethylene oxide
  $CO_2$ evolution 170
  effect of ethylene partial pressure 173
  effect of oxygen partial pressure 174
  effect of product partial pressure 172
  multiple selectivity-conversion plot 162
  operating conditions 163
  reaction kinetics and network 170–3
Ethylene glycol 161
Ethylene oxide 85
  production flowsheet 163
  production technology 162
  selectivity as function of oxygen coverage of silver catalysts 182–3
  *see also* Ethylene epoxidation to ethylene oxide
Exhaust gases 197, 216
Exothermic adsorption 37
Exothermic reactions 28
External diffusion 25–7

Fick's laws of diffusion 25
Fixed bed oxychlorination units 270
Fixed bed reactor 9–11, 13, 70, 136, 137, 272
Flammability limits 136, 245–7
  in air 21
  influence of inerts 24
  influence of pressure 24
  influence of temperature 21–2
  *see also* Lower flammability limits; Upper flammability limits
Flammability of hydrocarbon-oxidant mixtures 20–5
Fluidized bed reactor 246–9, 273, 276
Fluorene 110
Fluorenone 110
Formaldehyde, methanol oxidation to 74, 79, 88
Free radical species 77
Fuel NO 194
Furan 135, 153

Gas phase oxygen species 83–5
Gas-solid interface, adsorption at 29–44
Gibbs free energy
  as function of temperature 282
  of formation of metal oxides 227

Haber process 6
Halogens, poisoning by 286
Halsey equation 56
Heats of adsorption 33, 36–42, 85
  comparisons 37
  initial and final 38, 40–1
  isosteric 37
Heats of desorption 33
Heats of oxygen chemisorption 230
Heptane oxidation 215
Heterogeneous catalysis
  fundamental processes 14–15
  kinetic rate laws 46–7
  mode of activation of hydrocarbon and discriminating capacity of active sites 72–81
Heterogeneous catalysts 1, 2, 5
  dimensions important in 10
  macroscopic structure 9–14
  mechanical properties 13–14
  microscopic structure 9–14
  overview 8–64
  testing 15–25
Heterogeneous catalytic oxidation
  chlorinated hydrocarbons, by-product formation 290–5
  general principles 65–101
  generic features 66–72
  role and importance of 5–7

volatile organic compounds (VOCs) 285
Heterogeneous catalytic oxidation reactions
generic features 66–72
involving chlorine 264–300
Hexaaluminates 211–13
Hexachloroplatinic acid 204, 209
High resolution electron microscopy 94
Hydrocarbon-oxidant mixtures, flammability of 20–5
Hydrocarbons 89, 197
activation mode 72–7
saturated 84
Hydrogen peroxide 326
Hydrogen peroxide-TS-1 system 317, 324–6
Hydrogen sulfide 216
Hydrothermal synthesis of sodium aluminosilicates 312
Hydroxylamine reaction with cyclohexanone 262
Hydroxylation, phenol 320

Internal diffusion 27–9
Isopropylbenzene 77
Isosteric heats of adsorption 37
Isothermal reactions 28
Isotope distribution 116, 118
Isotopic exchange of oxygen 87

Kelvin radius 56
Ketones, ammoximation 261
Kinetic isotope effect (KIE) 72–7, 117
Kinetic rate laws
basic types 44
heterogeneous catalysis 46–7
Kinetics of oxidation reactions 44–9
Kisliuk equation 38, 42
Knudsen diffusion 27

$LaAlO_3$ 211
Langmuir adsorption isother 33, 34, 54
Langmuir-Hinshelwood mechanism 48–9, 172, 225, 226, 256–7
Lanthanum 211
Lanthanum chloride 275, 280
Lattice oxygen 86–90, 121–2, 124
Lean burn petrol engine 197–8
Le Chatelier's law 22
Lewis acid sites 304
Low energy ion scattering (LEIS) 147
Lower flammability limit (LFL) 21–2
influence of temperature 23
*see also* Flammability limits

Macropores 55
Maleic anhydride 1, 18, 74–5, 80, 108
annual production 133
main use in polymer industry 133
manufacture 133
principal uses 133
production flowsheet 137
production technologies 135–8
*see also* *n*-Butane oxidation to maleic anhydride
Mars-van Krevelen mechanism 44–5, 119
MCM-41 328, 330–6
MCM-48 328
MCM-50 328
Me-Cl bond 279
Mesopores 55, 56
Mesopororous materials
catalyst supports 335–6
structured 328–36
Metal oxides, Gibbs free energy of formation 227
Metallacycle mechanism 325
Methane 21, 75, 79, 88, 89
combustion in pulse mode 236
combustion over palladium catalyst 237
oxidation to ethane 79
platinum oxide in partial oxidation of 234
reaction with $PtO_x$ 232–3
Methanol oxidation to formaldehyde 74, 79, 88
Methionine 103
Methyl radical 89
$MgAl_2O_4$ 108
Micropores 55
Migration 39
Mixed oxides 109
$Mo_{12}Bi_1Co_8Fe_3{}^{16}O_x$ 121
Mo=NH surface species 259
Monodentate formate 223
Monsanto solvent collection system 137
$MoO_3$ 126, 128

Multiple selectivity-conversion plot 69, 81
Multiple selectivity-conversion plot, *n*-butane oxidation to maleic anhydride 80, 134, 149
Multiple selectivity-conversion plot
ethylene epoxidation to ethylene oxide 162
methane oxidation to ethane 79
methanol oxidation to formaldehyde 79
propane ammoxidation to acrylonitrile 69, 72, 242
propene ammoxidation to acrylonitrile 242
propene epoxidation with oxygen 179, 181
propene oxidation to acrolein 80, 104
VPO catalyst 155

Napthalenes (PCNs) 295–9
Natural gas 5
$NH_3$:hydrocarbon ratio 256
Ni(111) surfaces 221
NiO(111) surfaces 221
Nitrogen activation 240–63
Nitrogen insertion
in ammoxidation 254–60
in ammoximation 260–2
into organic compounds 240–63
Nitrogen oxides (NOx) 2–3, 197
catalytic combustion 192–6
formation during combustion 193
strategies to minimize emissions 194–6
*n*-Nonane 25
Nucleation, zeolites 313

Olefins
epoxidation with molecular oxygen 180–2
reactions with $CuCl_2$ on pumice 278
Organic chemicals 6
Oxidation catalysis, active oxygen species involved in 81–90
Oxidation catalysts
characterization 50–61
evaluation 15
macroscopic features 90–6
structural features 90–9
textural features 92–3
Oxidation processes 5
overview 4
Oxidation reactions 2
catalyst testing facility 16
kinetics of 44–9
Oxychlorination 2, 5–7
catalyst composition, preparation and characterization 275–6
ethylene 265, 270–5, 280
mechanism 276–84
principal uses 6
production capacity 6
reactions 1
vinyl chloride 266–8
Oxygen
activation, reaction energy profile 231
as oxidant 273–4
*see also* Lattice oxygen
Oxygen-$\sigma$-allyl species 117
Oxygen anions 90
Oxygen concentration of acrolein 125
Oxygen isotopes
exchange mechanisms 87
initial distribution in reaction products of *n*-butane 151
Oxygen-palladium metal interaction 35
Oxygen-palladium ratio 36
Oxygen species 81–90
adsorbed 85–6
gas phase 84–5
in combustion catalysis 231–8
interaction and interconversion on alkali metal promoted silver surfaces 177–9
on and under silver surface 174–9
reactivity on silver surfaces 179–88
*see also* Lattice oxygen
Oxygen transport within bismuth molybdate catalysts 124–6

Palladium 66
Palladium catalysts 229, 234, 236, 237
Palladium oxides 229–30, 234
decomposition 235
Palladium particles 11
Particle growth, surface hydroxyl groups in 209

Particle size measurement by X-ray diffraction (XRD) 51–3
$Pb_{1-x}Bi_{2x}\phi_x(MoO_4)$ 249
$Pd/Al_2O_3$ 218, 230, 234, 236
$PdO/Al_2O_3$ 203
$PdO/\alpha$-$Al_2O_3$ 207
$PdO_x$-Pd equilibrium 238
Peroxo complex 324–5
Petrochemicals industries, overview 4
Phase identification, X-ray diffraction (XRD) 50
Phenol hydroxylation 2, 320
Phthalic anhydride 91
Physicochemical techniques 57
Physisorption 29–30, 36
  characteristics 31
  oxygen onto palladium metal 35
Platinum 66, 231
Platinum-honeycomb catalyst 216
Platinum oxides 231–2
  in partial oxidation of methane 234
Platinum oxychloride 291
Platinum single crystals 96
PMPHP 327–8
Pollutant emissions 191
Pollutants, average human exposure 192
Polychlorinated aromatic compounds 296
Polychlorinated biphenyls (PCBs) 266, 295–9
Pore size distribution 55–7
Pore volume diffusion 27
Porosity 14
Potential energy diagram 30
Power law kinetic expressions, combustion catalysts 234–5
Power rate laws 49
Precious metal catalysts, active species 227–38
Precursor state 39
Pre-equilibrium state 39
Promoters 110
Prompt NOx 194
Propane-air-nitrogen system 24
Propane ammoxidation to acrylonitrile, multiple selectivity–conversion plot 69, 72, 242
Propane oxidation over metals or metal oxides 202
Propene 78–80, 88
  reaction with metal chlorides 278
Propene ammoxidation, mechanism 254
Propene ammoxidation to acrylonitrile 75, 241
  flow diagram 244
  multiple selectivity-conversion plot 242
  production technology 243–9
  summary information 242
Propene epoxidation with oxygen, multiple selectivity-conversion plot 179, 181
Propene oxidation 20
Propene oxidation to acrolein 74, 85, 102–31
  bismuth molybdate catalysts 105–16, 249
  multiple selectivity–conversion plot 80, 104
  overall reaction mechanism 126–8
  product distribution 126–7
  production flowsheet 104
  source of oxygen 119–23
  summary data 103
$Pt/Al_2O_3$ 96, 208, 215, 229, 290, 291
$Pt/\tau$-$Al_2O_3$ 209
$Pt(OH)_xCl_y$ 204
$PtO_xCl_y$ 204
Pt-$PdO/Al_2O_3$ 204
Pt-$PdO/\alpha$-$Al_2O_3$ 207
$Pt/SO_4^{2-}/ZrO_2$ 215
Pump-probe experiment 20
Pyrophosphate termination of (020) plane of $(VO)_2P_2O_7$ 157

Radical allyl species 128
Radical formation 77
Radical species 127
Raman band shifts 123
Rate of adsorption 33–4
Rate of desorption 33–4
Reaction energy profile, oxygen activation 231
Redox molecular sieves 313–28
Residence times 42, 43
R-H bond 73
R-H (D) bond 74
Rh-MCM-41, transmission electron micrographs 331

Rietveld analysis 51

Safety considerations 20–5
Sb:U ratio 98
Scanning electron micrographs, silver catalysts 166
Scherrer equation 52
Selective chemisorption
  applications 59
  continuous mode 57–9
  pulse mode 57–9
  techniques 57–9
Selective oxidation 1, 5–7, 76, 90
  kinetic isolation of reaction products 66–72
  model 68
  principal uses 6
  production capacity 6
  reaction network 69
  reaction profile 70–1
  thermodynamics of reactions 67
Selective oxidation catalysts, active sites 96–9
Selective oxidation product (SOP) 68–9
Selectivity-yield-conversion plot 69
Sequential reaction 20
Side-bonded hydroperoxide 324
Silicalite-1 320–1
Silver catalysts 163
  comparison of BET and silver metal surface areas 166–7
  influence of mean particle size and added caesium 168
  ISS and Auger data 167
  oxygen on pure silver surface 174–6
  oxygen under surface of silver 176
  preparation 164–5
  scanning electron micrographs 166
  selectivity to ethylene oxide as function of oxygen coverage 182–3
  surface chlorine content 168
Silver surfaces
  adsorbed molecular oxygen 174–5
  alkali metals 177–9
  atomic oxygen on 176
  chlorine on 185–6
  $^{16}O_2$ incorporation and $C_2H_4{}^{16}O$ formation over 183
SIMS analysis 213
Sn-V-Sb oxides
  unit cell parameters 52
  XRD patterns 50–1
Sodium aluminosilicates, hydrothermal synthesis 312
Sodium chloride 6
Solid solutions 107–8, 126
Steady state testing procedures 17
Sticking coefficients 38
  for oxygen on Pt surfaces 228
  for oxygen on silver surfaces 175
Sticking probability 39
  and surface coverage 42
Structure sensitivity 94–6
Structured mesoporous materials, transmission electron micrographs 330
Structured mesoporous silicates 328–36
  preparation 329
  synthesis and characterization 328–33
Sulfur dioxide 216
Supersaturation, zeolites 312–13
Surface allyl species 116–19
Surface area determination 53–5
Surface composition multicomponent bismuth molybdate catalysts 112–16
Surface coverage and sticking probability 42
Surface dehydroxylation 208, 210
Surface hydroxyl groups in particle growth 209
Surface residence times 39
Synthesis gas 5

Tammann temperature 203
TAP reactor 88, 121, 135, 150, 153, 231, 250, 259
TBHP 327–8
Temperature programmed desorption (TPD) 59–61, 176, 287, 293
Temperature programmed reaction 184–5
Temperature programming reduction (TPR) 59–61
Temporal analysis of products (TAP) 17, 19
Tetraethylorthosilicate (TES) 319
Tetraethyltitinate (TET) 319
2,2,3,3-Tetramethylpentane 25

Tetrapropylammonium hydroxide (TPAOH) 319
Thermal NO 193–4
Thermogravimetric analysis, $(VO_2)P_2O_7$ 146
Thermogravimetric cycle, PdO decomposition 235
Thiele modulus 28, 29
Three way catalysts 196–9
pollutants treated 197
reactions 198–9
Ti-beta 322–3, 334
Ti-MCM-41 333–5
Ti-Mordenite 322
Time-resolved sum-frequency generation (SFG) spectroscopy 221
$TiO_2$ 11
$TiO_2$ anatase 94
$TiO_2$ rutile 94
$TiO_2/SiO_2$ 315
Titanium-containing crystalline silicas 313–15
Titanium silicalite (TS-1) 91, 315–26
addition of alkali metals during preparation 320
oxidation of unsaturated alcohols 317–18
oxidation reactions 323–6
preparation 318–20
reactions catalysed by 316
structure 320–2
X-ray diffraction patterns 321
Titanium silicalite (TS-2) 322
Titanium silicate 315
TMPHP 327–8
Toluene 251–3, 256
Total oxidation catalysts 91
Total oxidation products ($CO_x$) 68–9
Total oxidation reactions 90, 91, 98, 223, 225
reaction profile 70–1
Transient analysis procedures 17
Transition aluminas 211
Transmission electron micrographs
Rh-MCM-41 331
structured mesoporous materials 330
Transmission electron microscopy 112
1,1,2-Trichloroethane 295–9
Tubular fixed bed reaction, temperature profile 67
Turnover frequency (TOF) 228, 230

UHV techniques 95
ULM-5 313, 314
Unsaturated alcohols, oxidation on TS-1 317–18
Upper flammability limit (UFL) 21
*see also* Flammability limits
Uranium oxide 99
$USb_3O_{10}$ 98, 99

$V_2O_5/TiO_2$ 91, 250
Vanadium antimony oxides 250
Vanadium oxidation state in working catalyst 144–5
Vanadium oxide 88, 89, 94
Vanadium phosphorus oxide (VPO) catalysts 133–6, 138–47
comparison of bulk and surface P:V ratios 146
crystalline structure 142
high resolution electron micrographs 148
influence of P:V ratio on performance 147–8
oxidation state of vanadium 144–5
phase composition 139–42
preparation 139–42
P:V ratio 145
role of distortion of particle shape on chemical properties 144
scanning electron micrographs 141
surface composition 147
surface texture 147
X-ray diffraction patterns 142–3
Vanadyl pyrophosphate 108–9
Vinyl chloride
oxychlorination 266–8
production facility 269
production flow diagram 268
production technology 268
$(VO)_2H_4P_2O_9$ 139, 141
$(VO)_2P_2O_7$ 140, 142, 154, 155, 250, 253
thermogravimetric analysis 146
X-ray diffraction patterns 145
Volatile organic compound (VOCs)
catalytic combustion 191
catalytic oxidation 285

Volatile organic compound (*cont.*)
concentrations in indoor and outdoor air 192
exhaust 216
multiple selectivity-conversion plots 155
oxidation 190

Weisz window 15, 68
$W/F$ function 17

X-ray diffraction (XRD) 50–3
line broadening 52–3
particle size measurement by 51–3
phase identification 50
X-ray diffraction (XRD) patterns 108
MCM-41 332
TS-1 321
$(VO_2)P_2O_7$ 145
VPO 142–3
X-ray photoelectron spectroscopy 112, 115
surface P:V ratio 146

*o*-Xylene 91

Zeolites 301–13
basic building unit 308
common codes 305
containing framework titanium 322–3
crystal growth 313
electron micrograph 303
formation 312–13
generation of Brønsted and Lewis acid sites 304
nomenclature 305
nucleation 313
preparation 308–12
primary and secondary structures 305
secondary building units 309
structural features 302–8
structural properties 306–7
supersaturation 312–13
synthetic 305, 308
ZSM-5 321